T0295298

Build Like It's the End of the World

Build Like It's the End of the World

A Practical Guide to Decarbonize Architecture, Engineering, and Construction

Sandeep Ahuja
Forbes 30 under 30, UN Speaker, Inc.
Top 100 Women CEO
CEO, cove.tool

Patrick Chopson
AIA, CPO, cove.tool

Published by John Wiley & Sons, Inc., Hoboken, New Jersey.
Published simultaneously in Canada.

Library of Congress Cataloging-in-Publication Data:

Names: Ahuja, Sandeep, author. | Chopson, Patrick, author.
Title: Build like it's the end of the world: a practical guide
to decarbonize architecture, engineering, and construction / Sandeep Ahuja, Patrick Chopson.
Description: Hoboken, New Jersey : Wiley, [2024] | Includes index.
Identifiers: LCCN 2023057795 (print) | LCCN 2023057796 (ebook) | ISBN 9781394179176 (cloth) | ISBN 9781394179190 (adobe pdf) | ISBN 9781394179183 (epub)
Subjects: LCSH: Construction industry–Environmental aspects. | Carbon dioxide mitigation. | Sustainable construction.
Classification: LCC TD428.C64 A48 2024 (print) | LCC TD428.C64 (ebook) | DDC 690.028/6–dc23/eng/20240325
LC record available at https://lccn.loc.gov/2023057795
LC ebook record available at https://lccn.loc.gov/2023057796

Cover Design and Image: © Patrick Chopson

Set in 9.5/12.5pt STIXTwoText by Straive, Chennai, India

SKY10081401_080724

To my dearest Aurora, who graced our lives as these pages were being woven together. Your birth is the embodiment of hope and urgency that this book speaks to—the necessity to act for the planet's future. With each day of your growth, you reinforce why our mission for carbon reduction is not just imperative but personal. "Build Like It's the End of the World" *is not merely a title but a dedication to you and the generations to follow, who deserve a world abundant in life and possibilities. May this book serve as a testament to our commitment to that future.*

Contents

Foreword

Our slow but certain descent from climate change to climate crisis indicates that we are indeed attempting to avert a path to the end of the world. Most of humanity continues to be numb to the facts. We are resigned from fighting planetary threats that are no longer in the distant future. There was a time before for theorizing and speculating, and there is a time now for getting practical. There is a hunger for fast, accurate, and reliable solutions to the biggest contributor to this dilemma: the built environment.

This book answers the following questions: "Why should you care, at all?" "Where do you start with carbon positive design?" "How do you manage decarbonized construction?" "How is what you built maintained after construction?" "How can your building be a minimal, yet positive, contributor to the fight against climate change?"

This publication also showcases the philanthropic perspective and the business opportunity that are presented as harmonious and necessary for our human nature.

In earlier chapters, you will be acquainted with the "why" with the intention of motivating you to learn the "how." You can use the presented arguments to empower your professional or academic engagements through historic and future-forward evidence of why we need immediate solutions. Then, you will learn about decarbonization (both in design and in construction) and how they come together, discovering why you cannot learn one without the other if you are building as if the world is ending. Practical approaches for how your decisions can be made early, impactful, and lasting through the many stages of construction are demonstrated. You, and your practice or class, will have a clear vision of what buildings of the future will not just look like but also feel and perform.

At the heart of any decision-making process is data. The management of data allows you to make informed decisions. You can start with minimal data to capitalize on low-hanging fruits. You can also empower your practice or scholarship with advances in data analytics to make effective technology-integrated decisions. The resources provided in this book come from an understanding of practice by engaging hundreds of firms and thousands of buildings across the globe.

The authors Sandeep Ahuja and Patrick Chopson are on a mission to save the world. This book is not one-tracked; it aims to address the entire architecture, engineering, and construction industry.

Knowing Sandeep and Patrick personally is seeing hope where there could not be. By reading this book and equipping yourself and others with the tools necessary to fight climate change in this profession, you become a major part of that hope. We can be heading toward the end of the world, or the end of a world experience that is now an alternative that we successfully evaded. You are now part of that revolution.

Dr. Tarek Rakha
Associate Professor, Director of the High Performance Building Lab and Program
School of Architecture, College of Design, Georgia Institute of Technology

Acknowledgments

Every book is a journey, and every journey is enriched by companions. In writing this book, I have been blessed with the support, wisdom, and encouragement of many. To them, I extend my sincere thanks and acknowledgment.

Co-Founder

Daniel Chopson, your dedication and vision have been instrumental in transforming the AEC space into practical, usable software and workflows. Your efforts have empowered thousands of architects, engineers, and contractors to make low-carbon decisions for countless buildings, creating benefits that will outlast us all. I am deeply grateful for your professional excellence, passion, and kindness.

Mentors and Advisors

In memory of Professor Godfried Augenbroe and Dr. John Haymaker, your mentorship and guidance remain the compass of our careers. To Dr. Tarek Rakha, Ed Akins II, Manole Razvan Voroneanu, Liz Martin, Louis Joyner, Greg Stephens, and Randy Deutsch, your insight and constructive critique have been instrumental in my professional growth.

Professional Associates

Natalie Terrill, Michael Beckerman, and Shaun Abrahamson, your contributions have significantly deepened the book's content. Tristram Carfrae, Dennis Shelden, Pablo La Roche, Chirag Mistry, Sarah Gudeman, Eric Borchers, Paul Mckeever, Brian Campa, and Kate Simonen, your pioneering spirits have pushed the boundaries of our field.

Research and Technical Assistance

The cove.tool research team, spanning the years 2017–2023, your unwavering dedication to climate action and sustainability has not only propelled our field forward but also greatly enriched this book.

Editorial and Publishing Support

Kalli Schultea, Ms. Annika Kraft, Ms. Caroline Windham, Ms. Nandhini Karuppiah, and Ms. Krystl Black—your collective expertise in navigating the publishing labyrinth has been nothing short of extraordinary. Your meticulous efforts have ensured the seamless realization of this book.

My deepest appreciation goes to each of you. Without your support, guidance, and invaluable contributions, this book would not have come to fruition.

1

A Call to Action – Climate Change and the Role of the Built Environment

> The greatest danger to our planet is the belief that someone else will save it.
>
> – Robert Swan, Polar Explorer

1.1 How Big Is the Problem of Climate Change?

This book is about decarbonization in the built environment. When one opens their phone and checks the news over morning coffee, the effects of climate change are increasingly shaping every aspect of our lives. But what can one do about it? It can feel like a problem too big to confront, and it is easy to "throw it over the wall" and hope someone else in the design and construction process will "handle the sustainability" bits. Yet, reducing carbon emissions from buildings is an urgent task, and as professionals in the architecture, engineering, and construction (AEC) industry, our daily choices have an impact far beyond that of the average citizen.

1.1.1 Historic Trends in Context

Is it not true that the earth has gone through warming and cooling phases in the past, long before humans were around? Unfortunately, there is unambiguous evidence that the current global warming is not caused by natural phenomenon. The primary driver of our current warming is a rapid increase in greenhouse gases in the Earth's atmosphere, which can only be explained by human activities. The burning of fossil fuels such as coal, oil, and natural gas releases large amounts of carbon dioxide and other greenhouse gases into the atmosphere. Greenhouse gases trap heat from the sun, causing the Earth's temperature to rise. By comparison, natural sources of greenhouse gases, such as volcanic eruptions and the decomposition of organic matter, are much smaller than human-caused emissions. In addition, the rate of increase in greenhouse gas concentrations in the atmosphere over the past century is much faster than natural processes could account for.

Build Like It's the End of the World: A Practical Guide to Decarbonize Architecture, Engineering, and Construction, First Edition. Sandeep Ahuja and Patrick Chopson.

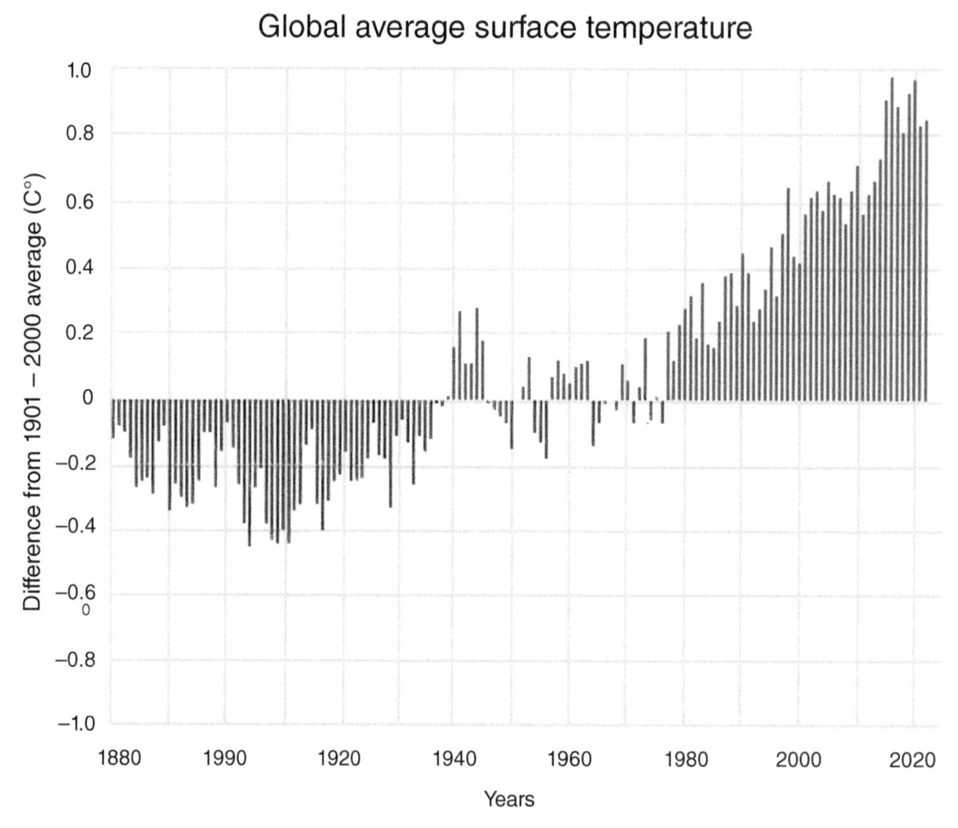

Synergistic mutual effects through the food web and environment between hybrid tilapia (*Oreochromis niloticus* × *Oreochromis aureus*) and the bottom feeder common carp (*Cyprinus carpio*). Source: REBECCA LINDSEY AND LUANN DAHLMAN / NOAA / https://www.climate.gov/news-features/understanding-climate/climate-change-global-temperature#:~:text=Earth's%20tempera[...]OAA's%20temperature%20data / last accessed on January 22, 2024 / Public Domain.

Buildings are responsible for a staggering 40% of global carbon emissions, and a small group of design and construction professionals makes the decisions about those emissions. In the United States, architects and engineers (design team) represent just 0.3% of the total population. According to the Intergovernmental Panel on Climate Change (IPCC), 91% of the carbon reductions required to stay under a "safe" amount of warming have not yet been made, although there is no safe limit of harm to our planet. Despite this alarming statistic, we still struggle to convince our colleagues and clients of the importance of designing low-carbon buildings. It is easy to become numb to the hard facts, but there is hope.

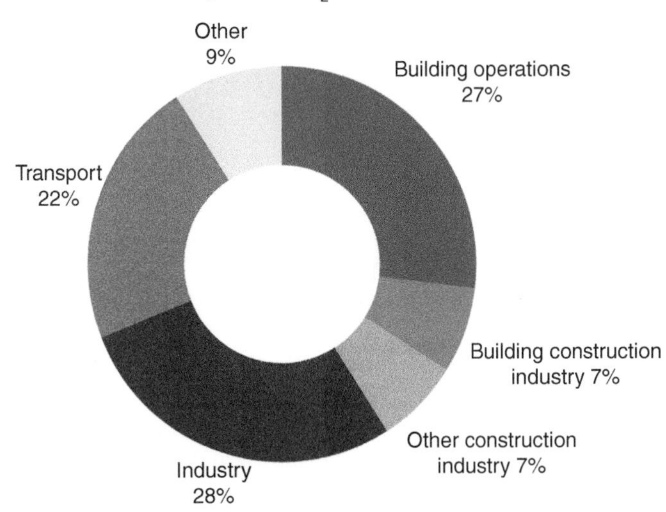

This graph showcases the total annual global CO_2 emissions, both direct and indirect energy and process emissions (36.3 GT), in a built environment. Source: Adapted from Architecture 2030, https://architecture2030.org/why-the-built-environment/.

In this book, we will be talking a lot about carbon and presenting data to back up our arguments, but it is important to remember that it is crucial to construct a narrative for ourselves and others around taking action. All the facts and simulations in the world will not change how people think, but stories are powerful. In the end, it comes down to the story we tell about the work we do and how we do it. The design and construction professions need to articulate stories about the way the world should be. These are stories that can help us recognize and adapt our work processes and profit motivations. It is time to take responsibility and inspire others and to have the courage to challenge conventional thinking. It is time to design and build like it is the end of the world.

Designing buildings in the 21st century requires a sense of urgency and an understanding of the consequences of inaction. Our choices have a significant impact on the planet and future generations, so it is important to take a long view and consider the uncertainties and challenges ahead. Ignoring them or blindly assuming certain events will not happen is not smart. Courage is required to challenge conventional thinking and present the facts in a way that helps people understand the urgency. While kindness is the key to helping people accept the hard truths that come with understanding this, ultimately, it is about transferring that courage to others so that they too can be armed with the data and narratives to create change.

Now for some facts. What is climate change exactly? Climate change refers to the long-term shifts in temperature, precipitation, winds, and other indicators of the Earth's climate. These shifts are unequivocally caused by humans burning fossil fuels, which release greenhouse gases into the atmosphere. These gases trap heat from the sun, causing the Earth's temperature to rise. The human-caused nature of climate change was first simulated by scientists Syukuro Manabe and Kirk Bryan in 1969. Since then, 50% of all carbon ever burned has been burned since 1985. Climate change is not a matter of belief, but a fact and it is happening right now. But this is not exactly the kind of conversation that makes one the life of the party.

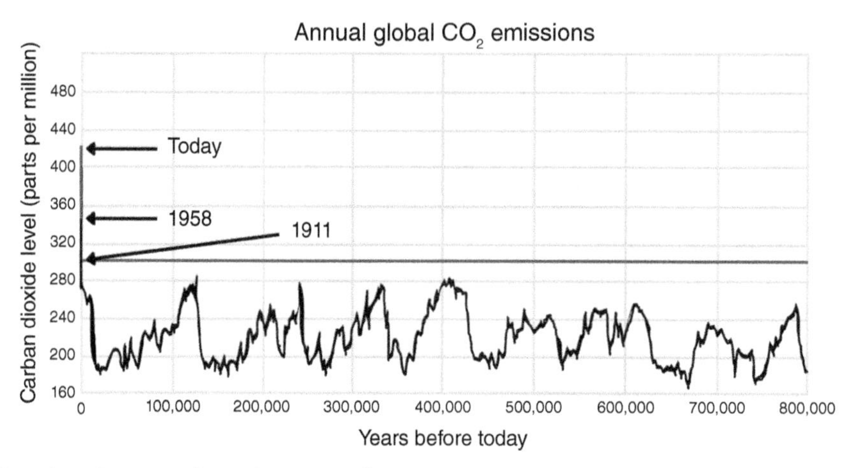

This graph, based on the comparison of atmospheric samples contained in ice cores and more recent direct measurements, provides evidence that atmospheric CO_2 has increased since the Industrial Revolution. Source: Luthi et al. (2008) / NASA / https://climate.nasa.gov/evidence/ Public Domain.

Digging deeper into the data, the IPCC, which includes more than 1300 scientists from the United States and other countries, forecasts a temperature rise of 1.1–5.4 °C (2–9.7 °F) over the next century. This will have numerous effects, including sea level rise of 0.3–2.4 m (1–8 feet) by 2100, stronger and more intense hurricanes, the likelihood of the Arctic becoming ice-free, more droughts and heatwaves, changes in precipitation patterns, and a lengthening of the frost-free season (and growing season). The effects of climate change are wide-ranging and affect lives and economies across the globe. Radical action is needed in the next decade if we are to stall the disastrous impacts of a spike in global temperatures.

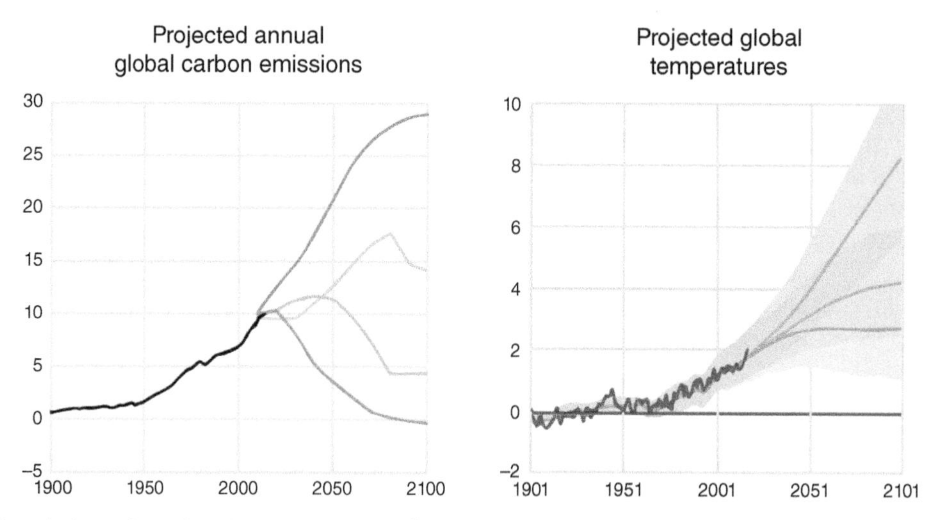

(Left) Speculative trajectories of carbon emissions ("representative concentration pathways" or RCPs) in the 21st century influenced by varying energy policies and economic growth trends. (Right) Anticipated temperature shifts relative to the 1901–1960 baseline, contingent upon our adoption of specific RCP scenarios. Source: REBECCA LINDSEY AND LUANN DAHLMAN / NOAA / https://www.climate.gov/news-features/understanding-climate/climate-change-global-temperature#:~:text=Earth's%20temperature%20has%20risen%20by,based%20on%20NOAA's%20temperature%20data / last accessed on JANUARY 18, 2024 / Public Domain.

As the largest contributor to the problem, the building industry has a significant role to play in shaping the crisis. Interestingly, buildings offer some of the most cost-efficient ways of reducing carbon emissions and combating climate change. This is because, unlike carbon capture or more efficient cars, buildings have a lot of room for improvement in their design and construction, which is often not even simulated or cost optimized. Due to advancements in technology and manufacturing, many new materials and efficient systems are already cheaper than conventional approaches. After all, most people rely on intuition and what worked on their last project when designing buildings.

Unfortunately, intuition and traditional thinking are the last things one needs when confronting a new challenge. Especially when we look at the short-term horizon of the next 10 years, the situation is dire. The graph below shows that if we do not act, the general collapse of civilization could begin sometime between 2030 and 2040. This prediction is based on an updated reassessment of the Limits of Growth computer model, which has accurately predicted these trends since it was developed in the 1970s. The model considers several factors such as population growth, resource depletion, and industrialization and shows how they can interact in a complex system. We must take drastic action in the next decade if we are to avoid the Business-as-Usual case 2 (BAU2) version of the simulation shown here. This is an outcome that should be avoided at all costs. Decarbonization is not a "nice to have," but necessary for the survival of civilization itself.

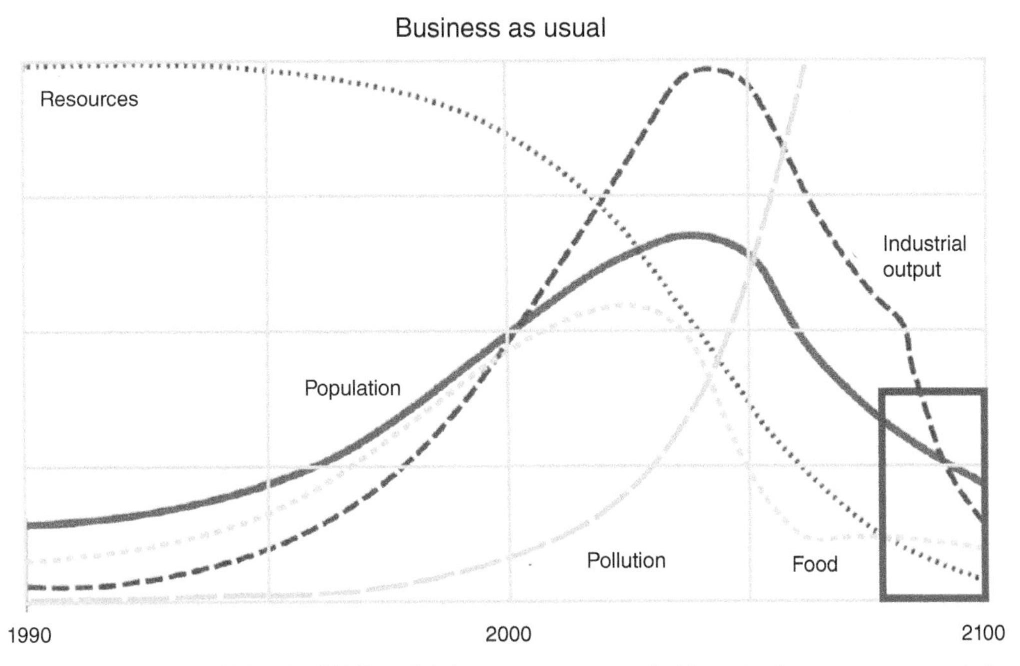

Limits of Growth model utilizing the BAU2 model showcases an unsustainable path when resources, population, pollution, food, and industrial output are misaligned. Source: Adapted from (Turner, 2014).

According to the Limits of Growth model, if we do our part, the comprehensive technology (CT) scenario shows a path forward that is survivable. Humanity has just over a decade to change course and avoid societal collapse. This can be achieved through a combination of technological progress and increased investments in public services, which would not only avoid the risk of collapse but lead to a new, stable, and prosperous civilization operating within planetary boundaries. As KPMG director Kate Herrington points out, we must adopt an "agnostic approach to growth" that focuses on other economic goals and priorities, rather than simply pursuing economic growth for its own sake. If we fail to act in the next decade, the consequences could be dire, with economic decline and societal collapse becoming more likely. However, by making the necessary investments and embracing technological progress, we can create a plausible scenario for a better world.

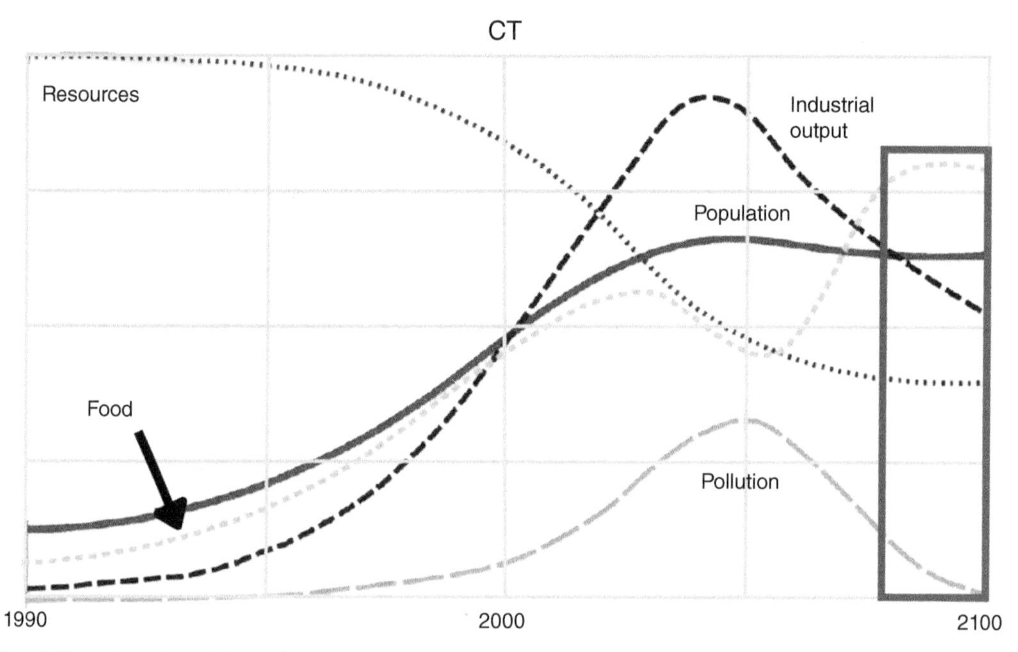

Limits of Growth model with the Comprehensive Technology (CT) showcases a survivable path when resources, population, pollution, food, and industrial output are all aligned. Source: Adapted from (Turner, 2014).

With the building materials and construction industry as one of the most significant contributors to greenhouse gas (GHG) emissions, it is exposed to carbon taxes such as the Inflation Reduction Act in the United States and the Carbon Border Adjustment Mechanism (CBAM) in Europe. Embodied carbon is the carbon used in building materials production, and operational carbon is from power and heat supply in buildings. In addition, the sector must also address infrastructure and sector decarbonization goals to combat climate change.

But the construction sector is not just creating risk for other people from our carbon emissions, it is also vulnerable to the physical impacts of climate change, like more extreme weather conditions on construction sites, water shortages, and other deteriorating environmental conditions like temperature increase and flooding. These physical risks can have a serious impact on our sector, and it is crucial that we take steps to address them.

So, what can the construction sector do to combat climate change? One key area is reducing carbon emissions and transition risks. This means lowering the carbon intensity of building materials in upstream production and implementing climate-smart, low, and clean energy consumption in the use phase of real estate and infrastructure. It also involves designing more recyclable materials and closed material flows in the refurbishment and demolition phases to increase the circularity of building materials.

Another critical area is building resilience against the environmental consequences of climate change. This includes measures like increasing the durability of materials against extreme weather conditions, overhauling heating/cooling, and insulation concepts, and revising water management toward more climate smart systems during the construction and use phases of buildings. It also involves lowering the potential negative effects of the construction sector on the environment, such as soil sealing and land use change.

1.1.2 How Is Human Health Affected?

Climate change has significant impacts on human health, both directly and indirectly. It affects the social and environmental determinants of health, such as clean air, safe drinking water, sufficient food, and secure shelter. Between 2030 and 2050, climate change is expected to cause approximately 250,000 additional deaths per year due to malnutrition, malaria, diarrhea, and heat stress. The direct damage costs to health, which exclude costs in health-determining sectors like agriculture, water, and sanitation, are estimated to be between US$2 and US$4 billion per year by 2030.

Climate Change Indicators: Health and Society

Heat-related deaths	Length of growing season
Heat-related illnesses	Growing degree days
Cold-related Deaths	Ragweed pollen season
Residential energy use	Lyme disease
West nile virus	Heating and cooling degree days

A synopsis of climate-related health hazards, their exposure routes, and factors of susceptibility. The influence of climate change on health, encompassing both direct and indirect effects, is intricately shaped by environmental, social, and public health variables. Source: U.S. Environmental Protection Agency / https://www.epa.gov/climate-indicators/health-society / Public Domain.

Areas with weak health infrastructure, particularly in developing countries, will be the least able to cope with the effects of climate change without assistance to prepare and respond. It is important to note that reducing emissions of GHGs through better transport, food, and energy-use choices can result in improved health, particularly through reduced air pollution.

Rising temperatures	More extreme weather	Rising sea levels	Increasing CO_2 levels
Environmental degradation Forced migration, civil conflict, mental health impacts	Extreme heat Heat-related illness and death, cardiovascular failure	Changes in vector ecology Malaria, dengue, encephalitis, hantavirus, rift valley fever, lyme disease, chikungunya, west nile virus	Increasing allergens Respiratory allergies, asthma
Water and food supply impacts Malnutrition, diarrheal disease	Severe weather Injuries, fatalities, mental health impacts	Air pollution Asthma, cardiovascular disease	Water quality impacts Cholera, cryptosporidiosis campylobacter, leptospirosis, harmful algae blooms

Comprehensive overview of climate change's influence on health outcomes. This illustration delineates key facets of climate change, including elevated temperatures, heightened weather extremities, rising sea levels, and escalating carbon dioxide levels. It also elucidates their impact on exposures and the resultant health effects stemming from these alterations in exposure patterns. Source: Centers for Disease Control and Prevention / https://www.cdc.gov/climateandhealth/docs/Health_Impacts_Climate_Change-508_final.pdf / Public Domain.

According to the 2022 report of the Lancet Countdown, climate change is undermining every aspect of global health that is monitored, increasing the fragility of the global systems that health depends on, and increasing the vulnerability of populations to coexisting geopolitical, energy, and cost-of-living crises. It is increasingly undermining global food security and exacerbating the effects of the COVID-19 pandemic, geopolitical, energy, and cost-of-living crises. A new analysis of 103 countries shows that days of extreme heat, which are increasing in frequency and intensity due to climate change, accounted for an estimated 98 million more people reporting moderate to severe food insecurity in 2020 compared to the average from 1981 to 2010.

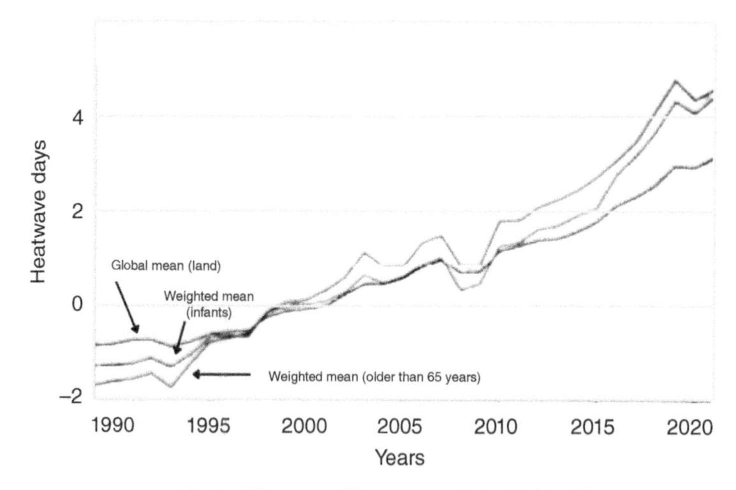

Contrasting heatwave occurrences with the 10-year rolling mean of the 1986–2005 baseline. Heatwave days are showcased through three different perspectives: weighted averages based on land surface area, infant population, and the population aged 65 years and older. Source: Adapted from (Romanello et al., 2022).

Well-prepared health systems are essential for protecting populations from the health impacts of climate change. However, global health systems have been significantly weakened by the effects of the COVID-19 pandemic, and funds available for climate action decreased in 30% of 798 cities. Health systems are also increasingly being affected by extreme weather events and supply chain disruptions.

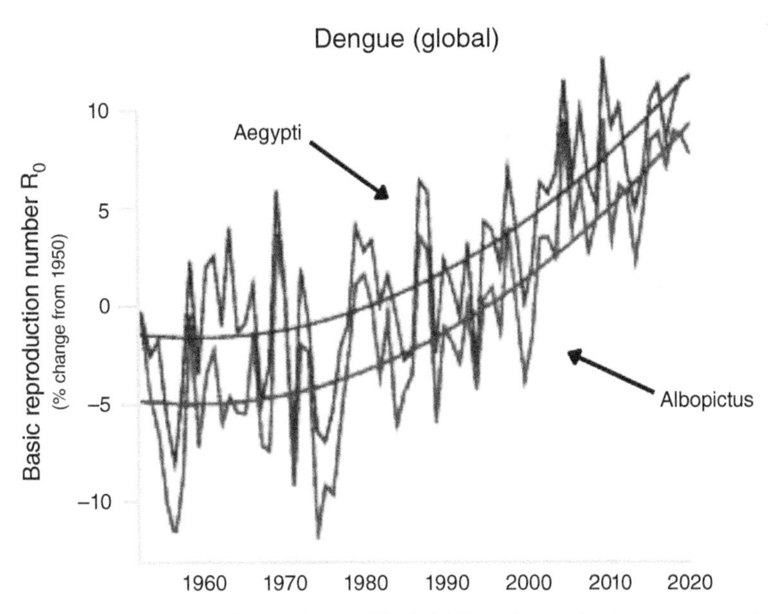

Narrow lines represent the year-to-year fluctuations, while bold lines depict the long-term trends starting from different years: 1951 for malaria and dengue, 1982 for *Vibrio bacteria*, and 2003 for *Vibrio cholerae*. HDI stands for the Human Development Index. Source: Adapted from (Romanello et al., 2022).

In the United States, each region experiences the impacts of climate change on health differently due to location-specific climate exposures and unique societal and demographic characteristics. In the United States, the CDC Climate and Health Program supports states, counties, cities, tribes, and territories to assess how climate change will affect their community, identify vulnerable populations, and implement adaptation and preparedness strategies to reduce the health effects of climate change. Most countries have some type of health risk assessment that provides striking conclusions that we ignore at our own peril. According to Dr. Daniel Bressler in his paper in *Nature*, there is a mortality cost of carbon. Every 4,434 metric tons of CO_2 in 2020 (equivalent to three to four people's lifetime emissions) causes one excess death globally in expectation between 2020 and 2100. Overall, climate change clearly has significant impacts on human health, and it is important that we act to mitigate these impacts and protect vulnerable populations. Changing the way we design and build buildings is one of those actions.

1.1.3 Climate Risk by Location

In 2019, Mozambique, Zimbabwe, and the Bahamas were the countries most affected by the impacts of extreme weather events. Between 2000 and 2019, Puerto Rico, Myanmar, and Haiti were the countries most affected by these impacts. In total, over 475,000 people (roughly the population of Toulouse, France) lost their lives and losses amounted to around $2.56 trillion in purchasing power parity (about $7,900 per person in the United States) due to more than 11,000 extreme weather events globally during this period.

Global climate risk index 2000–2019

Countries affected by extreme weather:

- Puerto rico
- Myanmar
- Hati
- Philippines
- Mozambique
- Mozambique
- The bahamas
- Bangladesh
- Pakistan
- Thailand

Identifying the primary victims of extreme weather incidents: weather-related loss events in 2019 and 2000–2019. The Global Climate Risk Index 2021 assesses the degree to which countries and regions have experienced the impacts of weather-related loss events, including storms, floods, and heatwaves. The analysis incorporates the most recent available data, encompassing the years 2019 and the period from 2000 to 2019. In 2019, the countries and regions most severely affected were Mozambique, Zimbabwe, and the Bahamas. Over the extended period from 2000 to 2019, Puerto Rico, Myanmar, and Haiti emerged as the top-ranking areas in terms of vulnerability to these events. Source: Adapted from Germanwatch (2021). Global Climate Risk Index 2021.

The 10 most affected countries in 2019

Ranking 2019	Country	CRI score	Fatalities	Fatalities per 100,00 inhabitants	Absolute losses (in million US$ PPP)
1	Mozambique	2.67	700	2.25	4930.08
2	Zimbabwe	6.17	347	2.33	1836.82
3	The bahamas	6.50	56	14.70	4758.21
4	Japan	14.50	290	0.23	28899.79
5	Malawi	15.17	95	0.47	452.14
6	Islamic republic of afghanistan	16.00	191	0.51	548.73
7	India	16.67	2267	0.17	68812.35
8	South sudan	17.33	185	1.38	85.86
9	Niger	18.17	117	0.50	219.58
10	Bolivia	19.67	33	0.29	798.91

Identifying the primary victims of extreme weather incidents: weather-related loss events in 2019 and 2000–2019. The Global Climate Risk Index 2021 assesses the degree to which countries and regions have experienced the impacts of weather-related loss events, including storms, floods, and heatwaves. The analysis incorporates the most recent available data, encompassing the years 2019 and the period from 2000 to 2019. In 2019, the countries and regions most severely affected were Mozambique, Zimbabwe, and the Bahamas. Over the extended period from 2000 to 2019, Puerto Rico, Myanmar, and Haiti emerged as the top-ranking areas in terms of vulnerability to these events. Source: Germanwatch, 2021. Global Climate Risk Index 2021.

Storms and their direct implications of precipitation, floods, and landslides were a major cause of losses and damage in 2019. Six out of the ten most affected countries in 2019 were hit by tropical cyclones, and recent science suggests that the number of severe tropical cyclones will increase with every tenth of a degree in global average temperature rise. In many cases, single exceptionally intense extreme weather events have such a strong impact that the countries and territories concerned also have a high ranking in the long-term index. In recent years, countries like Haiti, the Philippines, and Pakistan, which are recurrently affected by catastrophes, have consistently ranked among the most affected countries in both the long-term index and the index for the respective year.

Developing countries are particularly vulnerable to the impacts of climate change. They are hit hardest because they are more vulnerable to the damaging effects of a hazard but have lower coping capacity. Eight out of the ten countries most affected by the quantified impacts of extreme weather events in 2019 belong to the low- to lower-middle income category, and half of them are Least Developed Countries. While the impact on human life is higher in developing countries, the financial penalty of rebuilding is much higher in advanced economies with high-cost infrastructure and buildings. The global COVID-19 pandemic has highlighted the fact that risks and vulnerability are interconnected and systemic, which is why the Limits of Growth model continues to hold true.

Traditionally, when assessing risk, businesses and individuals tend to look back into the past to assess whether an event has happened before to predict when and at what frequency it will happen again. Now, with exponentially changing weather conditions, looking into the future using climate models is necessary. Not only that, one must also look at how climate risks can interact with previously known events like earthquakes. It is important to note that earthquakes are not related to climate change, but they can still pose a significant risk to buildings and infrastructure in certain locations when combined with a climate event. As such, it is important to consider earthquakes as part of the overall risk picture when designing and constructing buildings in areas that are prone to this type of risk. This may involve adopting building codes that require the use of earthquake-resistant materials and construction techniques, as well as investing in infrastructure and other measures that can help reduce the impacts of climate change on communities.

To bring this topic to life, let us journey into the world of architecture, weather, and climate, and unearth the specific risks and strategic design principles that must be considered when constructing a building. Imagine you are erecting a structure in Tornado Alley, USA, or on the hurricane-prone coastlines of Florida. Or, consider a setting in the Arctic Circle, where extreme cold is a constant companion. Every location poses unique challenges and threats, dictated by their individual climates, that could potentially impact the integrity of a building and the safety of its occupants.

Now, imagine the complexity of planning and constructing a building that can withstand the brute force of a tornado or the devastating floodwaters of a hurricane. For instance, the Fujita Scale, used to measure tornado intensity, categorizes an F5 tornado as having wind speeds over 200 mph, strong enough to level well-built houses and send cars flying through the air. In these regions, it is not just about creating a visually pleasing structure, but one that can stand up to the elements.

Billion-dollar weather and climate disaster cost

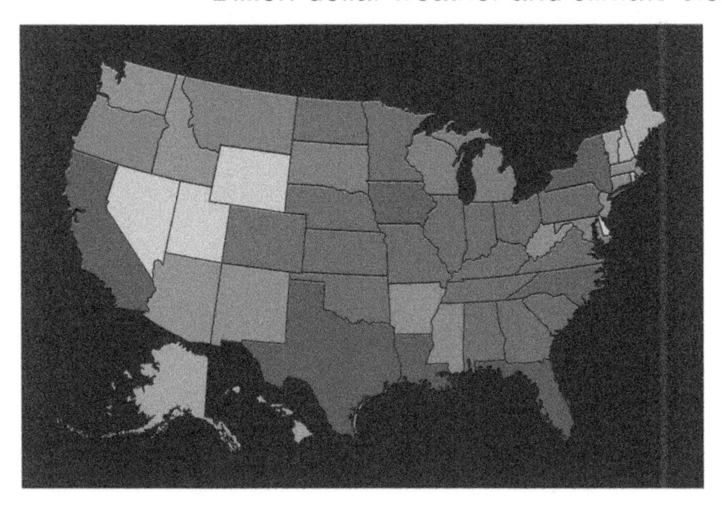

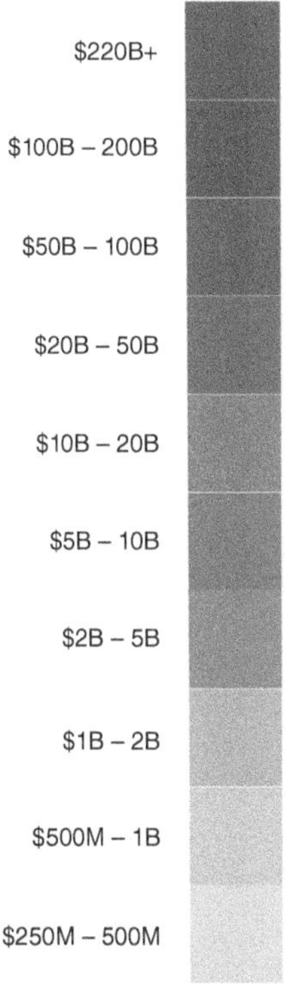

Mapping billion-dollar weather and climate disasters. From 1980 to 2023, the United States has witnessed 372 confirmed events with losses surpassing $1 billion (CPI-adjusted) each. These events encompassed various categories, including droughts, floods, freezes, severe storms, tropical cyclones, wildfires, and winter storms. While this map displays the cumulative number of billion-dollar events for each state affected, it does not imply that each state incurred at least $1 billion in losses for every event. NOAA National Centers for Environmental Information (NCEI) US Billion-Dollar Weather and Climate Disasters (2023).

Similarly, envision a building in the freezing temperatures of Alaska, where the permafrost can cause soil instability and compromise the building's foundation. Buildings in these regions must be constructed to withstand severe temperature fluctuations, and the design process must include an assessment of the physical risks posed by these extreme weather events as well as the potential for power and other service disruptions that could affect habitability.

Consider passive design principles by focusing on energy efficiency, minimal use of fossil fuels, and creating a comfortable living environment. This approach is a prime example of a resilience strategy used to protect occupants from climate risks like extreme heat and cold, air pollution, and other health risks. Buildings constructed following these principles are designed with airtight envelopes, super insulation, and advanced window technology to reduce energy use and thus limit the impact of external temperature fluctuations.

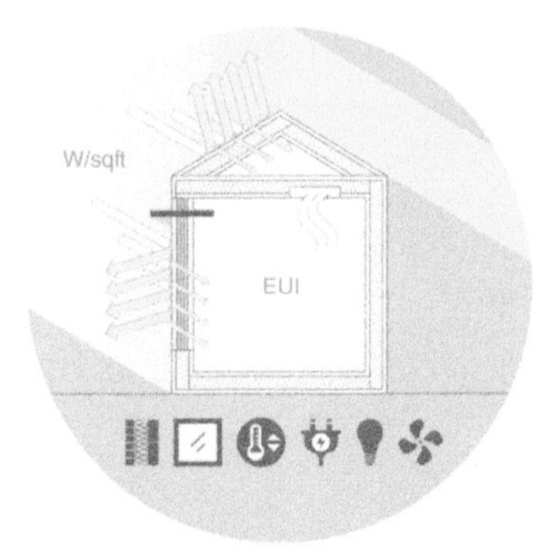

EUI model showing that reducing the SHGC value while allows finding acceptable visibility for occupants will help keep the heat out, lowering demand on the mechanical system.

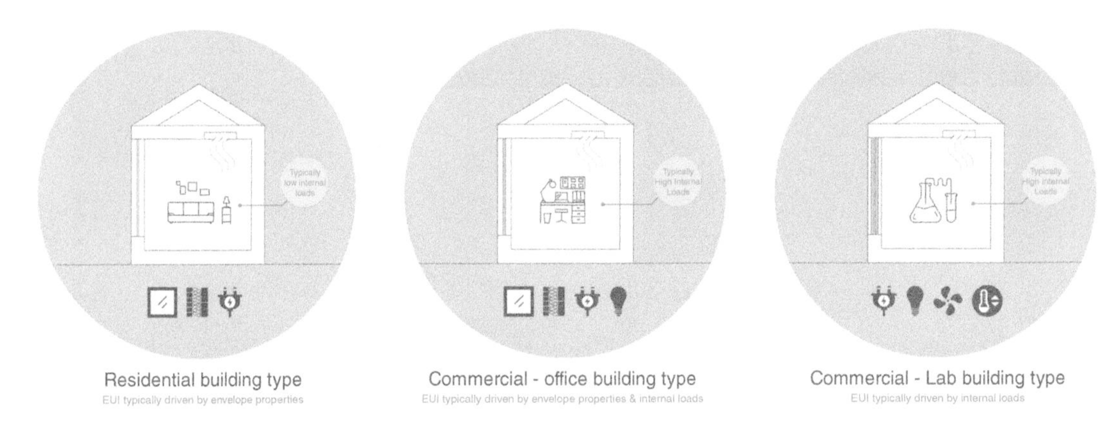

Shading design strategies based on building types.

Moreover, we must plan for the unexpected, such as a natural disaster like a hurricane or fire that might knock out power and water supplies. Look at the impacts of Hurricane Katrina, which left much of New Orleans without power or safe drinking water for weeks. Buildings designed with backup systems, such as generators or rainwater collection systems, can provide a lifeline in such times of crisis, especially for vulnerable occupants like the elderly or those with health conditions. It is not just about surviving the immediate event but also ensuring a level of comfort and normalcy in the aftermath.

We need to ensure that our buildings are adaptable, able to evolve to meet changing needs and climate conditions. Just as a chameleon changes its color to match its environment, buildings too must be able to adapt to ensure longevity. For instance, warehouses have been converted into residential lofts in urban areas experiencing population growth, demonstrating adaptability and resilience against obsolescence.

The frequency and intensity of extreme weather events like hurricanes, typhoons, tornadoes, and thunderstorms can cause significant damage to buildings, including wind damage, flooding, and landslides. Consider Hurricane Andrew, a Category 5 hurricane that hit Florida in 1992, which caused extensive damage due to high winds and flooding. Buildings in hurricane-prone areas are designed with these types of events in mind, using reinforced concrete and high-impact glass to withstand the wind force and materials that resist water damage to mitigate flooding.

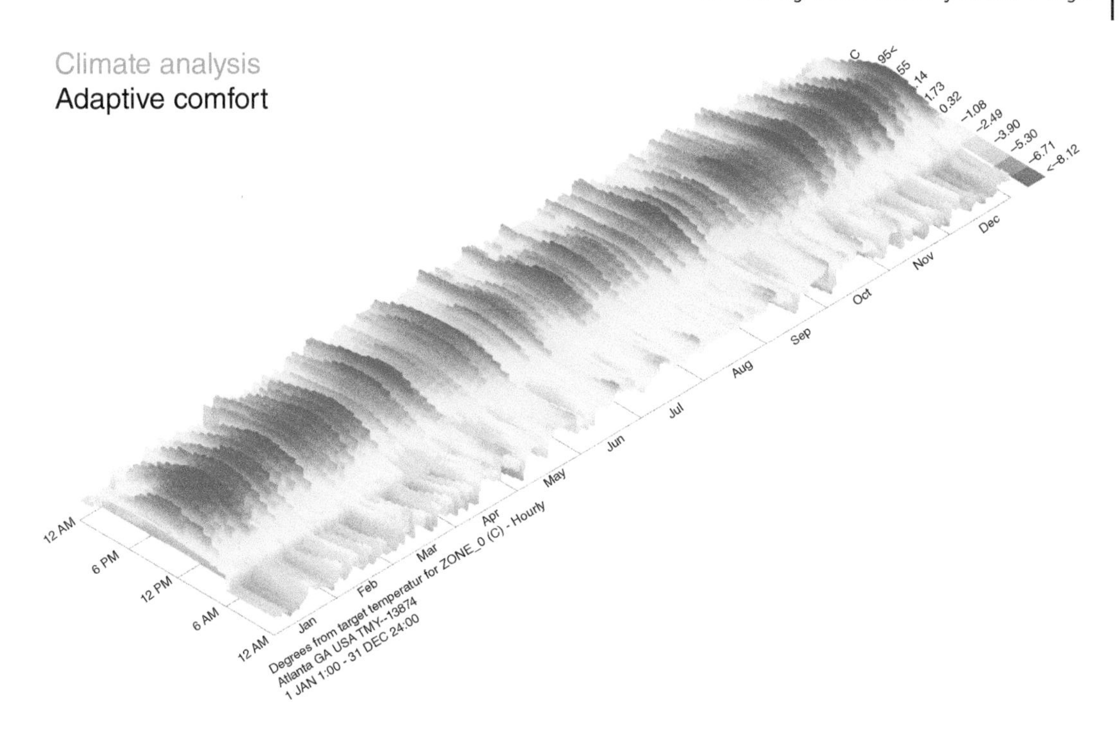

Adaptive comfort chart showing the time of day and time of year with the greatest human comfort for your location.

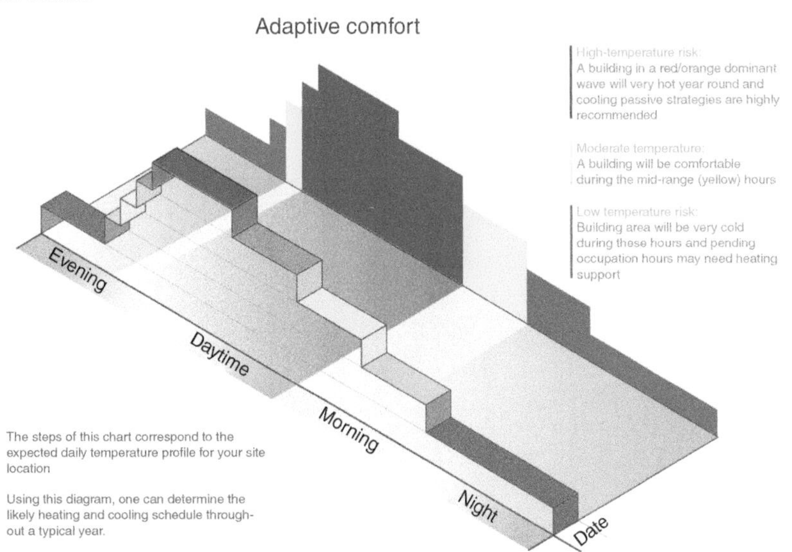

This Climate Analysis/Adaptive Comfort chart shows the time of day and time of year with the greatest human comfort. Source: cove.tool.

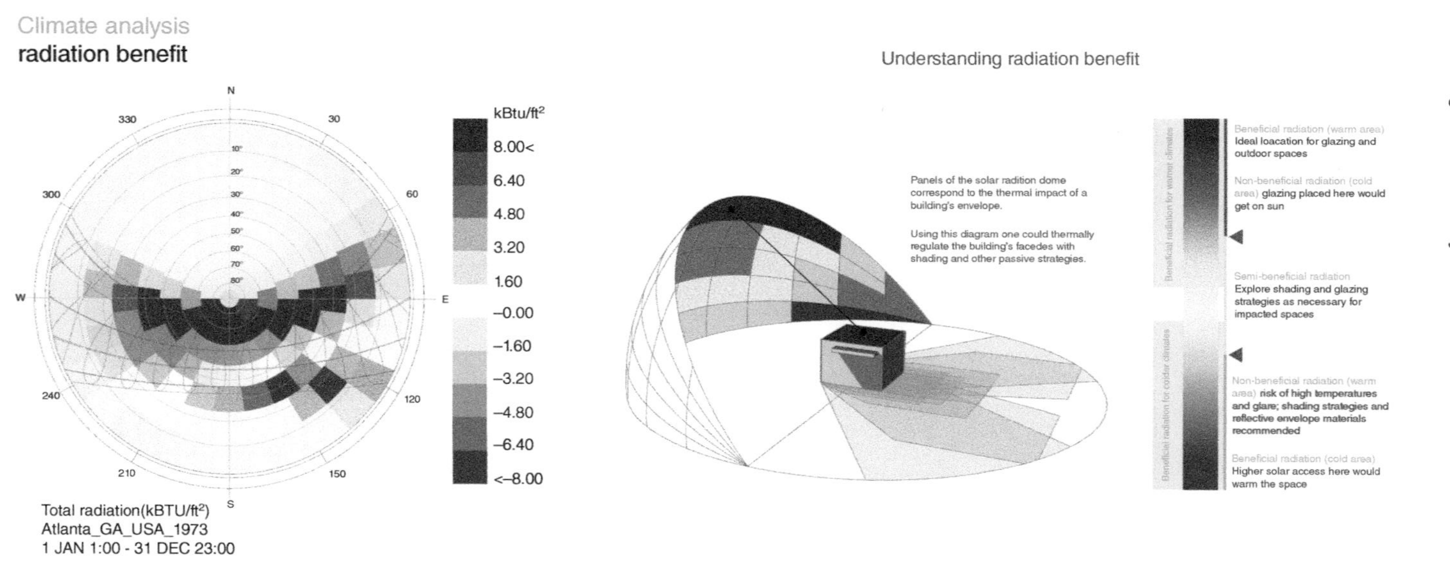

This Climate Analysis/Radiation Benefit chart provides visual representation of how the solar radiation dome corresponds to the thermal impact of a building's envelope. Using this diagram, one could thermally regulate the building's facades with shading and other passive strategies. Source: cove.tool.

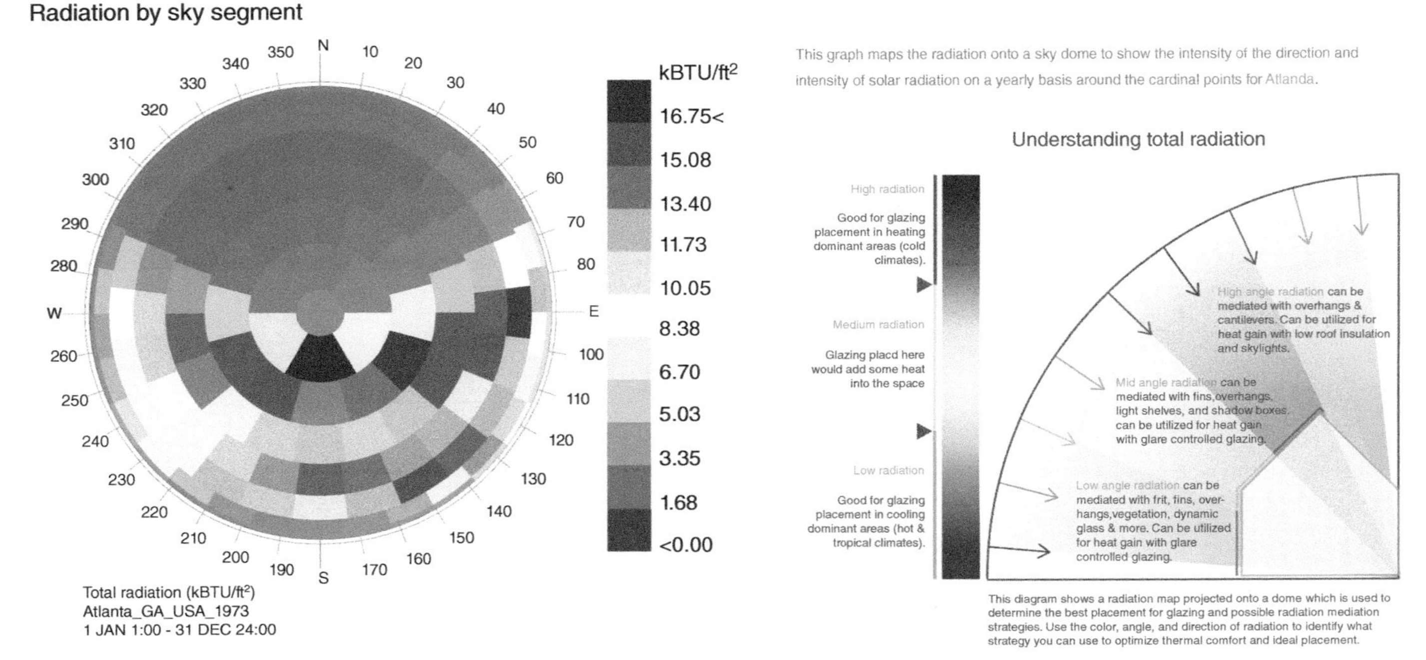

This Climate Analysis/Radiation by Sky Segment graph maps the radiation on to a sky dome to show the intensity of the direction and intensity of solar radiation on a yearly basis around the cardinal points for Atlanta, Georgia, USA. Source: cove.tool.

Extreme heat and cold also pose a significant threat to buildings. In areas prone to heatwaves, such as Death Valley in California, buildings need to be designed with excellent insulation and energy efficiency to maintain a comfortable indoor environment, while in cold areas, adequate insulation and heating systems are crucial. Reflective roofing materials can repel heat, while certain types of insulation can keep warmth inside during frigid winters.

Building in diverse geographical locations requires comprehensive planning to address the specific risks that may be encountered. Let us illustrate this with some vivid examples. In the desert, for example, where heat is an obvious risk, the right construction materials can make a world of difference. Take adobe, an age-old building material that works wonders in these arid climates. The thick walls made of adobe create a barrier between the scorching outside temperatures and the indoor environment. Furthermore, the adobe slowly releases heat absorbed during the day, providing warmth during chilly desert nights.

Insulation is another key consideration in such environments, helping to block the intense heat from entering the house and driving down energy costs by reducing the strain on air conditioning systems. A lighter color scheme for the house, like beige or grey, can also reflect the heat before it reaches the interior, like how wearing a light-colored shirt in the summer keeps us cool. These strategies are not just theoretical but have been applied in real-world desert construction.

Moreover, the sun's constant presence in desert environments makes solar panels an excellent building material choice for homes in desert regions. Solar panels not only lower energy costs but also increase a home's value, making them an attractive investment for homeowners.

But it is not just about the heat. Despite the scorching summers, desert regions also experience cold late-autumn and winter seasons. Trombe walls, designed in the 1970s, are a smart addition for desert homes, offering an energy-efficient way to trap daytime heat from the sun and gradually cool the interior during the colder nights.

Similarly, in locations prone to extreme weather events such as hurricanes and tornadoes, resilient design and construction techniques are paramount. Remember Hurricane Sandy in 2012? It illustrated the destructive potential of these extreme weather events and the importance of designing buildings to withstand them. This includes, for example, wind-resistant materials and construction techniques, along with backup power and water supply systems to ensure building resilience during service disruptions.

When it comes to extreme heat and cold, the design of buildings plays a vital role in ensuring a comfortable and safe indoor environment. Just as we layer our clothes to adapt to fluctuating outdoor temperatures, buildings too need layers of protection. For instance, well-insulated and energy-efficient buildings can help reduce heat gain during the summer months and retain warmth during winter. The use of shading, ventilation, and heat-reflective components, such as reflective roofing materials, can keep buildings cool during heatwaves. Meanwhile, operable windows and cross ventilation can reduce indoor air temperatures by 10°–15°, which can be lifesaving during extreme heating events, particularly if power for HVAC (Heating, Ventilation and Air Conditioning) systems is lost.

Similarly, in extreme cold, window orientation and the ability to accept heat from the sun is key to survivability. Insulation can help reduce the amount of heat that escapes the building during the winter event, making it easier to maintain a comfortable indoor temperature. The use of high-quality windows and doors designed to reduce heat loss and that are not thermally bridged can increase the time the building stays warm. While envelope tightness certainly helps, ventilation is necessary to ensure that adequate oxygen is provided. Switching from gas to electric heating removes the risk of carbon monoxide poisoning if the system fails, which sends over 50,000 people to the hospital and kills hundreds of people every year in the United States, according to the CDC.

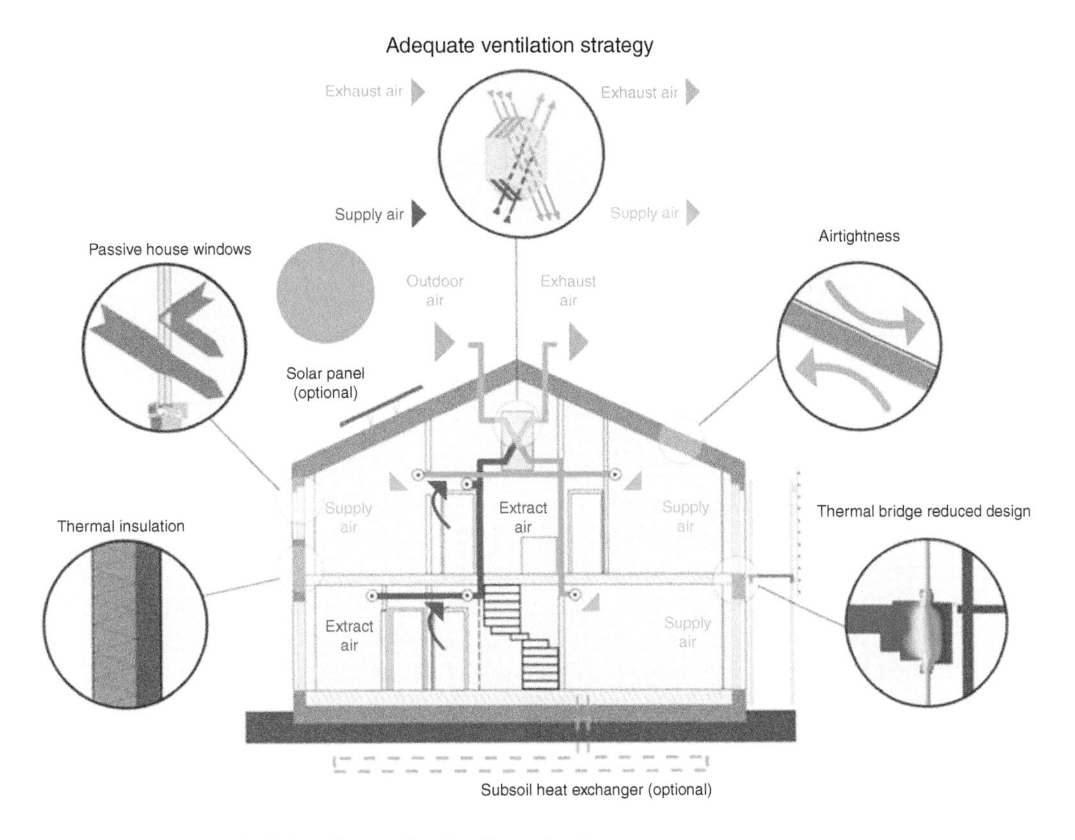

The five Passive House principles. Source: Passive House Institute.

Overall, it is important to carefully evaluate the major climate risks that may affect a building in any given location, and to design and construct the building in a way that is resistant to these risks and ensures the habitability and resilience of the structure for its occupants. By considering the potential loss of power and other services, as well as the physical risks posed by extreme heat, cold, and other temperature fluctuations, it is possible to create buildings that can withstand and recover from the impacts of climate change.

Sea level rise and coastal flooding are a major climate risk that we need to be prepared for. As sea levels rise, coastal communities are at increasing risk of flooding, which can damage buildings, disrupt the lives of occupants, and threaten the viability of entire neighborhoods. That is why it is so important to consider the potential for sea level rise in the design and construction of buildings in coastal areas.

When dealing with the threat of coastal flooding, it is essential to build with techniques that are resistant to damage. This could mean installing impact-resistant doors and windows for high winds, elevating the building above predicted flood levels, or incorporating breakaway flood-resistant features into the design. Adequate drainage systems and flood protection measures can also help ensure the habitability of a building during times of coastal flooding by directing water away from the structure and protecting it against water damage. One also needs to consider the potential for disruptions to essential services such as power and water during flooding. That is why it is a smart idea to design buildings with backup power and water supply systems above the worst-case flooding levels, as well as other measures to ensure the resilience of the building in the event of a power outage or other service disruption like solar backup and batteries. Because of the widespread nature of risk, coastal areas have the largest adaptation cost.

However, those who live inland have another risk to consider. Wildfires are one of the most terrifying and destructive climate risks that we face. With the right combination of dry, hot, and windy conditions, these raging infernos can threaten homes, businesses, and entire communities in an instant. That is why it is so important to consider the potential for wildfires in the design and construction of buildings in areas prone to this type of risk.

To protect ourselves against the threat of wildfire, it is essential to use materials and construction techniques that are resistant to fire. This could mean selecting noncombustible materials, installing fire-resistant roofing, or incorporating other fire-resistant features into the design of the building. Additionally, having a strong HVAC system in place can help ensure the habitability of a building during times of wildfire by filtering out smoke-filled air and fine particulate matter. By investing in high-quality air filters and other air filtration measures, one can help protect the indoor air quality of one's building and keep one's occupants safe and comfortable.

Wildfires not only damage buildings and natural habitats but also disrupt essential services like power and water. Take, for instance, the Camp Fire in 2018, one of the deadliest wildfires in California history, which left the town of Paradise virtually destroyed and without access to power or water for an extended period. It is a stark reminder that designing buildings with backup power and water supply systems is not just a luxury but a necessity in such high-risk areas. By preparing for the worst, we can ensure not just the safety and well-being of occupants, but also the long-term viability of our buildings.

Designing a building with resilience in mind is like crafting a ship to withstand the stormy seas. It is challenging, yes, but it ensures that our creations can endure and recover from the impacts of climate change, standing the test of time.

Economic impacts from climate events can be severe when damage is not fully repaired, causing people to move away from the affected area. Let us consider the devastating impacts of Hurricane Katrina in 2005 on New Orleans. The city's economy was crippled as people fled, and damaged buildings and infrastructure were left unrepaired. One key way to reduce such economic damage is to adopt stringent building codes that require the use of materials and construction techniques resistant to the specific types of risks inherent to each location. For example, in areas prone to hurricanes, building codes might require hurricane-resistant windows and wind-resistant roofs. Such codes not only reduce the risk of damage to buildings and infrastructure but also ensure the long-term viability and resilience of communities.

Investing in infrastructure and other measures to reduce the impacts of extreme events is another way to mitigate the economic damage from climate risks. Early warning systems, flood control measures, and other types of natural infrastructure often play an underappreciated role in mitigating climate risks. Fuel breaks and fire-resistant vegetation, for instance, can help to reduce the risk of wildfires, acting like buffers between flammable forests and buildings. For flooding, natural infrastructure like sand dunes, marshes, and mangroves can absorb floodwaters and protect communities from rising sea levels, much like a shield against the invading sea.

In places like Houston, Texas, which was severely affected by flooding in 2017 when Category 4 Hurricane Harvey hit, the lack of on-site water retention and zoning controls on impervious surfaces exacerbated the situation. If natural flood control measures such as floodplains, bioswales, and wetlands had been implemented, they could have mitigated the damage caused by the hurricane.

Urban forests, green roofs, and reflective roofs are the champions in mitigating the risks when we think about extreme heat in cities. The green roofs act like a natural air conditioner for buildings, reducing the heat absorbed by the roof. Urban forests not only provide shade but also reduce the urban heat island effect by releasing moisture into the air, like nature's own evaporative cooling system.

Greenbelts and urban forests also serve as protectors against both extreme heat and cold. In places prone to harsh winters, snow fences and windbreaks can reduce wind chill and protect against extreme cold. Urban forests can absorb excess rainfall, reducing the risk of flooding in areas prone to heavy precipitation, while also mitigating the heat island effect.

Managed retreat offers another tool in our arsenal against climate change. It involves strategically relocating buildings and infrastructure out of areas at substantial risk due to climate change. A classic example of this approach can be seen in the managed retreat of part of the Matatā township in New Zealand following two climate change-related landslides in 2005. Not only does managed retreat protect people and property from immediate harm, it also reduces the long-term costs of disaster recovery and rebuilding.

But, as mentioned earlier, all these measures can be expensive and complicated, often involving intricate negotiations among multiple stakeholders. There is a delicate balance between ensuring safety and maintaining the cultural, social, and economic fabric of a community. It is important to note that these approaches are not one-size-fits-all, but rather must be tailored to local conditions and the specific risks they face.

The future of resilient design is exciting and challenging. As designers and planners, we are not just creating buildings and cities but shaping the future of communities and, in the process, combating climate change. The task is monumental, but the potential rewards are immense, not just in terms of safety and security, but also in the promise of sustainable, resilient, and vibrant communities.

1.1.4 Changing Regulatory Environment

As the climate change crisis grows more severe, energy codes and regulations around the world are constantly evolving to meet more stringent targets for energy efficiency and GHG emissions reduction. These changes can have a significant impact on the design and construction of buildings, as they often require the use of innovative technologies and materials as well as more energy-efficient systems and appliances.

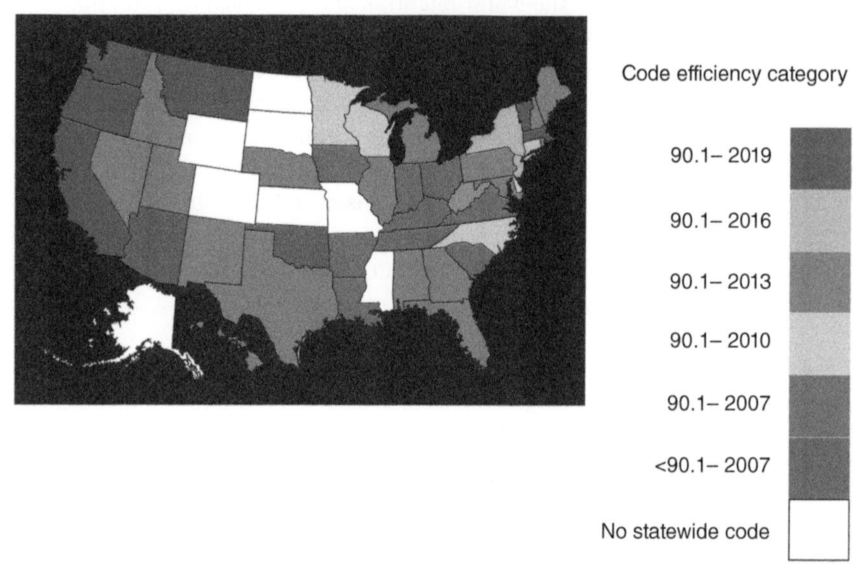

Assessment of the US states' adoption of energy codes in comparison to recent versions of ANSI/ASHRAE/IES Standard 90.1, the nation's model code for commercial structures. Source: Building Technologies Office. 2022 / Department of Energy (DOE) / Public Domain.

One example of these changes is the International Energy Conservation Code (IECC). It is a model building code that establishes minimum requirements for energy-efficient building design and construction. The IECC is developed by the International Code Council (ICC) and is updated every three years to reflect the latest energy-saving technologies and techniques.

One of the main sources of information and guidance used in the development of the IECC is the ASHRAE 90.1 standard, which is published by the American Society of Heating, Refrigerating, and Air-Conditioning Engineers (ASHRAE). The ASHRAE 90.1 standard provides detailed requirements for energy-efficient building design, including requirements for lighting, HVAC systems, and building envelope performance.

The IECC adopts the ASHRAE 90.1 standard as a reference document, and then states and jurisdictions can adopt the IECC model code with their own amendments. This allows states and jurisdictions to tailor the IECC to their specific climate and energy needs while still maintaining a baseline level of energy efficiency.

The IECC and ASHRAE 90.1 are both important tools for improving the energy efficiency of buildings, and they work together to set standards and guidelines that help reduce energy consumption and GHG emissions from the built environment. Many other countries also base their codes on ASHRAE 90.1 making each new version an incredibly influential document.

Another example is BREEAM, or Building Research Establishment Environmental Assessment Method, a sustainability assessment method for buildings in the United Kingdom. It is used to evaluate the environmental performance of buildings, infrastructure, and developments and is widely recognized as a leading sustainability assessment method in the construction industry.

It was developed by the Building Research Establishment (BRE), an independent research organization that focuses on sustainability and environmental performance in the built environment. The BREEAM assessment method covers a wide range of environmental and sustainability issues, including energy use, water use, materials, transport, land use, waste, pollution, health, and well-being. It is based on a set of criteria that are divided into categories, such as energy, transport, water, and materials. It evaluates the building's performance in each of the categories and then assigns it a rating, which can range from "Pass" to "Outstanding."

It is widely used in the United Kingdom and it is also recognized internationally. Thus, it has been adopted by a number of countries around the world, including the Netherlands, Germany, Poland, Sweden, and the United Arab Emirates. In the United Kingdom, BREEAM is often used as a benchmark for sustainability in the construction industry, and it is often a requirement for public sector building projects. Perhaps its most significant impact on the decarbonization of buildings comes from setting a high standard that encourages the construction industry to adopt more energy-efficient and low-carbon technologies and materials reducing the carbon footprint of buildings. It has also encouraged developers to consider the environmental impact of their projects and to adopt more sustainable design and construction practices.

Standards can take you only so far. A truly titanic shift has occurred in the United States with the passage of the Inflation Reduction Act (IRA) in 2022. It is a major piece of legislation aimed at reducing climate pollution and promoting clean energy technologies in the building sector. By investing over $50 billion into clean energy technologies and improvements, the IRA aims to lower energy bills; improve the health and safety of homes, workplaces, and schools; and prioritize the delivery of these benefits to low-income and environmental justice communities. These improvements include the use of efficient electric appliances, weatherized homes, rooftop solar panels, residential geothermal systems, low-carbon building materials, and more.

According to analysis by the Rocky Mountain Institute (RMI), the IRA could drive millions of retrofits, upgrades, and clean technology installations in the building sector. These estimates based on the IRA include uncapped building retrofit tax credits that extend for a decade, potentially enabling even more energy efficiency and electrification retrofits than currently envisioned. In addition, the law includes more than $30 billion in flexible GHG reduction spending for states and cities, which will allow the federal government, the largest real estate holder in the United States, to address the GHG footprint of its properties. Funding for embodied carbon labeling and Environmental Product Declaration (EPD) systems for construction materials will also help develop the market for lower embodied carbon in buildings nationwide.

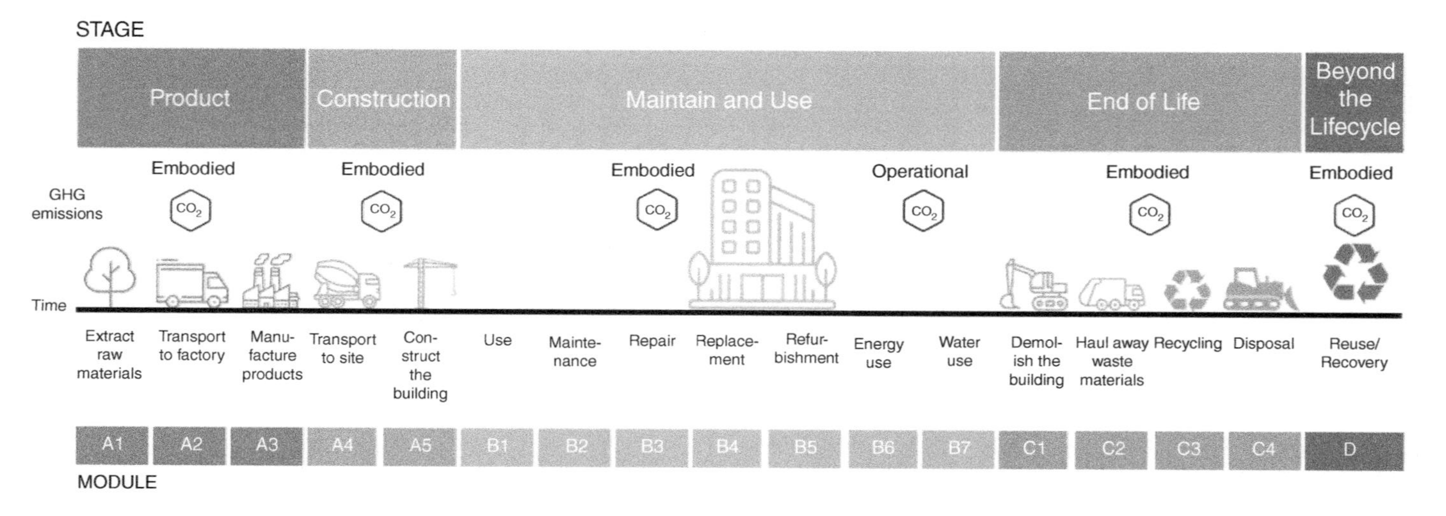

Lifecycle stages of building carbon. Data source: BS EN 15978:2011. Source: Adapted from Bowles, Cheslak, and Edelson 2022.

One of the key ways in which the IRA aims to transform the building sector is using financial incentives that encourage individuals and businesses to invest in clean, efficient alternatives for their homes and workplaces, rather than mandating these changes. These incentives include rebates of up to $14,000 for low-income households to switch from polluting gas to clean energy, as well as funding for the installation of efficient electric heat pumps, induction cooktops, insulation, windows, doors, and upgraded electrical panels and wiring.

The IRA could have a significant impact on reducing climate pollution in the building sector, with estimates suggesting that it could reduce emissions by anywhere from 33 to 100 million metric tons, equivalent to between 10% and 30% of the sector's 2030 goal to cut emissions in half. That would be a major achievement! However, the full impact of the IRA will only be seen over time as it is implemented and more data becomes available.

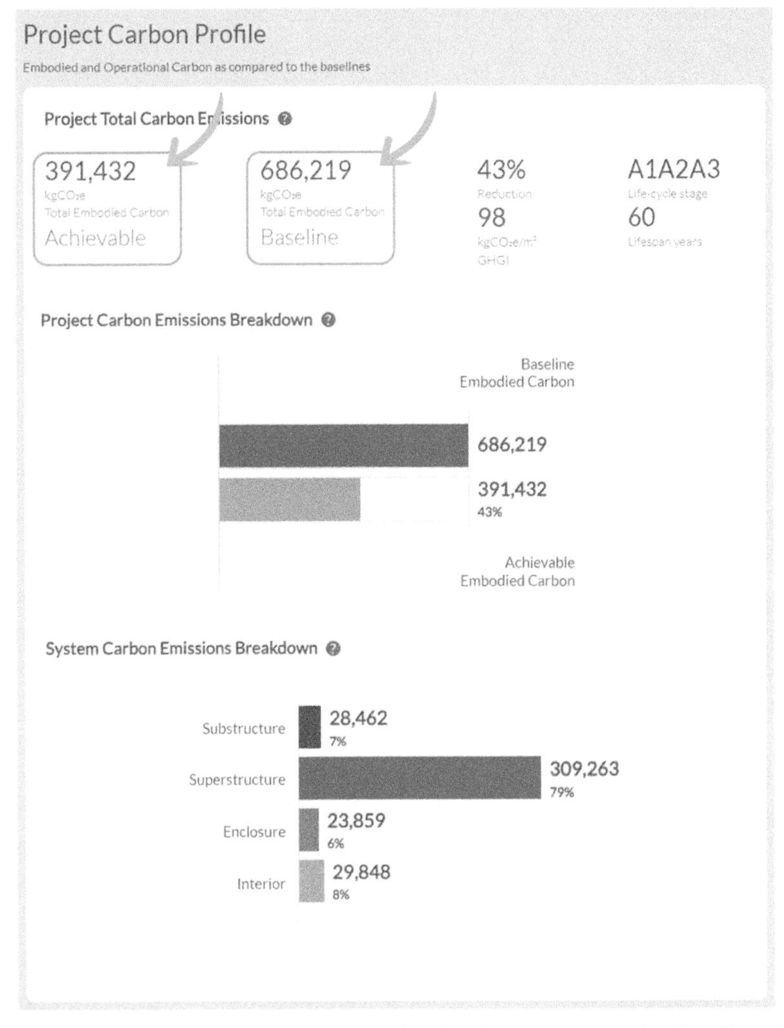

Project Carbon Profile – calculating embodied and operational carbon, as compared to baselines. Source: cove.tool.

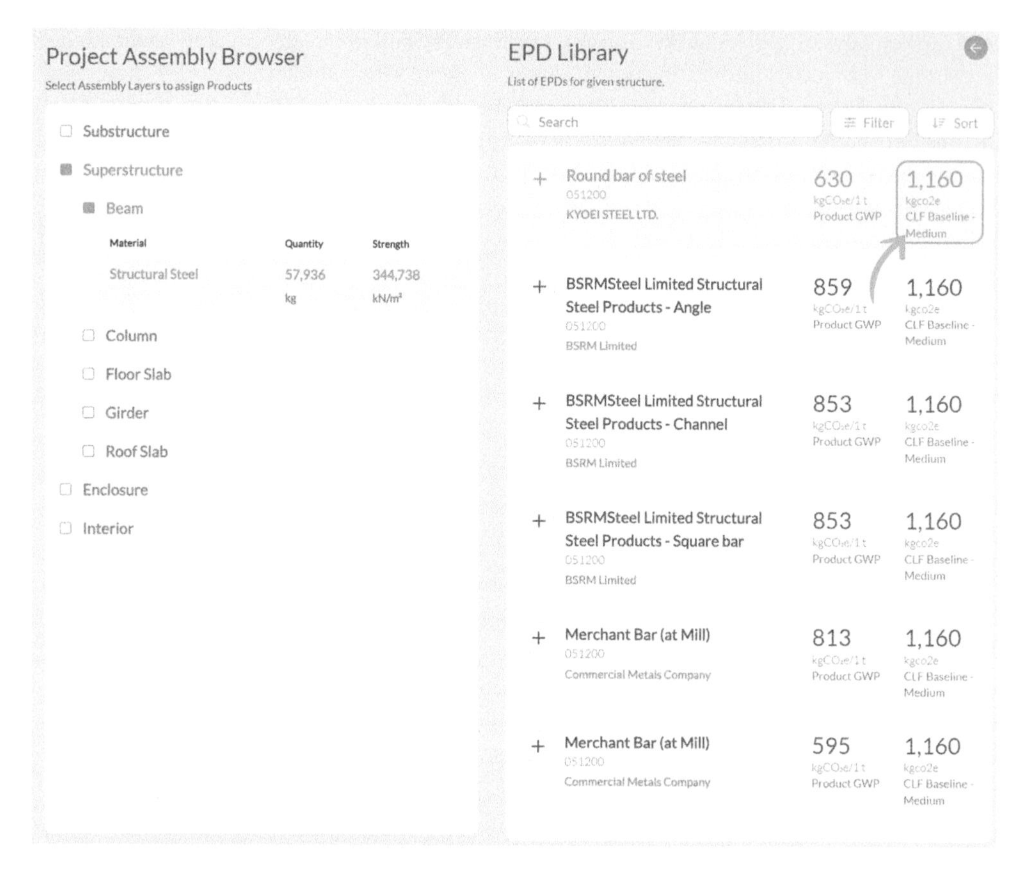

Environmental Protection Declaration (EPD) Library inside of the Project Assembly Builder. Source: cove.tool.

In Europe, the European Union (EU) has set ambitious targets for energy efficiency and renewable energy use and has implemented several policies and regulations to help achieve these goals. One of these is the Energy Performance of Buildings Directive (EPBD), which sets energy efficiency standards for new and existing buildings. The EPBD has been implemented in various forms in different countries within the EU and has helped drive the adoption of energy-efficient technologies and practices in the building sector.

Perhaps one of most monumental regulations starting in 2023 is the EU's Carbon Border Adjustment Mechanism (CBAM). It is a climate measure designed to prevent carbon leakage and support the EU's increased ambition on climate mitigation while also ensuring compliance with World Trade Organization (WTO) rules. Carbon leakage occurs when companies based in the EU move carbon-intensive production abroad to take advantage of laxer environmental standards, or when EU products are replaced by more carbon-intensive imports. This can shift emissions outside of Europe and undermine global climate efforts. The CBAM aims to equalize the price of carbon between domestic products and imports, encouraging producers in non-EU countries to green their production processes.

Initially, the CBAM will apply only to a select number of goods at high risk of carbon leakage: iron and steel, cement, fertilizers, aluminum, and electricity generation. A reporting system will be in place as of 2023 for these products, with importers beginning to pay a tax in 2026. The CBAM will be phased in gradually in hopes of providing businesses and especially manufacturers with time to adjust.

The CBAM has significant implications for the building design and construction industry, as many of the targeted goods are used in construction. The financial adjustment applied to imports may increase the cost of these materials, potentially affecting the overall cost of building projects. Additionally, the CBAM may incentivize the use of domestically produced materials, as they will not be subject to the same financial adjustment. This could lead to a shift toward using more locally sourced materials in building projects, potentially supporting the development of local industries and supply chains. It will also force global suppliers to modify their energy sources to remain competitive.

By equalizing the price of carbon between domestic and imported goods, it aims to ensure that the EU's climate objectives are not undermined by companies relocating to countries with less stringent environmental policies. The impact on the building design and construction industry, particularly through locally sourced materials, is just one aspect of this important policy.

Other countries around the world have also implemented their own energy codes and regulations, including Canada, Australia, and Japan. These codes and regulations often draw on best practices from around the world and are designed to help reduce energy use and GHG emissions in the building sector.

In addition to energy codes and regulations, there are also several voluntary programs and initiatives encouraging the adoption of energy-efficient practices in the building sector. One example of this is the Leadership in Energy and Environmental Design (LEED) program, which provides certification for buildings that meet certain standards for energy efficiency and sustainability. LEED has become a widely recognized symbol of green building and has helped drive the adoption of energy-efficient technologies and practices around the world.

1.1.5 Why People Do Not Take Action

As the world struggles to come to terms with the devastating realities of climate change, it is increasingly clear that a lack of action on this issue is a major barrier to progress. Despite overwhelming scientific evidence and consensus, many people are still failing to take the necessary steps to address this global crisis. So why is this the case?

Source: Adapted from World Green Building Council. Beyond the Business Case: Why You Can't Afford to Not Invest in a Sustainable Built Environment. P.15.

It is believed that near misses in games of chance, such as lotteries or slot machines, can encourage future play even when the probability of winning remains constant. This is because the occurrence of a near miss may be seen as an encouraging sign, confirming a player's strategy and raising hopes for future success (RL Reid, Psychology of the Near Miss). Nowhere is this more important than in the inability to take action on climate change. When individuals or communities experience "near misses" with the negative consequences of climate change, such as a narrow escape with extreme weather, it encourages a perversely false sense of security and a belief that the risk is not as severe as it is. This mistaken invincibility leads to a lack of action on climate change, as individuals or communities feel that they have "dodged a bullet" and are no longer in danger. Near misses discourage action as survivors rebuild and shift the baseline of what is "normal" weather, so we must understand that sustained climate action will not happen due to people experiencing it. Only through consistent, earnest, and prolonged persuasion can we address the issue and advocate for change in the face of seemingly successful avoidance of negative consequences.

To get people collectively moving, we must recognize that a lack of understanding of the magnitude or awareness of the issue is the primary driver of inaction in many communities. Some people may be unaware of the extent to which human activities contribute to climate change, or they may not understand the impacts that climate change will have on the environment, economy, and society. But the truth is that the longer we wait, the more difficult and costly it will be to address the problem. According to the IPCC, "Since systematic scientific assessments began in the 1970s, the influence of human activity on the warming of the climate system has evolved from theory to established fact." Scientific information taken from natural sources (such as ice cores, rocks, and tree rings) and from modern equipment (like satellites and instruments) all show the signs of a changing climate. From global temperature rise to melting ice sheets, evidence of a warming planet abounds.

Lack of political will is another cause that has historically been a problem as governments may be unwilling to implement policies or regulations that would address climate change. Due to a prioritization of economic or political interests over the environment, climate legislation was often relegated to the back burner. This lack of political will has made it difficult for individuals to act on climate change, as they feel their efforts will be insignificant in the face of larger systemic issues. Thankfully, this is becoming less common every year, as more people vote and select leaders that prioritize climate action. Building and manufacturing regulations have proven especially easy to pass into law given their low impact on individual citizens. Design and construction teams should expect this macro to only accelerate each year toward full decarbonization before 2030 as targets are moved forward to win elections and demonstrate action to people.

Another reason project teams may not act on climate change is a lack of personal or collective agency. Some people may feel that they do not have the power or resources to make a difference, or they may not see themselves as part of a larger collective that can take action. This can lead to a sense of helplessness or resignation, which can discourage people. On project and construction teams, it often takes just one person speaking up and asking questions to galvanize others. Large changes in carbon usually have low impact on the project budget and time. Even better, they can in many cases cost less than business as usual.

There are a variety of psychological and social factors that can discourage people from acting on climate change. For example, people may be more likely to do something about climate change if they feel a sense of personal responsibility for the issue, if they feel that their actions will have a meaningful impact, or if they feel that acting is consistent with their values and beliefs. Conversely, people may be less likely to act on climate change if they feel overwhelmed by the scale of the problem, if they feel that their actions will be insignificant, or if they feel that taking action is inconvenient or costly.

So, what can you do to change hearts and minds? It can be challenging to persuade someone to act on climate change if they hold beliefs about the issue contrary to scientific evidence. In the context of a building design and construction project, it may be necessary to address these beliefs to move things forward. One approach that usually works well is to focus on the practical benefits of taking action on climate change, rather than getting bogged down in debates about the science. Do not use politically charged words like "sustainability" or "green" to describe the strategy. For example, you could emphasize the financial savings that can be achieved through energy-efficient design. Try not to talk about why those extreme events might be happening.

In the past, we have found success in appealing to common values or beliefs that the person holds. If the person is religious, you could highlight the moral implications of climate change by discussing the importance of stewardship of the earth. If the person is politically conservative, you could emphasize the importance of individual responsibility and self-sufficiency which meshes well with resiliency strategies. Ultimately, it may not be possible to change someone's deeply held beliefs about climate change. In these cases, it is necessary to find common ground and look for ways to move forward that are acceptable to both parties. This could include focusing on more practical considerations, such as the cost-effectiveness of unique design options or the potential benefits to the community.

One cause of the inaction is the common misconception that decarbonization is too expensive or too difficult to achieve. Many people believe that it would require significant financial sacrifices, disruptions to the project time, and that the costs of retrofitting or rebuilding existing buildings would outweigh the benefits. The reality is that decarbonization can save money in both the short and long run by reducing energy costs and increasing the value of properties. Both market forces and regulations are rapidly moving in this direction, so it is vital for firms, contractors, and manufacturers to get aligned on true costs and adopt new tools and processes. Those that adapt stand to gain dramatic financial benefits in the decarbonized economy.

1.2 What is the Business Opportunity for the AEC Industry?

1.2.1 Why Should Building Owners Care?

	Number of companies with assets at high physical risk on at least one indicator	Market cap of companies with assets at high risk $US trillion	Revenue of companies with assets at high risk $US trillion
S&P/TOPIX 150	40 (27%)	$1.3 (42%)	$2.1 (55%)
S&P/ASX 200	88 (44%)	$0.9 (89%)	$0.4 (82%)
S&P GLOBAL 1200	521 (43%)	$27.3 (66%)	$16.6 (66%)
S&P EUROPE 350	101 (29%)	$4.3 (52%)	$3.2 (48%)
S&P 500	297 (59%)	$18.0 (72%)	$9.0 (74%)

Climate change is a global crisis that poses significant risks to the environment and economy. As a result, building owners are increasingly recognizing the importance of developing, owning, and operating low-carbon buildings. There are several key reasons why building owners care about this issue.

Primarily, there is a growing awareness that climate risk is financial risk. As the impacts of climate change become more severe, the cost of repairing damage caused by extreme weather events and other natural disasters is likely to increase. This is a major concern for building owners, as these costs can have a significant impact on their bottom line. In addition, financial institutions are increasingly taking climate risk into account when deciding whether to provide loans for construction projects. This means that loans for low-carbon buildings are likely to be more readily available and have more favorable terms than those for buildings with higher carbon emissions.

Another reason building owners care about low-carbon buildings is that they can help reduce the overall carbon footprint of their building portfolios. By investing in energy efficiency measures and renewable energy technologies, building owners can significantly reduce their carbon emissions and make a meaningful contribution to the fight against climate change. In addition, low-carbon buildings are often more resilient to the impacts of extreme weather events, which can help protect the value of these assets over the long term.

Typically, building owners care about low-carbon buildings because they are good for business. In today's socially conscious world, consumers and tenants are increasingly looking for companies and buildings that are committed to sustainability. By investing in low-carbon buildings, building owners can not only attract and retain customers and tenants, but they can also enhance their brand reputation and differentiation in a crowded market.

In addition to these financial and business considerations, building owners are also influenced by a variety of economic, social, and political factors. For example, the adoption of stricter energy codes and regulations around the world is driving the demand for low-carbon buildings. In the United Kingdom, for example, the BREEAM sets stringent sustainability standards for new construction projects and major renovations. Similarly, the IECC in the United States sets energy efficiency standards for new buildings and major renovations and is updated every three years to reflect the latest best practices. In addition, building owners can also leverage the economic, social, and political macro-trends in favor of low-carbon buildings. For example, as governments around the world increase their carbon reduction targets and implement policies to incentivize low-carbon building projects, building owners can take advantage of these policies to secure funding and other forms of support for their projects. Additionally, building owners can also tap into the growing demand for low-carbon buildings from tenants and buyers, who are increasingly conscious of the environmental impacts of the buildings they occupy or purchase. By positioning their buildings as low carbon, building owners can differentiate themselves from competitors and potentially command a higher price or rental rate.

Furthermore, building owners can also use their influence and leadership to advocate for low-carbon policies and practices at the local, regional, and national levels. By working with policy makers, industry organizations, and other stakeholders, building owners can help shape the regulatory and market conditions that support the transition to low-carbon buildings. This can include supporting the development of energy efficiency standards and regulations and participating in advocacy efforts to raise awareness about the benefits of low-carbon buildings.

Overall, there are many compelling reasons why building owners should care about developing, owning, and operating low-carbon buildings. By doing so, they can not only mitigate the financial risks associated with climate change, but also take advantage of the economic, social, and political opportunities that are emerging as the world transitions to a low-carbon future. By embracing these trends, building owners can not only contribute to the fight against climate change but also create value for their businesses and the communities they serve.

1.2.2 How Are Architects a Part of the Solution?

Climate responsive design and performance-based approaches in architecture are key to addressing the challenges of global warming and climate change. They represent the first step in the building process and can give momentum to this new approach. It focuses on designing high performance buildings. However, some architects have been critical of this approach, arguing that they turn architects from designers into engineers and that there is a tension between beauty and rationality in architecture theory.

One of the key reasons for the tension between beauty and rationality in architecture theory can be traced back to the Enlightenment period, where architects were grappling with the question of what constitutes architecture versus just building. Just like today, architects were also influenced by the ideas of the Scientific Revolution and the Enlightenment, which placed a heavy emphasis on reason and logic. This eventually led to a focus on functionality and efficiency in architectural design, rather than aesthetics.

This culminated in the modern movement in architecture that emerged in the early 20th century, which was characterized by a focus on functionality, simplicity, and the use of new technologies and materials. Architects of the modern movement rejected historical and ornamental elements in favor of a sleek, minimalist aesthetic. They sought to create a new architecture that was rational, objective, and universal, and that would reflect the technological and social advancements of the time.

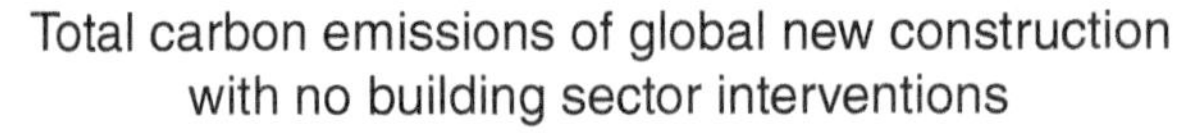

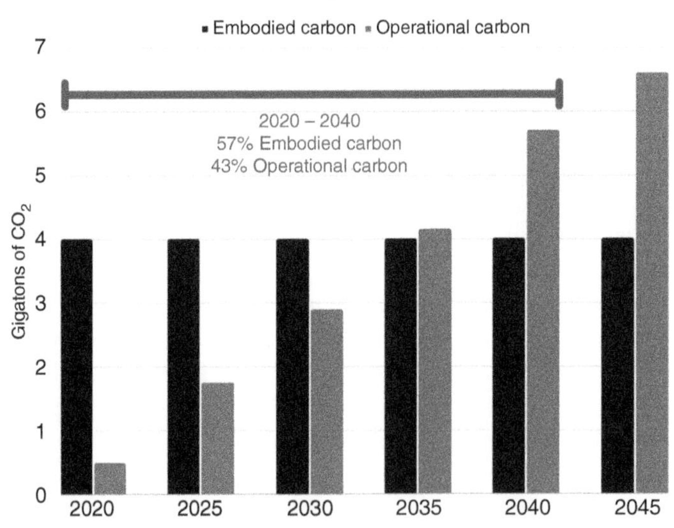

"Total carbon emissions of global new construction with no building sector interventions". Source: Adapted from Architecture 2030.

However, by the 1950s and 1960s, Modernism in architecture began to face criticism for its lack of attention to social and cultural context, and for its failure to address the needs of the inhabitants. Writers like Jane Jacobs criticized the modernism's disregard for the historical and cultural context of cities and the way they designed buildings with little consideration for the needs of the community. Additionally, many modernist buildings had poor performance in terms of energy efficiency, comfort, and livability since building performance simulations were impossible to do when they were designed and built given the limitations of pre-digital workflows.

The failure of Modernism in architecture was also driven by the changing social and political context of the time. The post-war period was marked by a growing sense of disillusionment with the technological and social progress promised by Modernism. Additionally, the civil rights movement and the rise of feminist and environmental movements brought attention to the ways in which architecture had been used to maintain social and economic hierarchies, calling into question a rational design philosophy.

According to Alberto Perez-Gomez in his book "Architecture and the Crisis of Modern Science," during the Enlightenment, architects faced a similar crisis of identity as they tried to reconcile the ideals of reason and beauty in architecture. They struggled to find a balance between the functional requirements of a building and the aesthetic aspirations of architecture. As a result, architecture began to be viewed as a mechanical art focused on the technical aspects of building design rather than the symbolic and expressive aspects of art. In some ways, we could view the end of Modernism as a natural swing of this same ideological pendulum.

Starting from the 1960s, architects began rejecting the rational underpinnings of design in favor of a "postmodern" ideology. Postmodernism in architecture, as in other fields, emphasizes self-referentiality, a theory of knowledge based on moral relativism, pluralism, and irreverence. Postmodern architects focus on the effects of ideology, society, and history on culture and reject the idea of objective reality, morality, truth, human nature, reason, language, and social progress. They also reject universalist ideas of architecture, such as function over form, and instead focus on the building's expressive and symbolic qualities.

Even today many architects are still educated in Postmodernist ideologies and thus reflexively reject high performance or climate responsive designs. Is decarbonization rejected because it represents a worldview associated with Enlightenment ideas of objective reality? Despite rationality being the basis of architectural theory and practice for centuries, most architects today are educated to be skeptical of solutions that are valid based on the context and, instead, focus on narratives. These truths could be based on groups, cultures, traditions, or races. While it is important to focus on individuals and avoid the excesses of Modernism, a mental construct that rejects the possibility of building science guided approaches will not be our best bet in the fight to slow down climate change.

Another reason architects might hesitate against climate responsive design approaches is that they prioritize formal design over function architects fear that performance-based processes could lead to a homogenization of architectural styles. This is a valid criticism if designers do not simulate but rely on rules of thumb. Poorly realized projects with a performance focus often obsess over the technical aspects of building design, such as insulation, natural ventilation, and daylighting, rather than balancing them against the aesthetic qualities of the building. This can lead to buildings that are highly energy-efficient but lack the aesthetic appeal and sense of place that is so important in architecture.

It should go without saying that "high-tech" Postmodernist architecture, which emerged in the 1970s, is not the answer to climate action. High-tech is characterized by its use of advanced technology and materials, such as steel and glass, to create a sleek and minimalist aesthetic. This movement was heavily influenced by the beautiful work of architects such as Richard Rogers, Norman Foster, and Renzo Piano, who were interested in using technology to create buildings that look efficient, functional, and expressive. However, high-tech architecture has been criticized for being more focused on the aesthetic aspect of the "machine-like" appearance of the buildings rather than their performance. This is because the high-tech aesthetic was often used as a decorative element rather than being integrated into the building's design as a functional component. While they may look technologically advanced, they lack actual metric-based performance. The aesthetic suggested highlighting the difference between the approach of high-tech postmodern architecture and high-performance buildings that can take any form. A data-driven high performance approach emphasizes the integration of technology, materials, and design in a holistic and functional way, rather than using them as mere aesthetic elements.

Does performance-based design limit creativity and experimentation in architectural design? It is important to note that throughout history, architects have been able to successfully incorporate other meaningful ideas such as building codes, ADA access requirements, and fire codes into their work without compromising the design quality. These regulations have become a part of the design process and architects have found ways to integrate them into their design while still maintaining creative freedom. The same can be done with decarbonization strategies, by finding ways to incorporate renewable energy sources, energy-efficient design, and sustainable building materials into their designs in a way that enhances the aesthetic appeal and sense of place of the building. With the right approach and perspective, the push for decarbonization is an opportunity. Architects can push the boundaries of what is possible and experiment with new forms, materials, and technologies while imbuing them with values of our time.

Indeed, it is important to remember that climate responsive design and performance-based approaches do not have to be at odds with aesthetic considerations. These approaches can be incorporated into the design process in a functional and beautiful way. For example, architects can use passive cooling and heating techniques, such as shading devices and thermal mass, to create buildings that are both energy-efficient and visually appealing. They can also incorporate natural materials such as wood and stone into their designs to create a sense of connection and belonging between the building and its surroundings.

Moreover, architects should look outside of ideas of globalized architecture "culture" to the traditional designs of indigenous peoples and the vernacular architecture, which often seamlessly integrate form, beauty, and function to respond to the local climate. The vast and diverse array of building traditions from around the world, developed by different cultures over thousands of years, are finely balanced with the local environment. These building traditions are not only a rational response to the local climate but also an artistic expression of the society's culture, making it incredibly important to be humble and learn alternative strategies. Indigenous architecture is often characterized by its use of natural and locally sourced materials and its close connection to the natural environment.

One example is the Iroquois long houses, traditional homes of the Haudenosaunee people of the north-eastern region of North America. These houses were constructed using bark and wood, and they were designed to be highly efficient, with a central fireplace for heating, and openings at the top of the walls to allow for natural ventilation. Additionally, these long houses were built to accommodate several families and were a symbol of the community's values of sharing and cooperation.

Cross-section of a longhouse built by Haudenosaunee ("People of the Long House"), also known as the Iroquois. Source: New York State Museum / Public domain.

Another example is the Indian Mughal architecture, which developed during the Mughal Empire in the Indian subcontinent. This architectural style is characterized by its use of intricate geometric patterns, arches, and domes, and it often incorporates elements of Islamic, Persian, and Indian architecture. Mughal architecture is known for its use of gardens, water features, and other elements that helped to create a microclimate within the building, making it more comfortable.

Indian landmark Gadi Sagar located in Jaisalmer, Rajasthan, India. Source: Dmitry Rukhlenko/Adobe Stock Photos.

Vernacular architecture is the architecture of the common people, built by local traditions, materials, and techniques. It is the architecture of a place and a time, born out of the specific needs and resources of a community. Vernacular architecture often seamlessly integrates form and function to respond to the local climate and culture. It is an architecture that emerges organically, through the process of trial and error, rather than being designed by trained architects.

Burano island canal filled with colorful houses in Venice, Italy. Source: stevanzz/Adobe Stock Photos.

In contrast, high design by trained architects often prioritizes aesthetics over function and can be less adaptable to the specific needs of a place and its climate. According to Stewart Brand in his book "How Buildings Learn," vernacular architecture is better suited for the long term because it has been shaped by the forces of trial and error over time. These buildings are more adaptable to change, and they can evolve and adapt over time to meet the changing needs of the community. One of the key characteristics of vernacular architecture is its local adaptation to climate. Vernacular buildings are often designed to take full advantage of natural cooling and heating systems, such as natural ventilation, shading, thermal mass, and insulation. This results in buildings that are not only energy-efficient but also comfortable and healthy for the inhabitants. Additionally, vernacular architecture often incorporates local materials and building techniques, further increasing its adaptability and sustainability.

Added to that, high design by trained architects may not take into account the specific climate and local materials, leading to buildings that are less energy-efficient and less adaptable to the local context. Architects should not ignore these traditions as a source of inspiration for creating buildings that are not only energy-efficient and sustainable but also beautiful and expressive. These techniques and styles seamlessly integrate form and function to respond to the local climate and culture, and they are a testament to the ingenuity and resourcefulness of people who have been building for centuries in a specific location. Architects can learn from these examples and apply these principles to create buildings that are not only functional but also reflective of the society's culture and values.

It is a balance that can be achieved by taking inspiration from traditional and vernacular architecture and being mindful of the current state of climate change and the role of architecture in society. Rational and climate-responsive designs do not have to be at odds with aesthetic considerations. Historically, there are alternative ideologies that also support highly energy-efficient and sustainable outcomes. One of them is championed by architects like Michael Sorkin and Buckminster Fuller; it consists of a system-of-systems approach that emphasizes the interdependence of the building with its surroundings, the broader community, and the environment. When thought of this way, architects have a strong design theory framework to rely on since this approach allows architects to create buildings that are not only energy-efficient and sustainable but also responsive to the specific needs and context of the community. The systems method emphasizes flexibility and adaptability over time, as the building and its surroundings evolve, making it theoretically perfect for our changing planet.

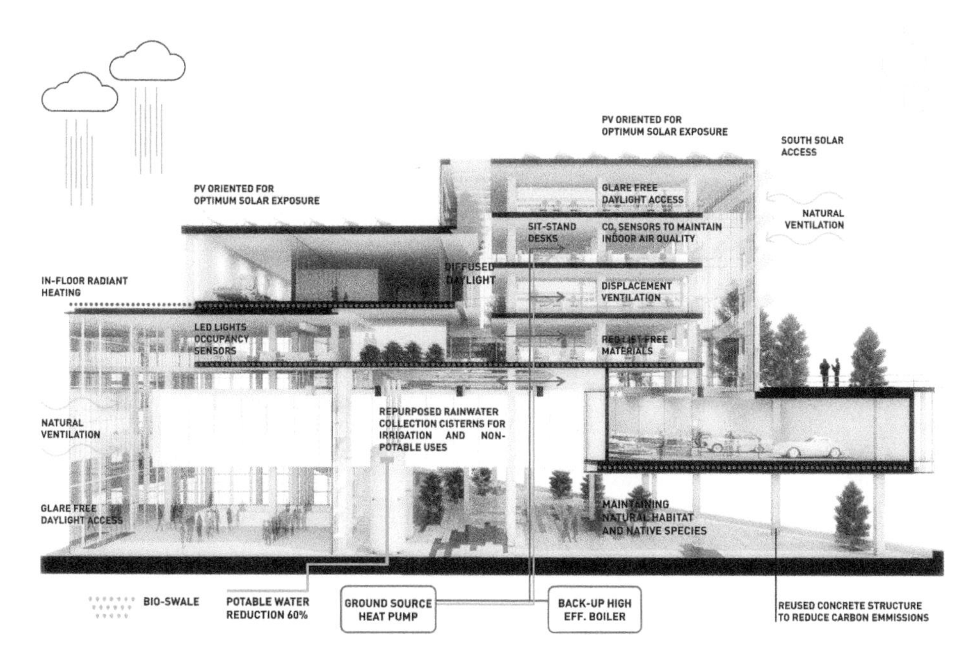

Building massing energy infographic Source: cove.tool.

It is worth noting that the current state of climate change calls for a new level of environmental responsibility in architectural design, and this responsibility can be a source of inspiration and meaning as well as a moral imperative. Architects have the opportunity to create buildings that not only address the urgent need for sustainability and energy-efficiency but also reflect the cultural values of our society. This innovative approach to design allows architects to push the boundaries of what is possible, experiment with new forms, materials, and technologies, and create buildings that are not only functional, but also beautiful and meaningful. The integration of environmental responsibility in the design process can elevate architecture to a higher purpose and give it a role in shaping a more sustainable future. Architects should embrace this responsibility as an opportunity to create buildings that are not only environmentally friendly but also inspiring for the communities they serve.

1.2.3 What Can MEP Engineers Do?

Mechanical, electrical, and plumbing (MEP) engineers are not just technical experts; they are the virtuosos who bring the concept of a "machine for living" to life. When they step into a building, they do not just see the physical structures; they see a dynamic system that's almost organic in its complexity. These engineers perceive each building as a living, breathing entity, complete with its own circulatory (plumbing), nervous (electrical), and respiratory (HVAC) systems.

In the spirit of envisioning buildings as machines for living, MEP engineers go beyond mere utility. They aim for a harmonious integration of form and function, where every mechanical system is not just installed but elegantly orchestrated. The focus is not merely on providing solutions for heating, lighting, or water supply, but on how these systems interact and how they can be optimized to enhance the quality of life for the inhabitants. The result is a living space that is more than the sum of its parts, a harmonious ensemble where mechanical efficiency meets human comfort, and where the built environment attains a seamless symbiosis with human needs and activities. Just as in a well-conducted orchestra where no single instrument dominates but each contributes to an overall harmonious sound, the MEP systems in a well-designed building contribute to a balanced and comfortable living environment.

The 20th century marked a pivotal era for mechanical engineering in the context of modern buildings, a revolution ignited by Willis Carrier's invention of modern air conditioning in 1902. This was more than a feat of engineering; it was a societal game-changer that fundamentally altered our relationship with indoor spaces. Before Carrier, buildings were largely at the mercy of the environment, with design considerations often dictated by the need to manage heat and circulate air naturally. The advent of modern air conditioning shifted this paradigm, granting architects and engineers unprecedented control over the indoor climate. This opened the floodgates for a torrent of innovations—high-efficiency HVAC systems, building automation, and eventually, sustainable marvels like geothermal and heat recovery systems. These technological advancements, armed with sensors and controls, operate like a building's brain, making real-time decisions to optimize energy usage.

However, the meteoric rise of mechanical engineering capabilities in buildings also introduced a complex dilemma—the tension between function and aesthetics. The newfound ability to control every aspect of a building's internal environment led to an increasing focus on optimizing for measurable, utilitarian outcomes, often at the risk of overshadowing the unquantifiable aspects of human experience within these spaces. Engineers must elevate their work by seeing it as part of a whole building effort to reduce carbon. Assemblages of advanced technologies are not necessarily soulful spaces that engage our senses, celebrate our humanity, and also tread lightly on the Earth. The engineer is armed with the data and means to solve many of the complex trade-offs faced by the design team.

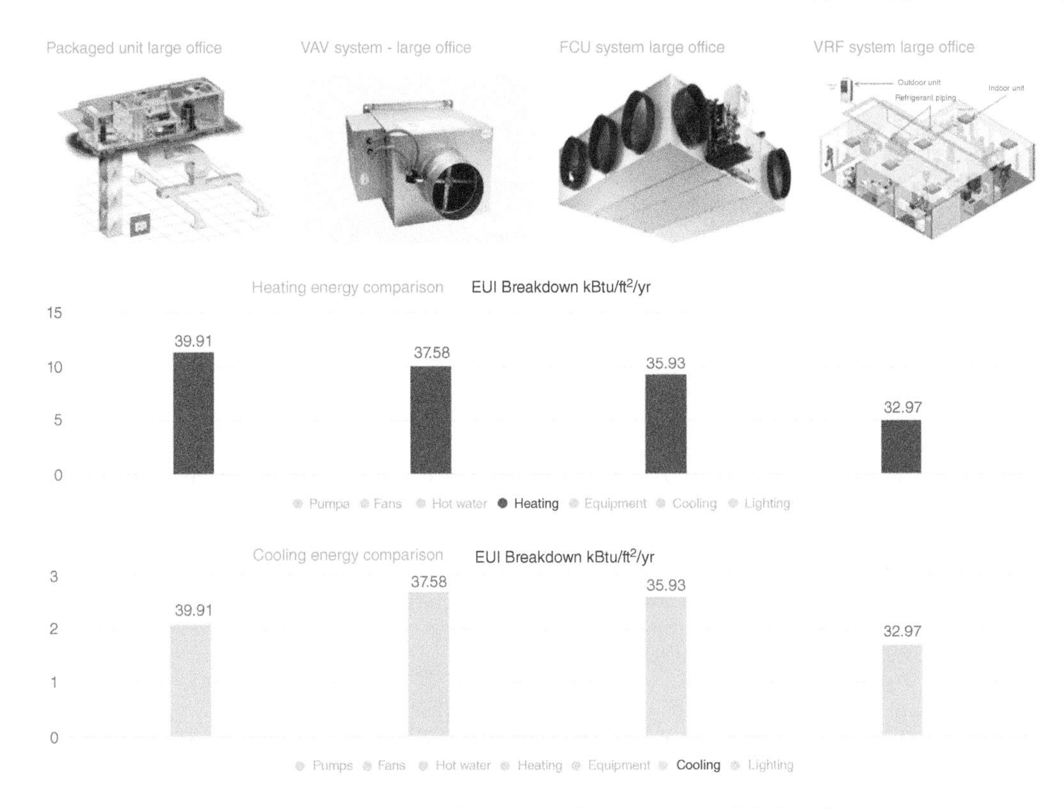

Comparison of different mechanical systems using cove.tool in early stages of design. Source: cove.tool.

The integration of renewable energy sources and resiliency against natural disasters has become a focal point in the evolution of mechanical systems for buildings. Solar and wind energy, for instance, are now being seamlessly integrated into HVAC systems, enabling buildings to generate a portion of their own energy. These renewable sources not only reduce a building's carbon footprint but also offer practical advantages. In the event of grid failures, especially during natural disasters like hurricanes or wildfires, these integrated systems can switch to locally generated power, ensuring that critical services within the building remain operational. This is not just sustainability; it is resilience built into the very mechanical systems that breathe life into modern structures.

Advancements in building automation systems are also contributing to disaster resilience. Smart sensors can now detect a range of environmental factors—from temperature and humidity to smoke and chemical levels—in real time. In case of an emergency, these systems can automatically adjust ventilation, close fire doors, or even alert emergency services, providing an extra layer of safety and resilience. Moreover, the development of energy-efficient microgrids within buildings is another leap forward. These microgrids can operate independently of the central grid, utilizing a mix of renewable energy sources and advanced battery storage solutions. In disaster-prone areas, this ensures not only energy efficiency but also uninterrupted power supply when it is needed most. As we look toward the future, the role of mechanical systems in buildings is clear: they must serve as both stewards of sustainability and guardians of resilience, ensuring that our built environment is not only in harmony with the natural world but also robustly prepared for its uncertainties.

A geothermal mechanical room; various geothermal technologies are showcased, providing contractors with the opportunity to gain knowledge, comprehension, and the ability to communicate the advantages of these technologies to their clients. Source: stevanzz/Adobe Stock Photos.

Embodied carbon has emerged as a crucial frontier in the fight against climate change, and MEP engineers find themselves at the epicenter of this battle. While buildings are often scrutinized for their operational energy use, which accounts for approximately 30% of annual global GHG emissions, the embodied carbon—the emissions associated with construction materials and processes—constitutes an additional 10%. This number becomes even more significant when you consider that MEP systems alone contribute to around 10% of a building's embodied carbon. Materials commonly used in these systems, such as galvanized steel, aluminum, copper, and plastic, are high in embodied carbon content, making the role of MEP engineers pivotal in mitigating a building's overall carbon footprint.

The first step in this critical journey is understanding the comprehensive life cycle impact of the systems being installed. While the Architecture 2030 initiative challenges architects to create more energy-efficient designs, MEP engineers must ensure that the mechanical systems, equipment delivery methods, choice of refrigerants, and even the overlooked aspect of refrigerant leaks are accounted for in life cycle analysis studies. The absence of these considerations often leads to an underestimation of embodied carbon, obscuring the full environmental impact of a building.

One practical approach lies in reducing heat loads within buildings. By doing so, the size of mechanical equipment can be minimized, which not only reduces the structural needs of the building but also leads to lower operational energy usage and, consequently, fewer carbon emissions. The strategic placement of equipment also plays a role. For example, placing heavy equipment on grade can reduce the building's structural load, which in turn decreases its embodied carbon. Moreover, MEP engineers have the opportunity to guide architects toward better product and design choices through simulation tools, making a significant dent in the building's overall embodied carbon emissions.

In the realm of materials, understanding EPDs becomes critical. These declarations provide transparent information about the life cycle environmental impact of products. Coupled with embodied carbon studies, MEP engineers can make data-driven decisions, optimizing not just for cost and energy efficiency but also for lower embodied carbon. The choice of refrigerants is another crucial area of focus, given that they contribute to 2% of global GHGs, opting for low global warming potential alternatives can have a substantial impact.

MEP engineers are no longer just the sub-consultants of the building industry, they are its environmental stewards. By incorporating life cycle analysis into their workflows, optimizing designs for both operational and embodied carbon, and making informed choices about materials and equipment, they have a unique and vital role to play. As the industry moves toward a more sustainable future, the MEP engineer's toolkit must expand to include better simulation software, EPDs, and a deep understanding of environmental impact. This holistic approach is not just good for the planet, it is also becoming a professional imperative.

In the rapidly evolving landscape of sustainable building design, speed and accuracy are of the essence, and this is where the role of simulation in the MEP engineer's toolkit becomes invaluable. Simulation and modeling tools serve as a linchpin in the design and decision-making process, acting like a computational crystal ball that allows engineers to foresee the environmental impact of their choices before they are physically implemented. These advanced software platforms can swiftly evaluate a myriad of variables—ranging from materials and equipment types to system configurations—and calculate their corresponding embodied and operational carbon emissions. This expedited evaluation is not just a time-saver, it is a catalyst for iterative design, enabling architects, engineers, and other stakeholders to quickly refine their strategies based on real-time data.

By offering a dynamic feedback loop, simulation tools empower the design team to explore multiple scenarios, balance trade-offs between energy efficiency and embodied carbon, and ultimately converge on solutions that meet both sustainability goals and performance criteria. Where more "experiments" can be conducted, the design teams can converge faster on the optimal design that minimizes carbon footprint while maximizing building performance. Simulation is not merely an add-on but an integral component in the modern MEP engineer's approach to effectively deliver a project.

Despite the importance of embodied carbon in the building industry, some MEP engineers may be resistant to changing their tools and workflows. The narrative surrounding innovation and decarbonization often assumes a universal eagerness to embrace change. Yet, from a business model perspective, the hesitance among MEP engineers to pivot toward greener technologies and practices is more nuanced. For many engineering firms, their value proposition has long been rooted in proven methodologies and longstanding client relationships, often built around traditional design paradigms. Pivoting toward a focus on sustainability might raise concerns about alienating a core clientele who prioritize cost-effective, tried-and-true solutions over eco-friendly innovations. There's also the matter of market differentiation; in an industry where many firms have already jumped onto the "green" bandwagon, some may question the business advantage of joining a now-crowded field.

But perhaps the most compelling reason lies in the financial mechanics of the traditional engineering business model. Firms are often contracted for specific phases of a project, with a predefined scope and budget that offer little room for experimentation with new, greener technologies. When billing is done on an hourly basis, time spent on researching and implementing novel solutions can be perceived as a financial risk, especially when the long-term ROI of such innovations is not immediately clear or quantifiable in the project's budget. Furthermore, the capital expenditure for new software and training can be significant, and in a business model sensitive to cash flow, such upfront costs may be hard to justify without guaranteed immediate returns.

However, the landscape is changing, and with it the long-term financial equations. Client demand for sustainable buildings is not just a trend but an emerging standard, one that comes with its own set of financial incentives, from tax breaks to premium pricing in the marketplace. Engineers who upskill now are not just meeting a market demand but are positioning themselves to capitalize on a growing niche. In addition, the integration of sustainable practices and technologies can open doors to new revenue streams, such as consulting services for green certifications or long-term partnerships focused on building sustainability and maintenance.

Moreover, innovations like BIM tools are transforming not just the design process but also the business model. By enhancing collaboration and reducing errors, these tools can lead to more efficient project timelines, potentially offsetting the upfront investment costs. They also allow engineers to demonstrate the long-term cost benefits of sustainable features to clients in a compelling, data-driven way, making it easier to sell these features as part of the project scope.

So, while the inertia against change in MEP engineering may be grounded in valid business concerns, the evolving market dynamics make a compelling case for re-evaluating the risks and rewards. Those who proactively adapt their business models to incorporate sustainability and innovation are likely to find themselves at the forefront of an industry in transition, with a competitive edge that is both financially rewarding and socially responsible.

1.2.4 What Can Structural Engineers Do?

Structural engineers are the unsung heroes behind every towering skyscraper, resilient bridge, and sustainable habitat. They are the great enablers of human civilization's progress, from the ancient Roman arches to today's data-driven design algorithms. Take a stroll down history lane, and you will find their fingerprints on every innovation that has made buildings safer, more efficient, and aesthetically appealing.

Imagine the grandeur of ancient Rome—the Colosseum, the Pantheon. Now picture the intricate labyrinths of a Gothic cathedral. What made these architectural feats possible? Arches and domes. These engineering marvels allowed for taller and wider structures, all while efficiently distributing weight and minimizing the need for bulky walls. The result? Not only awe-inspiring aesthetics but also an early nod to sustainability by conserving materials and maximizing natural light.

Fast forward to the industrial age. Enter steel and reinforced concrete—the building blocks of modernity. These materials made our cities reach for the sky, dotting landscapes with skyscrapers and intricate bridges. But it was not just about scaling heights; these materials offered newfound resilience against natural calamities like earthquakes and high winds. The dual benefits of safety and sustainability reduced the overall maintenance and repair costs, a fine balancing act that few could have envisioned.

The 21st century is witnessing another revolution—mass timber and cross-laminated timber. Forget the concrete jungles; think of living, breathing forests transformed into architectural marvels. Timber is not just renewable but also excels in embodying less carbon compared to traditional construction materials like concrete and steel. Add to that its thermal mass properties, and you have got an energy-efficient building that is as green as it is grand.

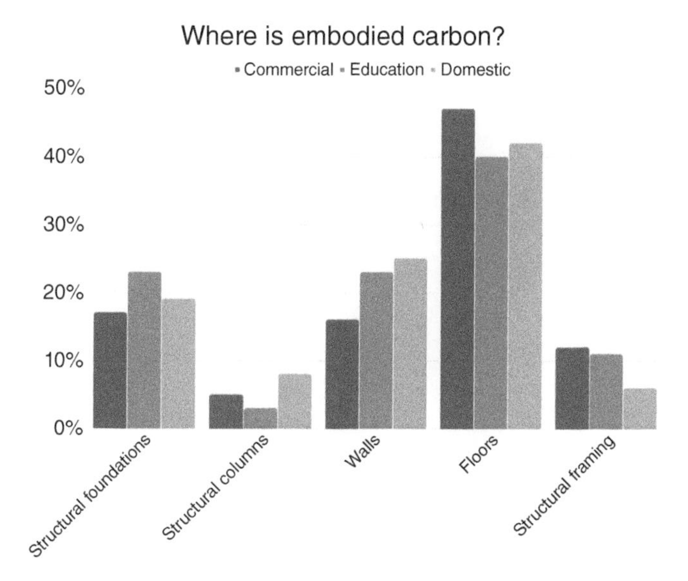

On average, floors account for the largest portion of embodied carbon from structural systems ('Where is the Embodied Carbon?' graph, Thornton Tomasetti).

Recent innovations in structural engineering also include the use of base isolation and energy dissipation systems that reduce the impact of earthquakes and other natural hazards on buildings. These systems use advanced materials and technologies to reduce the amount of force transferred to the building, reducing the risk of damage and increasing the resilience of the structure.

These innovations have not only improved the safety and aesthetic of the building but also made it more sustainable by reducing the amount of materials needed and increasing the resistance to natural hazards. With the current climate change crisis, it is more important than ever for structural engineers to continue to innovate and find new ways to reduce the carbon footprint of buildings and improve their resilience against the increasing extreme weather events. By looking to the past and learning from the successes of earlier generations, engineers can help pave the way for a more sustainable and safe future.

However, despite these innovations, there is a hesitance among many structural engineers to fully embrace the future. Why? A blend of tradition, unfamiliarity with new tools and, quite frankly, a lack of urgency. For some engineers, their outlook on structure as an exercise in just strength and serviceability is honed over decades and steeped in tradition. Adopting new tools and technologies feels like betraying their mentors. They question: "If it ain't broke, why fix it?" This mindset can hamper the adoption of new methodologies designed to mitigate climate change, leading to a cycle of reluctance with the architect's request feeling like an unwanted intrusion into the process.

It is human nature to stick to the familiar, even if it is to our detriment. Engineers are no exception. The lack of exposure to new software and techniques creates a self-imposed barrier. It is not just about learning curves, it is about trust. If you are not comfortable with a tool, you are less likely to see its utility, less likely to invest time in it, and certainly less likely to advocate for its adoption. If there is anything a structural engineer hates, it is a black box. One math error can lead to a collapsed building, so it is important that structural engineers be very risk adverse.

Ignorance is not bliss when it comes to climate change. Many engineers are simply unaware of the embodied carbon footprint in their designs. They may not have easy access to life cycle assessment studies that could enlighten them about the broader ecological impact of their work.

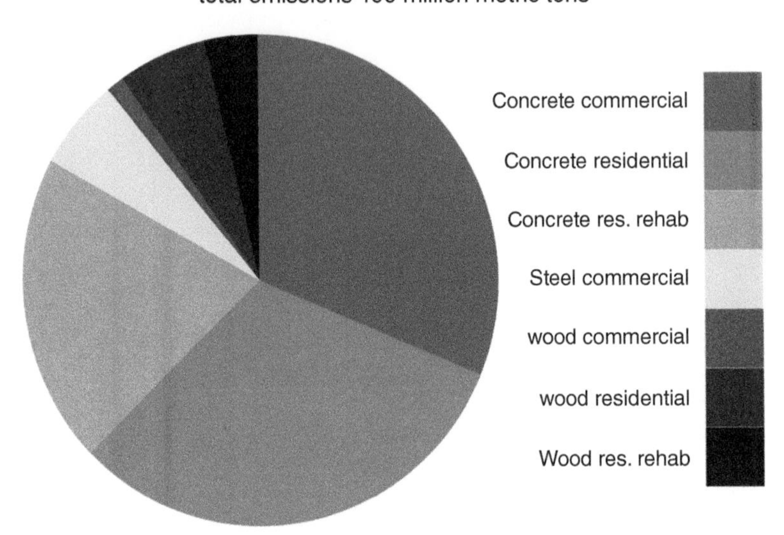

Annual CO_2 emissions associated with structural materials used in new construction in the United States by building sector, adapted for text.

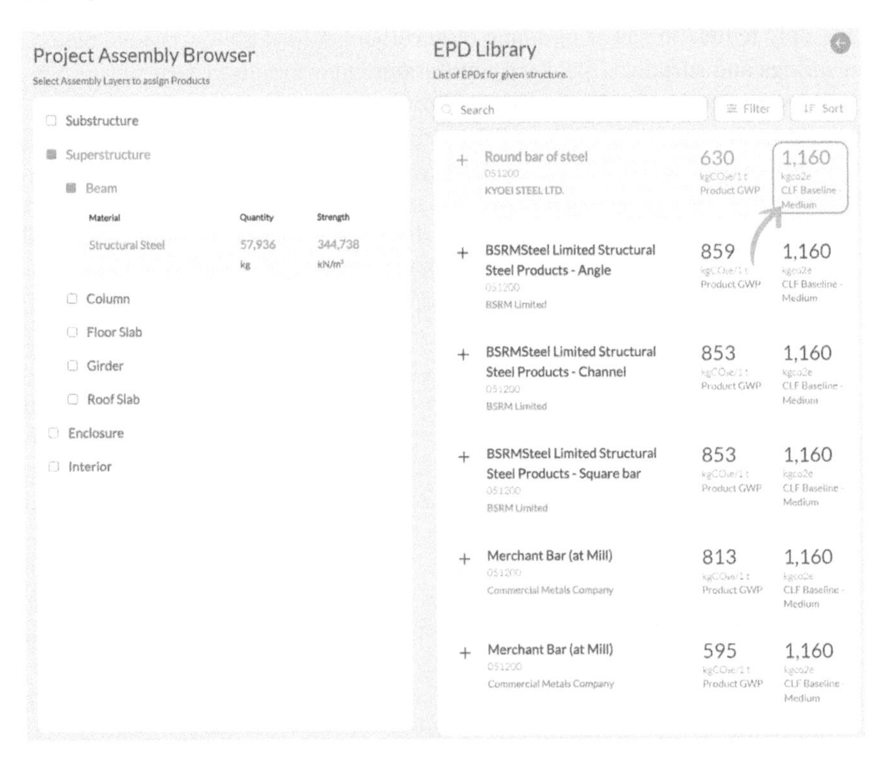

Environmental Protection Declaration (EPD) Library inside the Project Assembly Builder. Source: cove.tool.

The climate crisis is not merely a challenge for future generations; it is an existential threat that demands immediate attention. For structural engineers, this moral imperative to act is intertwined with a compelling business case for sustainable practices. Consider this—by reducing the embodied carbon in your projects, you are not only lessening your environmental impact but also potentially lowering material costs. This creates a win-win scenario where both the planet and the project budget benefit. For example, the use of recycled steel or sustainably sourced timber can bring down both carbon footprint and material expenses. These savings can be a major selling point for clients and could even be the competitive edge that wins your team a project over another firm.

Moreover, the financial gains from sustainable practices are not limited to material costs. By adopting advanced simulation tools and software that can accurately predict the environmental impact of different materials and designs, engineers can optimize their plans from the outset. This cuts down on expensive modifications and overruns down the line. Imagine the cost savings when you can avoid the domino effect of changes that often occur during the design phase. By running multiple scenarios through a simulation tool, you can foresee and address issues before they become expensive problems. This kind of proactive approach not only saves money but also boosts your reputation as a reliable, forward-thinking engineer.

One should not underestimate the power of reputation in a world increasingly focused on decarbonization. In today's market, a track record of eco-friendly projects is more than a badge of honor; it is a form of social currency. Companies and clients are more willing than ever to collaborate with firms that prioritize sustainability, often viewing such commitments as indicative of overall quality and reliability. For instance, if you were to implement a rainwater harvesting system or solar panels into your design, that would not only reduce the building's long-term carbon footprint but also attract clients interested in long-term savings and sustainability. Such innovations can serve as a cornerstone of your portfolio, drawing in business and fostering partnerships that could be lucrative for years to come.

Efficiency has also emerged as a key financial driver in engineering practices. The days of endless design revisions and costly delays are numbered, thanks to advancements in software and project management tools. Many of these technologies integrate seamlessly with existing workflows, making it easier for engineers to collaborate with architects, contractors, and other stakeholders. For example, BIM allows for real-time updates and shared access to project data. This kind of integration minimizes the risk of errors and miscommunications that can lead to costly delays. It is no longer just about being good at what you do; it is about being efficient, and in the business world, time is indeed money.

The financial benefits extend even to client relationships. The modern client is well-informed and often seeks more than just a structure; they are looking for a sustainable, long-term investment. By leveraging data analytics and artificial intelligence to predict maintenance costs, energy usage, and even the potential for future retrofitting, engineers can provide clients with valuable insights that go beyond the build itself. This value-added service can be a significant differentiator in the market, setting you apart from competitors who offer only traditional engineering services.

The role of the structural engineer is evolving at a pace we have never seen before, driven by both moral obligation and financial incentives. By embracing new materials, leveraging cutting-edge technologies, and integrating sustainability into every aspect of their work, engineers are not just shaping structures, they are shaping the future. It is an exciting time to be in this field, full of opportunities for those willing to innovate and adapt. Far from being a burden, the challenges of today's world offer engineers a chance to lead in ways that will leave a lasting impact on both the built environment and society at large.

1.3 A Roadmap for Contractors

Bringing the contractors to the conversation. Source: danr13/Adobe Stock Photos.

The shift from wood and stone to iron and steel didn't just change the skyline; it redefined what was possible. Consider the Empire State Building, constructed in 1931 by Starrett Brothers and Eken. The 102-story marvel was unimaginable in the era of wood and masonry. More than just a marvel of engineering, it became a symbol of what contractors could achieve with new materials—greater heights, more floors for tenants, and increased profitability. The steel framework provided durability and fire resistance, changing the risk profile and insurance calculus for decades to come.

The 20th century brought us prefabrication, an innovation that transformed the way buildings were conceived and executed. Skanska, a global leader in construction, leveraged prefabrication techniques in the Bertschi School project in Seattle. Let us break down the impact: a 20% reduction in costs is no small feat. It equates to quicker ROI and increased profit margins. The prefabricated components also ensured quality control, reducing the chances of costly on-site errors. It is a method that encapsulates efficiency, cost-saving, and quality—cornerstones for any successful contractor.

Enter the 21st century, and technology takes center stage with BIM. Gensler, the architect behind the Shanghai Tower, used BIM to bring to life this marvel of modern construction. But why is this significant for contractors? Gensler identified over 400 potential issues before actual construction began, resulting in an estimated saving of $11 million. For contractors, BIM is not just a fancy tool for architects; it is a risk mitigation strategy and a budget optimizer. It is like having a crystal ball that predicts the future of a building, helping to avoid pitfalls and seize opportunities for cost-saving.

Sustainability is no longer a buzzword; it is a business model. Turner Construction exemplified this in the Bullitt Center project in Seattle. The building uses 83% less energy than a typical office space and even generates its own electricity. Let us quantify this: the reduced utility costs translate into an attractive proposition for tenants, which, in turn, enables the building owners to charge premium rents. For contractors, specializing in such sustainable projects is like acquiring a golden ticket to a niche but rapidly growing market.

Contractors have benefited greatly from innovation throughout history. From the use of new materials to prefabrication and modular construction, to technology such as BIM, and green building techniques. These innovations have allowed contractors to take on larger and more complex projects, complete them more quickly and efficiently, and ultimately, increase their profits. As the industry continues to evolve, it is likely that new innovations will emerge that will further benefit contractors and the construction industry as a whole.

Collaborating with the construction team. Source: NDABCREATIVITY/Adobestock.

1.3.1 The Roadblocks for Contractors on Climate Action

But if the benefits are so palpable, why the hesitation? The first challenge is the upfront cost. Webcor Builders faced this dilemma head-on with the California Academy of Sciences project. The building features a living roof, recycled denim insulation, and an array of sustainable technologies. The initial investment was formidable. However, it is crucial to consider that sustainability often translates to long-term cost-saving. The building's energy efficiency and low maintenance costs offer a compelling counternarrative to the upfront expenditure.

Let us talk about the information gap. Bechtel, one of the largest construction companies in the world, emphasizes the importance of data-driven decisions. The lack of comprehensive research and case studies on sustainable technologies leaves many contractors in a quandary. It is akin to setting sail without a map; the unknown waters of sustainability are rife with both opportunities and risks. However, with more projects like the Bullitt Center offering tangible metrics on the benefits of going green, this gap is slowly closing.

There is the issue of regulatory ambiguity. Unlike sectors like automotive and pharmaceuticals, construction often lacks stringent guidelines for sustainability. This vacuum creates an environment of uncertainty. Contractors are hesitant to invest in technologies and methods that may or may not meet future regulatory standards. It is a bit like playing darts in the dark; you might hit the bullseye, but you are more likely to miss the mark.

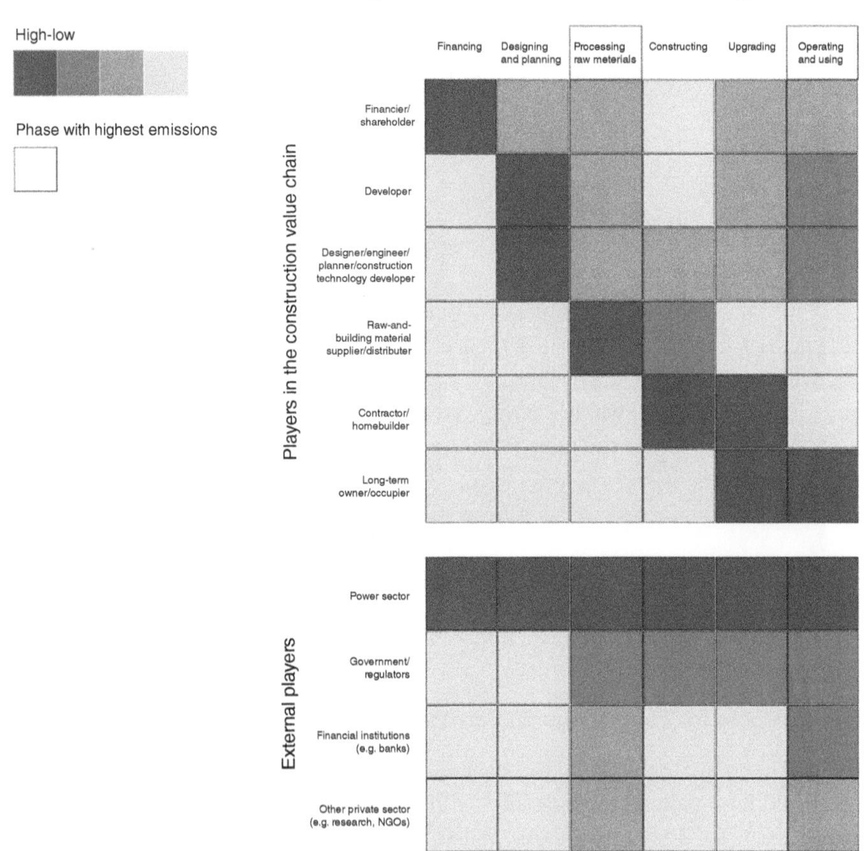

Impact and influence on emissions at each stage of the construction life cycle. Adapted for text. Source: Adapted from IEA CO, Emissions from Fuel Combustion 2018.

1.3.2 The Financial Upside of Climate Action

Despite these challenges, the financial benefits of sustainable construction are hard to ignore. Skanska's Kungsbrohuset building in Stockholm is a case study in innovation and profitability. The building uses the body heat from its occupants and an adjacent train station to reduce energy costs. This led to a 30% reduction in operating expenses, making it a compelling case for how ingenuity can drive profitability.

In an industry where reputation often precedes contracts, a commitment to sustainability can be a game-changer. Lendlease, an international construction company, has built its brand around sustainable practices. Their high-profile project, Barangaroo South in Sydney, aims to be Australia's first large-scale carbon-neutral community. The project has not only garnered media attention but has also attracted a new client base willing to pay a premium for sustainable construction.

The construction industry stands at a crossroads. One path leads to a continuation of traditional practices with their inherent limitations. The other path veers toward a future filled with innovation, sustainability, and unprecedented profitability. The examples of Starrett Brothers and Eken, Skanska, Gensler, Turner Construction, Webcor, Bechtel, and Lendlease serve as waypoints on this journey toward a greener and more profitable horizon.

For contractors willing to make the leap the rewards could be monumental, both in terms of profitability and contribution to a sustainable future. As the industry evolves, those who adapt will not just survive, they will thrive. The choice is clear. Will you be a spectator or a pioneer?

1.4 Conclusion

As we will see throughout this book, there are a lot of easy wins. The AEC industry has a crucial role to play in combating the climate crisis. Some of the techniques and processes we will dig into include things like how design teams can contribute by joining the commitments, adopting new technology, using better materials, and incorporating renewable energy systems. We will explore the role of educators and how to structure a curriculum to prepare the next generation to tackle the climate crisis as well.

2

What is Carbon Positive Design?

2.1 Introduction

As the specter of climate change looms ever larger, architects, engineers, and builders transition from being mere creators of functional and aesthetic spaces to active stewards of a sustainable future. The gravity of the situation cannot be overstated: By 2050, the global building stock is on track to double, resulting in an environmental impact akin to constructing a "new New York City" every month for four decades. This poses a critical challenge but also a unique opportunity to revolutionize the way we approach building design and construction.

To navigate this formidable landscape, this chapter offers a five-step framework: Setting Decarbonization and Project Goals, Evaluating Design and Construction Processes, Developing Targeted Solutions, Creating an Implementation Plan, and the final crucial step, Validating and Verifying the Finished Design. Think of these as the pillars supporting a new paradigm in construction. Setting goals serves as the blueprint, defining the metrics and targets that will gauge our efforts. Evaluation functions like an X-ray, diagnosing the current state of our design and construction approaches. Developing solutions is our toolkit, featuring an array of methods from cutting-edge technologies like building information modeling (BIM) and artificial intelligence (AI) to traditional techniques like passive design. Implementation is where we roll up our sleeves; it is the construction site where all planning takes life. Finally, validation and verification are the quality checks that ensure our edifice of sustainability stands firm.

1	Setting decarbonization and project goals
2	Evaluating design and construction processes
3	Developing targeted solutions
4	Creating an implementation plan
5	Validating and verifying the finished design

Construction project workflow template.

Build Like It's the End of the World: A Practical Guide to Decarbonize Architecture, Engineering, and Construction, First Edition. Sandeep Ahuja and Patrick Chopson.

In the pages that follow, we will dissect each of these pillars in detail, providing a comprehensive guide replete with actionable insights. From the intricacies of material selection to the utilization of advanced technologies and compliance measures, the chapter aims to be your go-to manual for low-carbon building projects. It is not just about planning for a sustainable future, but actively constructing it. The choices we make today will indelibly influence the health of our planet and the sustainability of our cities for generations. Let us build a legacy worth leaving.

2.2 Step 1: Set Decarbonization and Project Goals

In the world of carbon emissions, ignorance is not bliss—it is a missed opportunity for meaningful change. To effectively curb emissions, you need to know what you are measuring, much like a doctor needs to diagnose a condition before prescribing treatment. There are two headline metrics that should be on every design team's radar: embodied carbon and operational carbon. Embodied carbon refers to the emissions associated with the sourcing, manufacturing, and transportation of building materials, as well as the construction process itself. Operational carbon, on the other hand, pertains to the emissions produced during the building's lifecycle, such as heating, cooling, and electricity use.

The Tower at PNC Plaza in Pittsburgh. Source: Cbaile19/Wikimedia Commons.

Take, for instance, The Tower at PNC Plaza in Pittsburgh. This skyscraper did not merely aim to be another addition to the city's skyline; it aspired to set a precedent in sustainable design. By integrating features like a solar chimney and natural ventilation systems, the building achieved a staggering 50% reduction in energy consumption. Here operational carbon was scrutinized, turning a construction project into a landmark case study in emissions reduction.

2.2.1 Consult Experts to Identify Emissions Reduction Opportunities

Navigating the maze of emissions reduction requires more than just good intentions; it demands specialized knowledge that only experts in sustainable design and construction can provide. But how do you distinguish a true authority from a well-meaning amateur? Look for individuals with a proven track record—those who have not just talked the talk, but walked the walk. A credible expert will have tangible successes, like contributing to certified sustainable projects or pioneering innovative techniques that have moved the needle in carbon reduction. Their portfolio should include empirical evidence, whether it is substantial energy savings or documented material optimizations. Their expertise acts as a turbocharger to your emissions reduction vehicle, propelling your strategies from aspirational to actionable.

You do not have to be a guru with white hair to embody the principles of sustainable design. You can become the very expert that Fortune 500 companies consult by digesting and applying the comprehensive strategies and tools outlined in this book. It is important to not focus on getting more credentials but positioning yourself as a leader in the field of sustainable design and construction by learning to think. Take the example of Google's Charleston East campus. Although Google consulted external experts, imagine being the in-house authority who could drive such transformative change.

2.2.2 Define Specific, Measurable Emissions Reduction Goals

While it is tempting to make broad proclamations about "going green" or "being more sustainable," such generalities run the risk of evaporating into the ether of good intentions. The key to transformational change lies in the specificity of your goals. This is where the time-honored SMART framework—Specific, Measurable, Achievable, Relevant, Time-bound comes into play. By providing a structured approach to goal-setting, SMART turns lofty ambitions into concrete plans.

- **Specificity** in goal-setting for emissions reduction cannot be overemphasized. A vague goal like "we want to be more sustainable" is as ineffective as a ship without a compass; you may move, but you will lack direction. Instead, aim for granularity. For instance, if your focus is on reducing embodied carbon, specify which materials or construction phases you will target. Will you substitute traditional concrete with low-carbon alternatives? Will you aim to reduce the carbon footprint of your HVAC installations?

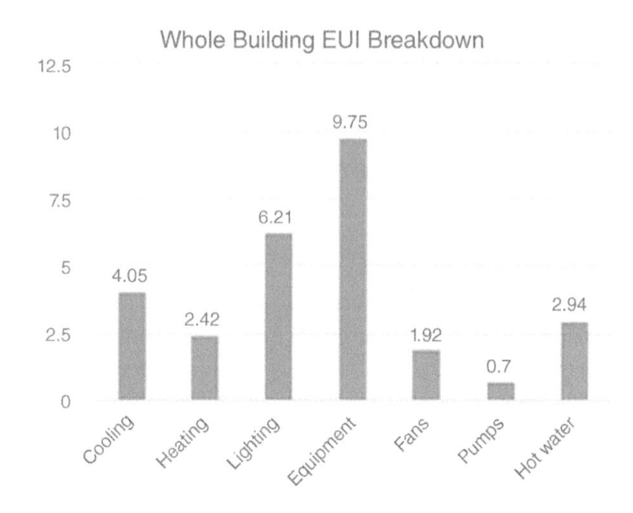

Whole building EUI breakdown created in cove.tool.

- **Measurability** is your yardstick for success. Without metrics, you are sailing blind. Using well-established indicators, such as kilograms of CO_2 per square meter or per square foot, allows you to track your progress and make adjustments as needed. Your project should not just aim for "better energy efficiency", it must target a specific reduction in energy consumption. Most buildings can easily pick up a 40% reduction in energy use and simultaneously discover a 2–3% reduction in construction cost.

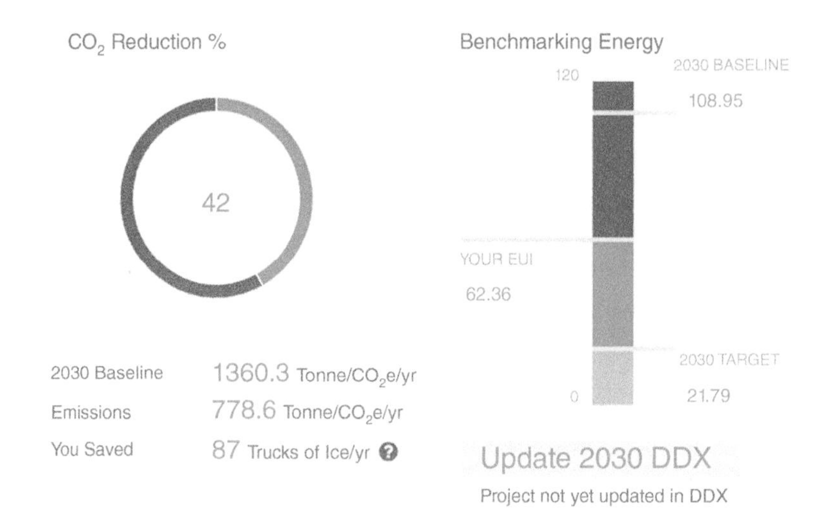

Model created by cove.tool to indicate potential reduction in carbon emissions against the 2030 Baseline.

- **Achievability** is about grounding your goals in reality. While it is exciting to shoot for the moon, your goals should be rooted in what is technically and financially feasible. This is where consultation with experts and rigorous research become invaluable, helping you gauge the attainability of your objectives.
- **Relevance** ensures your goals align with broader organizational or project objectives. If your project aims to be a model of affordable housing, then your sustainability goals should not drive costs to a point where affordability is compromised. The goal is to create a symbiosis between sustainability and the project's primary objectives.
- **Time-bound** parameters add urgency and focus to your goal-setting. Saying you will reduce operational carbon "someday" is an open invitation to procrastination. Instead, set deadlines. Whether it is a 20% reduction in embodied carbon by the end of your next project or a 10% decrease in water usage within the first year of operation, a timeline transforms your goal from a wish into a commitment.

By adhering to the SMART framework, your goals transform from nebulous ideas into a rigorous, actionable strategy. Whether you are targeting embodied carbon, operational carbon, energy use intensity (EUI), or any other metric, specificity is your linchpin, converting good intentions into impactful, real-world change.

2.2.3 Set Short-term and Long-term Emissions Reduction Targets

Navigating the path to a sustainable future requires a nuanced strategy, one that balances the urgency of immediate action with the foresight of long-term planning. It is a delicate equilibrium, much like a runner who must manage their speed and endurance to complete a marathon successfully. The goal is to build a framework where short-term targets act as stepping stones leading to your ultimate long-term objectives.

Short-term goals serve as much more than just preliminary milestones; they offer the psychological benefit of immediate gratification, a critical component in maintaining stakeholder engagement and team morale. Achieving these "quick wins" delivers a dopamine rush of success, making the longer journey ahead feel more achievable. This was evident in the Empire State Building's retrofit, where one of the initial steps involved upgrading the lighting systems—an action that yielded prompt energy reductions and served to energize the team for the challenges that lay ahead.

In contrast, long-term objectives are your legacy goals. These are the ambitious targets that will define the project's lasting impact on both the environment and the field of sustainable design. For the Empire State Building, the long-term goal was a hefty 38% reduction in energy consumption. Achieving this required significant alterations, like overhauling the HVAC system—a complex, resource-intensive task that would be daunting without the momentum generated by earlier, short-term successes.

Source: Casper Sørensen/Wirestock Creators/Adobe Stock Photos.

Emissions reduction is a dynamic field, constantly evolving with advancements in technology, materials science, and best practices. This fluidity necessitates an adaptive approach to goal-setting. While your long-term objectives may remain relatively stable, your short-term targets can—and should—be adjusted to capitalize on new opportunities or address unforeseen challenges. Think of it as a GPS for your sustainability journey, constantly recalculating the best route to your final destination.

Regular reviews of your short-term and long-term goals serve as "heartbeat checks" for your project. These reviews are critical points where you assess whether you are on track to meet your objectives and what course corrections might be needed. Such review sessions also provide an opportunity to celebrate achievements, however small, reinforcing the positive psychology that fuels long-term commitment.

By diligently setting and managing both short-term and long-term emissions reduction targets, you create a robust strategy that not only delivers immediate results but also builds a legacy of meaningful, lasting impact. This dual approach ensures that you are not just sprinting to temporary victories but running a marathon designed for sustained success.

2.2.4 Achieve Stakeholder Alignment

Getting everyone on the same page is more than a logistical necessity; it is the bedrock of a project's success, especially when it comes to something as complex and multifaceted as sustainability. Achieving stakeholder alignment is not just a matter of communication; it is about crafting a compelling narrative that turns passive listeners into active participants.

	Construction	Use stage		End of life
	Upfront carbon	Operational carbon	Use stage embodied carbon	End of life carbon
Architects	High	Low	Medium	High
Engineers	Low	High	Medium	Low
Cities,states, and Regional authorities	Low	High	Low	Low
Construction product Manufacturers	High	Low	Low	High
Contractors	High	Low	Low	Low
Property developers[4]	High	Low	Low	Medium
Property owners and investors[4]	Low	High	High	Low
Property managers agents and advisors[4]		Medium	High	
Tenants/occupiers		High	Medium	

Influence of Building Stakeholder across Building life cycle.

- **The Importance of a Shared Vision**

  The first step in securing stakeholder alignment is establishing a shared vision. This vision acts as a North Star for the project, guiding all decisions and actions toward a common objective. In the case of the BedZED development in London, the One Planet Living framework provided a well-defined roadmap for a zero-carbon community. This framework was not just a list of technical specifications; it was a story of what sustainable living could look like, capturing imaginations and inspiring commitment.

- **Crafting the Presentation: Facts, Stories, and Aspirations**

  A well-orchestrated presentation can be a game-changer in achieving stakeholder alignment. This presentation should be a blend of hard data, compelling narratives, and a glimpse into what the future could hold. Use facts and figures to lay down the urgency of the matter; climate change is not waiting, and neither can we. Incorporate stories or case studies that exemplify the positive impact of similar sustainability efforts. And do not shy away from painting a vivid picture of the project's aspirational goals—what could we achieve together if all goes according to plan?

- **Stakeholder-Specific Messaging**

  Not all stakeholders are created equal; each comes with their own set of priorities and concerns. Tailoring your presentation to address these specific interests can make the difference between lukewarm reception and enthusiastic endorsement. For financial stakeholders, focus on the long-term cost savings and potential for increased property value. For architects and designers, highlight the opportunity to be part of an industry-changing project. For local communities, emphasize the environmental and health benefits.

The author in a sustainability charette highlighting the value of shared vision.

- **The Role of Feedback Loops**

  Alignment is not a one-off event; it is an ongoing process. Creating feedback loops where stakeholders can voice concerns or offer suggestions is vital. It is a way to gauge the temperature of the room periodically and make course corrections as needed. After all, even the best-laid plans may need adjustments, and who better to provide those insights than the people who have invested their time, money, or expertise in the project?

- **Finalizing Commitments and Next Steps**

The ultimate goal of achieving stakeholder alignment is to turn a collective vision into collective action. The final part of the presentation should clearly outline the next steps, from immediate actions to long-term milestones. This is also the time to formalize commitments, whether they are financial contributions, resource allocations, or public endorsements.

By approaching stakeholder alignment as more than just a box to tick off, you turn it into a powerful catalyst for change. When all stakeholders are aligned in vision and commitment, the path from ambitious goals to transformative results becomes not just feasible, but inevitable.

2.2.5 Defining Goals for Embodied Carbon Reduction

Embodied carbon is the 'silent partner' in the sustainability equation, often overshadowed by the more immediate concerns of operational carbon. However, its role is anything but minor; it encompasses the carbon footprint of materials from extraction to manufacturing, transport, and installation. Setting explicit, aggressive goals for reducing embodied carbon can be a game-changer in your sustainable design strategy.

Project Carbon Profile generated using the Carbon Feature in cove.tool.

Choosing low-carbon materials is the most direct way to reduce embodied carbon. Consider specifying materials that have lower energy requirements in their production or that are sourced from recycled or renewable resources. Concrete, for example, is a significant contributor to embodied carbon; replacing it with engineered wood or low-carbon concrete can yield considerable reductions.

The Brock Environmental Center in Virginia Beach serves as a vivid example of how sourcing materials locally can drastically cut embodied carbon. By setting a goal to source recycled its materials from within a 500-km radius, the center not only achieved significant embodied carbon reductions but also boosted local economies and promoted sustainable supply chains.

The Brock Environmental Center & Mobile Oyster, Virginia Restoration Center, Virginia Beach, VA. Source: Jwallace72/Wikimedia Commons.

Embodied carbon is not just about the here and now; it encompasses the entire lifecycle of materials, from cradle to grave. Performing a lifecycle analysis can provide a more comprehensive understanding of a material's carbon footprint, helping you make informed choices that stand the test of time.

2.2.6 Defining Goals for Operational Carbon Reduction

Operational carbon is the long-haul trucker of the building world, racking up emissions over a building's entire operational life. This long-term impact makes it a cornerstone of any sustainable design project, and setting precise goals for its reduction is crucial. Melbourne's Council House 2 (CH2) showcases the power of innovation in reducing operational carbon. Targeting a six-star rating from Australia's Green Building Council, the building integrated cutting-edge technologies like phase-change materials for thermal regulation and co-generation systems to recycle waste heat. These were not bells and whistles but core elements of the building's design, aimed at minimizing operational carbon emissions.

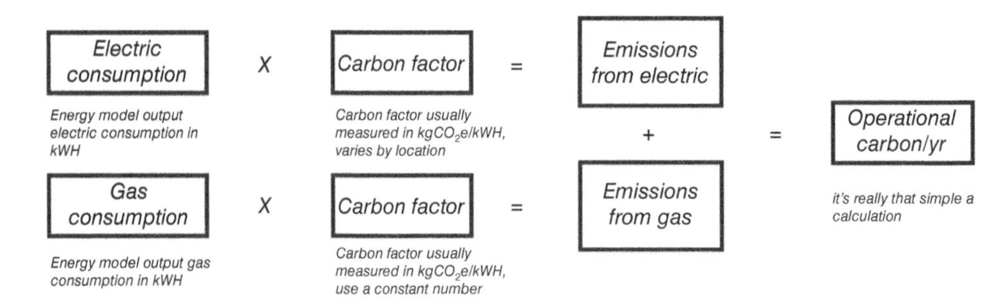

Calculating the operational carbon impact of your building.

EUI is the metric that measures operational carbon since the grid is typically powered by a mix of energy sources. The EUI provides a snapshot of a building's energy efficiency, expressed as the energy used per square foot of floor area. The New York Times Building aimed for a 50% reduction in EUI compared to typical Manhattan office buildings. Thanks to a carefully designed curtain wall that maximizes natural light and an innovative underfloor air distribution system, the building met this ambitious goal. Reducing EUI is not just about cutting electricity usage; but about improving thermal efficiency. Well-insulated buildings require less energy for heating and cooling, contributing to a lower EUI. Setting goals around thermal performance can lead to innovative solutions, which also reduce the structural load of the building and embodied carbon. Reducing operational carbon is not a 'set it and forget it' proposition. Ongoing monitoring and adjustment are essential. Advances in building management systems and Internet of Things (IoT) sensors now allow for real-time monitoring of energy usage, enabling dynamic adjustments to operational settings to minimize carbon emissions continually.

By setting focused and ambitious goals for both embodied and operational carbon reductions, you are committing to a holistic approach to sustainability that addresses the entire lifecycle of a building. It is not just about constructing a building that is less harmful to the environment; it is about creating a structure that actively contributes to a more sustainable and responsible world.

2.2.7 Set Goals for Increasing the Use of Renewable Energy

The shift toward renewable energy has evolved from a desirable feature to an imperative in sustainable architecture. Whether it is through onsite solutions like solar panels, wind turbines, or geothermal systems, or offsite strategies like purchasing renewable energy credits, setting specific goals for renewable energy usage is crucial. These are not merely aspirational targets but can provide resiliency when the electrical grid goes down.

Renewable energy resources, solar panels and wind turbines. Source: Lina/Adobe Stock Photos.

The beauty of renewable energy lies in its versatility. Onsite options offer the most direct control, allowing architects to integrate renewable energy systems seamlessly into the building's design and function. Solar rooftops, for example, can double as design elements, while onsite wind turbines can become iconic features of the architectural landscape. Offsite options, on the other hand, offer scalability and can be a more feasible choice for projects with space or budget constraints. The key is to find the right balance between onsite and offsite solutions, tailored to the unique needs and possibilities of each project.

The choice to incorporate renewable energy can be not merely a technical decision but a powerful architectural statement. It reflects a commitment to innovation and sustainability that resonates with stakeholders and the community at large. Renewable energy features can become defining elements of a building's identity, much like the wind turbines of the Bahrain World Trade Center became a symbol of sustainable innovation. By setting clear, measurable goals for renewable energy use, architects have the opportunity to create not just buildings, but landmarks in the transition to a more sustainable world.

2.2.8 The Daylight and Glare Balance

In the quest for sustainable architecture, the nuanced interplay between daylight and glare presents both an opportunity and a challenge. While natural light can significantly reduce energy consumption by

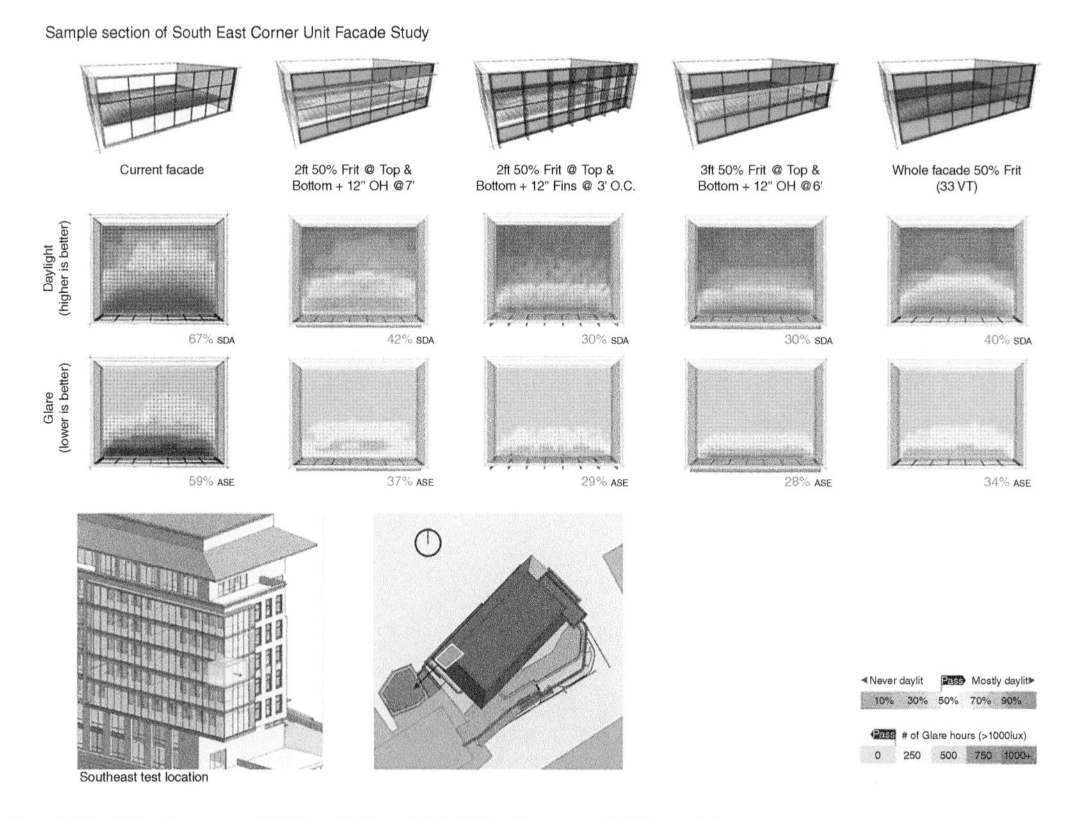

Spatial Daylight Autonomy (sDA) and Annual Sunlight Exposure (ASE) models.

decreasing the need for artificial lighting, unmanaged glare can compromise occupant comfort and even necessitate energy-intensive cooling solutions. Therefore, striking the right balance is not just a matterof visual aesthetics or comfort—it is also a critical factor in a building's overall energy efficiency. The use of quantitative metrics like Spatial Daylight Autonomy (sDA) and Annual Sunlight Exposure (ASE) can offer architects a robust framework to achieve this delicate balance, setting specific targets that guide design choices and assess performance.

sDA quantifies the availability of natural light in interior spaces, providing a more structured approach to setting daylight targets. Daylight helps architects and designers measure how much of a space receives sufficient daylight over a year. Setting daylight targets allows project teams to quantify their goals, making them specific and measurable. For instance, aiming for an sDA value of 55% would mean that at least 55% of the space should receive adequate natural light for at least half of the annual occupied hours. This numerical target serves as both a design guide and a benchmark for assessing performance.

While maximizing daylight is important, it is equally crucial to manage the risks of glare. This is where the ASE metric comes into play. ASE measures the amount of direct sunlight a space would receive and can help identify areas where glare could be an issue. By setting ASE targets, designers can proactively address glare concerns in the design stage itself. For example, a low ASE target might prompt the inclusion of automated shading solutions or the use of specialized glazing techniques to diffuse direct sunlight. By utilizing both sDA and ASE metrics, architects can achieve a finely-tuned balance between ample daylight and minimal glare, resulting in spaces that are both energy-efficient and comfortable for occupants.

San Francisco Public Utilities Commission Headquarters.

Water is another critical metric often overlooked in the emissions equation. The San Francisco Public Utilities Commission Headquarters set a goal to reduce potable water use by 60%. Through rainwater harvesting, greywater recycling, and low-flow fixtures, the building not only met but exceeded this goal. This initiative serves as a compelling example of how setting specific targets for water consumption can lead to innovative solutions that benefit both the environment and the building's operational costs.

Total Water Use Model. Source: Cove Tool, Inc.

As we navigate the complexities of building a sustainable future, the adage "What gets measured gets managed" rings particularly true. The landscape of emissions reduction and sustainable architecture is not just a moral or aesthetic endeavor—it is a scientific one, demanding precision, verification, and continual adaptation. From the granularity of embodied and operational carbon goals to the innovation of renewable energy and water conservation, the key to transformative change lies in the specificity and rigor of your objectives.

Detail of modern sustainable academy building. Source: creativenature.nl/Adobe Stock Photos.

In this multifaceted journey, each component—be it carbon emissions, energy use, water conservation, or even the balanced interplay of daylight and glare—serves as a puzzle piece. When these pieces come together under the umbrella of clearly defined, measurable, and achievable goals, they form a mosaic of sustainable design that is both impactful and accountable. The SMART framework is not just a tool but a philosophy that guides you from nebulous intentions to concrete achievements. It helps crystallize the intangibles of goodwill and ambition into metrics that can be tracked, analyzed, and celebrated.

2.3 Step 2: Evaluate Design and Construction Processes

In the journey toward sustainable building, evaluation is not merely a pit stop—it is the control tower guiding the entire flight. This step focuses on a rigorous evaluation of both design and construction processes to ensure they are not only aligned with sustainability goals but are also actionable and effective. Utilizing a blend of advanced simulation tools and data analytics, this step aims to transform raw numbers into strategic insights. This helps set the north star for ownership, design and construction teams to know if they are on the right path.

2.3.1 A Foundational Element, Not a Checkbox

While tempting to view the initial carbon audit as a mere compliance activity or an environmental token, it is actually much more than that. This audit is the linchpin that holds the project's sustainability ambitions together. It offers a data-rich snapshot of the project's current environmental impact, from energy usage to waste production. It is not about ticking boxes; it is about establishing a baseline.

Can an all glass building be sustainable? Evaluation is key to work with data. Source: Artinun/Adobe Stock Photos.

Armed with this baseline data, the design team can set realistic but ambitious targets that stretch the limits of what is possible. For example, if the audit reveals that the building's heating, ventilation and air conditioning (HVAC) system is a significant carbon emitter, the design team might aim to replace it with a more efficient system, perhaps even pushing the boundaries by exploring cutting-edge, energy-neutral options.

Think of the initial audit as the opening scene in a movie or the first chapter in a book. It sets the tone and lays the groundwork for the narrative that will unfold. It provides the context within which the design team will operate, similar to how a director uses the opening scene to introduce key elements that will be developed throughout the film.

2.3.2 The Role of Advanced Tools

Advanced tools like cove.tool serve as the backbone for informed decision-making in the complex realm of sustainable building design. Far from being mere calculators that churn out numbers, software programs function more like mission control centers, providing a comprehensive overview that includes multiple variables such as energy usage, material choices, and daylight. The role of software is not just to quantify, but helping architects and designers discern between tradeoff of various options and their respective impacts. In short, it is about making data actionable and choices insightful.

Taking the Georgia Tech Student Center as a case in point illuminates the power of such advanced tools. When planning their new building, the team used cove.tool to simulate different scenarios, including the ramifications of demolishing or preserving the existing structure. Not a simple "yes or no" questions, the team found many variables interacted with each other like embodied carbon of the existing student center structure, location of the campus loop, and the project goals for full daylighting of student spaces. Software offered a way to navigate this complexity, ultimately revealing that preserving the existing structure would not only make good financial sense but also offer substantial environmental savings.

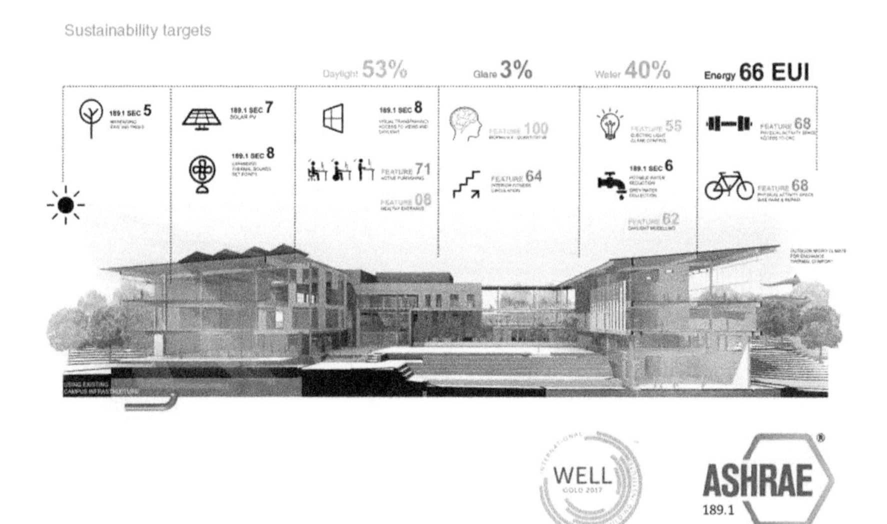

In image.

John Lewis Student Center, Georgia Institute of Technology, Atlanta, GA, renovation project completed by Cooper Carry.

This revelation led to an innovative shift in the project's design strategy. Instead of starting from scratch, the team chose to preserve and repurpose the existing concrete structure. This decision had a ripple effect, influencing subsequent choices about materials, energy systems, and even space utilization. It is an excellent example of how data, when interpreted and applied wisely, can lead to design choices that are both innovative and sustainable.

The role of advanced tools extends far beyond mere computation. They serve as an essential nexus where data meets design, facilitating choices that align with both aesthetic and environmental imperatives. In this way, they elevate the entire design process, transforming it from a series of educated guesses into a finely tuned orchestra of decisions, each contributing to the ultimate goal of sustainability.

2.3.3 Roadmap for Design Teams

The data gathered during the audit does more than just populate spreadsheets. Using a data-driven process helps the design team make choices that are in sync with both aesthetic and sustainability goals. For instance, if the data indicates that a particular type of window glass can significantly reduce energy costs, that becomes a material of choice for the architects. In essence, the baseline audit acts as a project's "North Star," providing a fixed point of reference for all subsequent decisions. It enables a more harmonious blend of form and function, steering the project in a direction that maximizes both aesthetic appeal and environmental responsibility.

The roadmap ensures that architects and engineers do not have to choose between creating a visually stunning building and an environmentally sustainable one. Instead, it shows them how to accomplish both, providing them with data-supported strategies to achieve a balanced outcome that satisfies both ethical and aesthetic objectives. By adopting these focused approaches, baseline audits and advanced tools become more than administrative or technical tasks. Using data and simulation are strategic activities to inform and guide the project from conception to completion. The aim must be to make sustainability not an add-on, but an integral part of the design DNA.

2.3.4 Data-Driven Construction Planning

When we talk about a data-driven construction strategy, we are not just talking about keeping tabs on costs and timelines, although those are important. Instead, we are talking about a more holistic approach that also incorporates environmental metrics.

Exterior view of the southwest corner of the Seattle Central Library building designed by the Dutch architect Rem Koolhaas. Source: Sergii Figurnyi/Adobe Stock.

The Seattle Central Library offers a great example, where data collection covered a broad spectrum, from energy consumption in HVAC systems to the environmental impact of material transportation. Just like a GPS system that is not just guiding you on the fastest route but also pointing out the scenic views and roadblocks you should be aware of.

Collecting data is only the first step; the real magic happens when that data is translated into actionable insights. In the context of the Seattle Central Library, data was not just an end but a means to an end. They did not merely compile spreadsheets; they used data as a lever to shift their approach. For example, if data indicated that the transportation of construction materials was a carbon-intensive process, they pivoted to local suppliers. It is like having a fitness tracker that not only tells you how many calories you have burned but also suggests specific exercises to improve your performance.

Data should function as a compass for construction teams, offering a directional sense that goes beyond mere number-crunching. Think of it as the team's playbook, providing a framework within which every decision can be made. Whether it is choosing an energy-efficient HVAC system or a sustainable material supplier, this data-driven compass ensures that all choices are made in harmony with the project's sustainability objectives. It is about what should be done to align with both immediate and long-term goals.

One of the most powerful aspects of this data-rich approach is the capability for targeted interventions. Instead of a scattergun approach, where efforts are spread thinly across many areas, data allows for a laser focus on the most impactful actions. If a particular material is identified as a major contributor to the carbon footprint, that becomes the priority for re-evaluation and change. It is the difference between a general practitioner and a specialist; both are important, but the specialist can go deep into a problem and offer tailored solutions.

By incorporating a meticulous, data-driven approach into both design and construction planning, we are not just ticking off boxes in a sustainability checklist. We are revolutionizing the way we conceive, plan, and execute building projects. It is about setting new benchmarks for what is possible in sustainable construction. This is not mere best practice; it is a clarion call for a fundamental shift in the industry's approach to building the future.

2.4 Step 3: Develop Targeted Solutions

It is by collaborating together in a clear strategy that we can all assemble behind a low carbon design solution that respects the design aesthetics, the client needs, the cost, and all the other parameters. It requires the team to trust the process, and the data to be ambitious. This ambition is no longer a 'nice to have', it is a 'must have'.

Whether it is the incorporation of low-carbon materials or the deployment of AI for design optimization, this is the stage where dreams are engineered into reality. Welcome to the mission control center of sustainable building design where every decision, large or small, is a coordinated step toward a more sustainable orbit.

2.4.1 Research Data Sources

Pioneers in sustainable building design offer more than just inspiration; they provide a concrete roadmap for action.

Bullitt Center, Seattle, WA, USA. Architect: Miller Hull Partnership, 2013. Elevation with angular photovoltaic panel canopy. Source: Another Believer/Wikimedia Commons.

Take the Bullitt Center in Seattle, often dubbed the "greenest commercial building in the world." Its radical approach to reducing both embodied and operational carbon offers a goldmine of information. Firms that study this building can uncover a multitude of practical steps—from optimizing energy systems to selecting low-carbon materials—that can be replicated. The Bullitt Center serves as an operational manual for architects, illustrating how bold visions can be translated into reality while adhering to stringent sustainability metrics.

In the age of information, databases are invaluable tools for architects targeting sustainability. For example, the Embodied Carbon in Construction Calculator (EC3) provides granular data on materials' carbon footprints. Projects like the Manetti Shrem Museum at the University of California, Davis, have capitalized on this resource to make informed decisions (https://www.usgbc.org/projects/uc-davis-manetti-shrem-museum-art). By selecting low-carbon materials, the museum achieved a reduction in embodied carbon, essentially rewriting the playbook on what is possible for museum constructions. In this way, databases act as an evolving resource guide, constantly updated to reflect the latest advancements in sustainable materials.

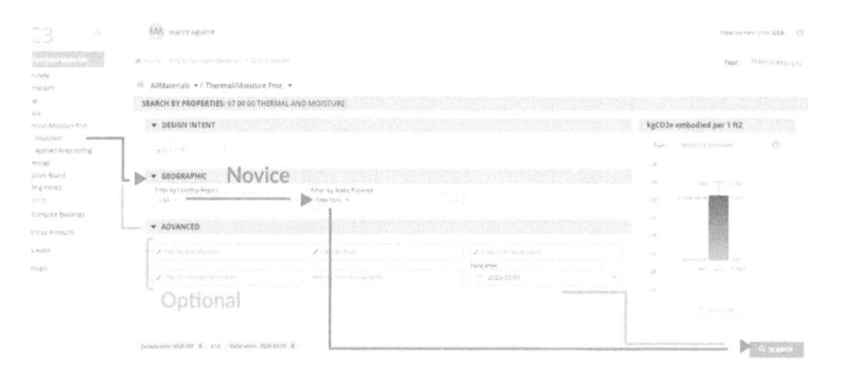

Workflow of building transparency's EC3 integration. Source: Cove Tool, Inc.

2.4.2 Engaging the Supply Chain: More Than Just Sourcing Materials

Engaging with the supply chain is akin to assembling the best engineers, pilots, and scientists for a groundbreaking space mission. Each member of this team—or in this case, each supplier and manufacturer—plays a crucial role in achieving sustainability goals. The Apple Park Campus in Cupertino serves as an exemplar of this integrated approach. Apple's strategy across their supply chain is not merely to source materials; it was to establish a collaborative dialogue with suppliers. They ensured that every piece used in construction is either recycled or sustainably sourced.

This is not just about meeting current sustainability standards; it is about pushing the envelope. Apple's engagement with its suppliers not only led to a significant reduction in the building's embodied carbon but also established new benchmarks for the industry. Apple effectively challenged its suppliers to think beyond the usual scope of their responsibilities, fostering a culture of environmental stewardship that extended throughout the supply chain. While by no means the greenest building, the goals should target what can be done.

The use of biobased and recycled materials is akin to using advanced recycling systems aboard a spaceship—turning waste into valuable resources.

Many manufacturers are leaning into this shift as well by making information more readily available. GAF Roofing, the leading manufacturers of roofing, have done just that. They have invested to create products that are more sustainable with their initiatives like product circularity, which is a material takeback program for recycled asphalt shingles with the goal of diverting 1 million additional tons of

Embodied carbon savings

The impact of embodied carbon is crucial to consider while designing and selecting building materials and products. By comparing options against industry standard baseline teams can understand how their solution is better than standard parctice.

01 /

Roof assembly 01

Roof assembly 1 consists of three new layers of the following products:
- EverGuard white smooth TPO 80 mil membrane
- Energy guard NH polylso 2.6 in ×2
- HD Polylso 0.5 in

43,569

Option embodied carbon, kg CO_2e

108,052

Industry baseline, kgCO_2e

40%

Lower embodied carbon with roof assembly 01 as compared to the baseline

02 /

Roof assembly 02

Roof assembly 1 consists of three new layers of the following products:
- EverGuard white smooth TPO 80 mil membrane
- Enery guard NH polylso 1.5 in
- HD Polylso 0.5 in

25,044

Option embodied carbon, kg CO_2e

42,349

Industry baseline, kgCO_2e

59%

Lower embodied carbon with roof assembly 02 as compared to the baseline

An example of GAF leveraging technology to showcase the impact of their focus on carbon reduction on a project that the architect is working on.

roofing materials from landfill annually by 2030 and establishing a circular roofing economy. They have leaned into carbon emissions reductions as well by reducing their operational carbon emissions by improving energy efficiency, introducing renewable energy sources and exploring and/or innovating new technologies to remove carbon from the environment. They also focus on operational waste diversion with a goal to achieve 80% waste diversion rates across all their manufacturing plants.

Centre for Interactive Research on Sustainability, University of British Columbia. Completed in 2011. 2260 West Mall, Vancouver, Canada. Source: Xicotencatl/Wikimedia Commons.

A striking example is the CIRS building at the University of British Columbia. Visitors to the building are often intrigued by its unique construction, providing an opportunity to engage them in dialogues about sustainability.

Environmental impact potentials for unfaced sustainable batt insulation (North America)

Impact category	Unit	Raw material A1	Raw material transport A2	Manufacture A3	Final product shipping A4	Installation A5	End of life transport C2	End of life disposal C4	Additional facing
GWP(T)	kg CO_2 eq	2.02E-01	1.69E-02	7.21E-01	6.42E-02	1.11E-02	2.92E-03	2.18E-02	5.19E-01
ODP(T)	kg CPC 11 eq	2.15E-09	1.41E-03	1.22E-10	5.69E-13	2.27E-11	2.58E-14	3.34E-13	2.45E-09
AP(T)	kg SO_2 eq	9.69E-09	2.18E-04	1.61E-03	2.99E-04	3.14E-05	1.36E-05	9.98E-05	1.94E-03
EP(T)	kg M eq	1.85E-04	1.03E-05	9.63E-05	2.46E-05	3.12E-06	1.12E-06	5.07E-06	2.25E-04
POCP(T)	kg CO eq	1.03E-02	5.04E-03	3.36E-02	9.90E-03	5.40E-04	4.50E-04	1.97E-03	3.32E-02
ADP fossil (T)	MJ	3.38E-01	3.12E-02	1.53E+00	1.22E-01	2.23E-02	5.55E-03	4.29E-02	1.01E-00

Life Cycle Impact Assessment Chart | TRACI environmental impact potentials for unfaced sustainable batt insulation (North America).

But the supply chain extends beyond suppliers; it also includes manufacturers who create the building materials. Architects and builders must demand information about the sustainability of materials—everything from their source to their manufacturing process. Requesting evidence-based simulations can reveal how a particular material will perform in terms of its carbon footprint over its lifecycle, from cradle to grave. This creates demand and drive investment decisions from manufacturers.

Simulations are invaluable tools in this context. They act as virtual test flights, revealing how materials will perform under different conditions and timelines. By leveraging sophisticated building simulation software, architects can ask manufacturers to provide data-backed evidence of their materials' environmental impact. This turns the process into a data-driven exercise, where claims about sustainability are substantiated through rigorous testing.

The relationship with contractors and manufacturers should not be transactional but rather a continuous loop of feedback and improvement. As architects receive data and test it through simulations, they can feed these insights back into the supply chain. This iterative process ensures that both suppliers and manufacturers remain committed to a shared mission of sustainability.

By taking a holistic approach to supply chain management, we are not just optimizing a single project for sustainability. We are transforming the very ecosystem that produces our buildings, steering the industry toward a trajectory of continuous environmental responsibility. This is the kind of comprehensive strategy that elevates the mission control center of sustainable building design into a hub for change, influencing practices across the architectural community.

2.4.3 The Art of Trade-offs: Balancing Decarbonization Goals Against Project Requirements

The evaluation of embodied carbon in construction materials is not an isolated task; it is a complex web of interconnected building systems. High-strength but lighter steel, for example, may seem like an excellent choice for reducing embodied carbon. However, its lighter weight can also influence the size and

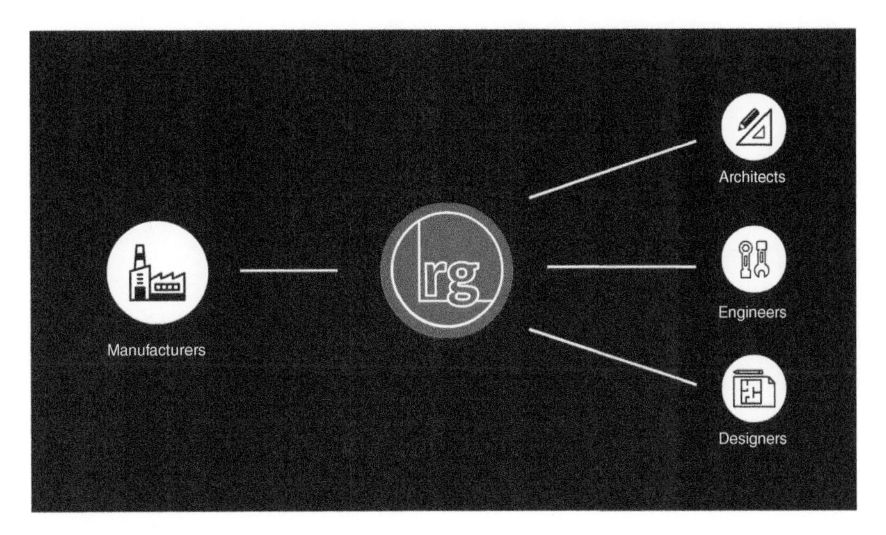

cove.tool's revgen.tool workflow.

composition of the building's foundation, which in turn reduces the project's overall embodied carbon footprint. In the realm of sustainable construction, each material choice initiates a ripple effect, impacting other systems and, consequently, the project's total carbon emissions.

To navigate these complex interactions, data-driven decision-making becomes indispensable. Building simulations can act as a virtual sandbox, allowing architects and engineers to test different scenarios and understand the cascading effects of their material choices. For instance, using insulated panels instead of conventional tilt-up construction might reduce the building's weight, thereby requiring less structural support. This leads to a lighter building with a lower embodied carbon footprint. Simulations offer a platform to quantify these interactions, providing empirical evidence that can inform material selection, ensuring optimal design and resource allocation.

Understanding and evaluating these trade-offs does more than optimize a single project; it advances the entire field of sustainable construction. By making informed decisions about embodied carbon, architects and builders contribute to a growing body of knowledge that challenges traditional practices. It sets a precedent for what is achievable in terms of both structural integrity and sustainability. These informed choices serve as case studies for future projects, illustrating how sustainability can be integrated into every aspect of building design, from material selection to structural requirements.

2.4.4 Passive Approaches

Orientation and shading are not just design elements; they are variables in the equation of building sustainability. The orientation and shading of a building must be meticulously planned to reduce operational carbon. The aim is to leverage natural elements—such as sunlight—to the building's advantage. A well-oriented building that captures natural light minimizes the need for artificial lighting, thus reducing electrical consumption.

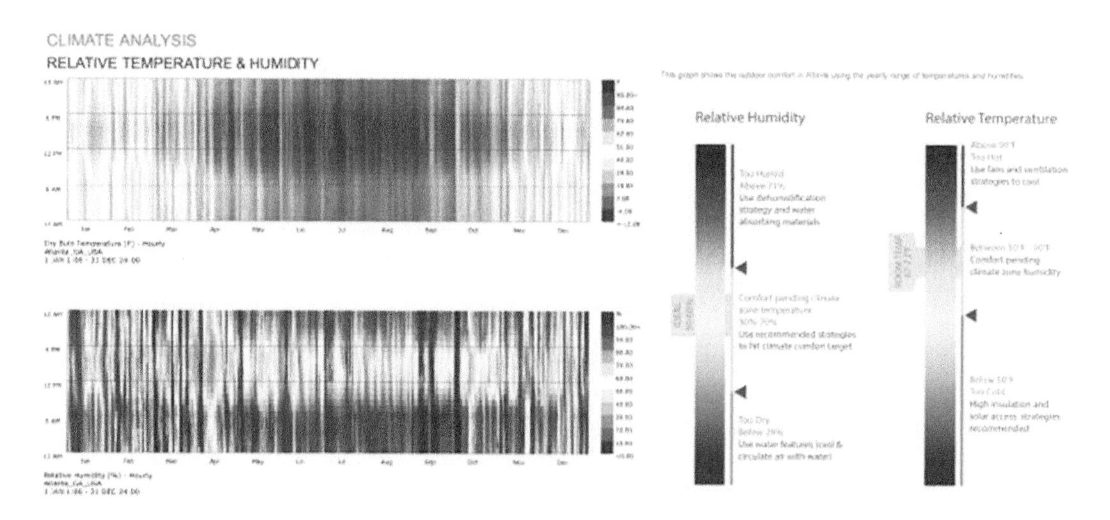

Climate analysis created in cove.tool.

Climate analysis created in cove.tool.

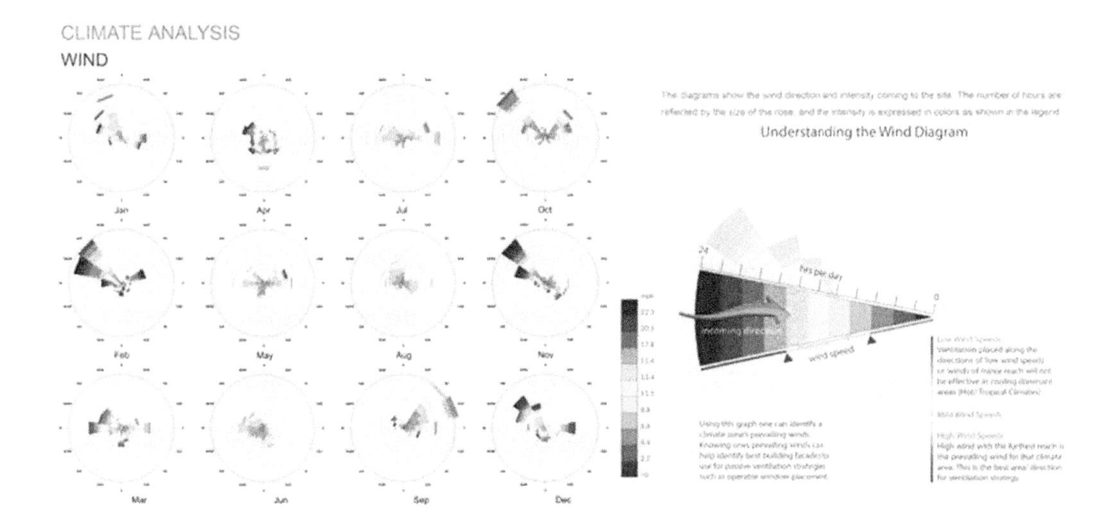

Climate analysis created in cove.tool.

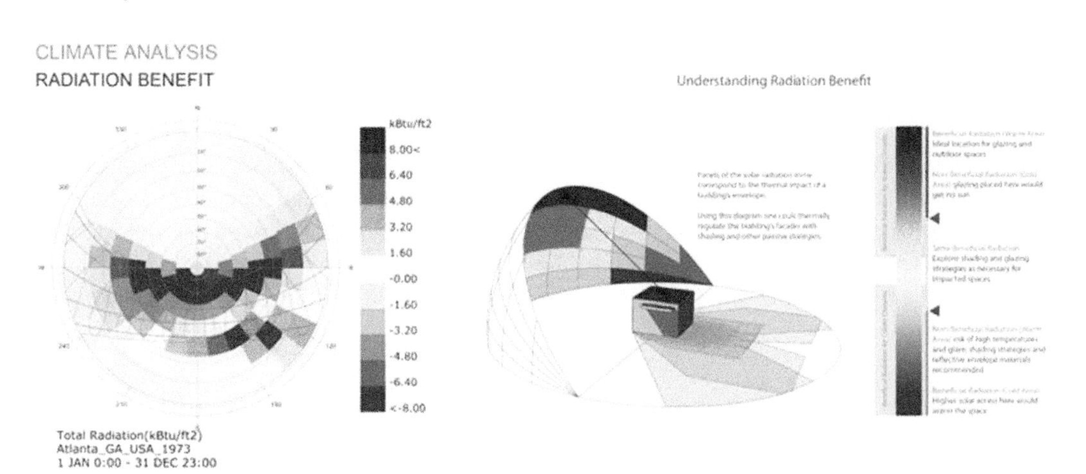

Climate analysis created in cove.tool.

Conversely, strategically-placed shading elements can shield interiors from excessive heat, decreasing the reliance on energy-intensive cooling systems. These passive design steps are not aesthetic choices made in isolation; they must be simulated in synergy with other building systems to reduce operational carbon.

In the complex puzzle of sustainable building design, technology serves as the compass guiding architects toward the most effective solutions. Advanced modeling and simulation tools have become indispensable in this context.

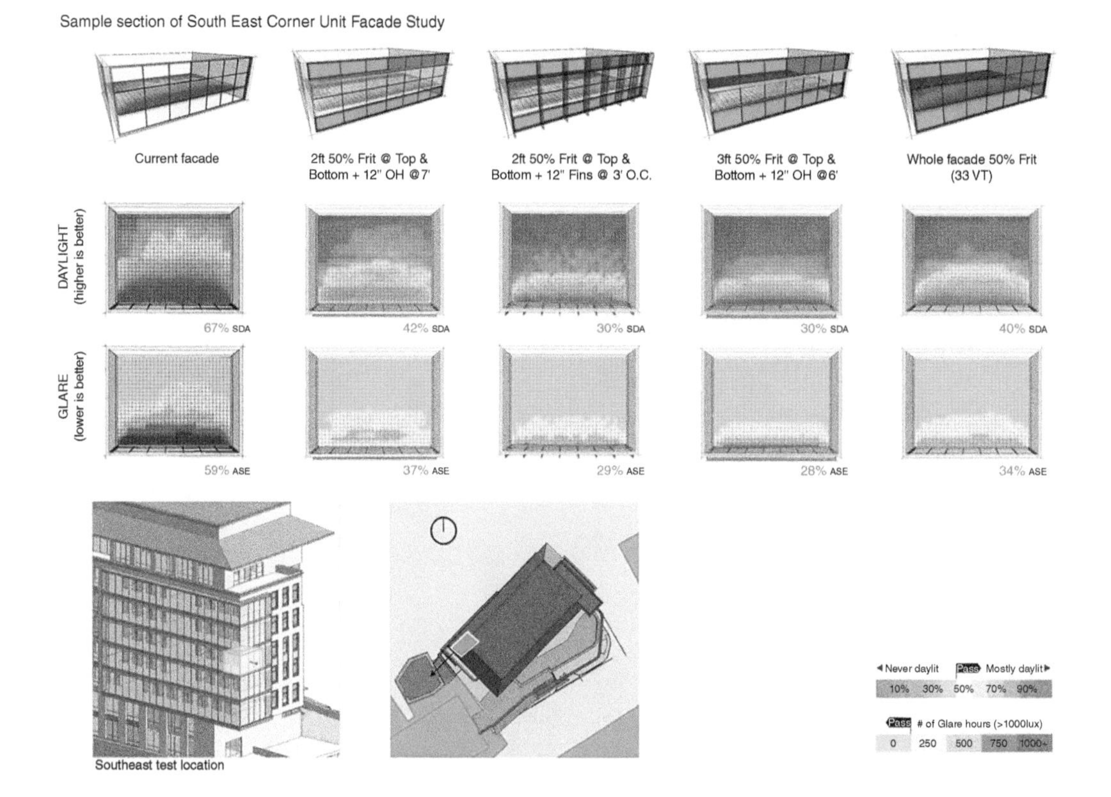

Simulation of facade options to find trade offs.

They allow architects to virtually manipulate building orientation and shading elements, providing real-time data on how these changes impact energy consumption. These simulations serve as preparatory exercises. By employing these high-tech tools, design teams can make data-driven decisions that align both with environmental sustainability and operational efficiency.

The importance of orientation and shading extends beyond individual buildings; it has the potential to influence community-wide sustainability initiatives. When these passive design elements are optimized, they not only benefit the building but also contribute to broader goals like reducing the urban heat island effect or lowering community-wide energy consumption. This is the broader canvas on which these design elements play a role. They are not just about making a single building sustainable; but about contributing to a collective effort to mitigate climate change.

2.4.5 Designing High-Efficiency Active Systems

HVAC and lighting systems serve as the active support systems of a building. These systems are often the largest consumers of energy in a building, making their efficiency critical for sustainability. High-efficiency HVAC systems, often leveraging technologies like geothermal heating and cooling, go beyond mere temperature control; they contribute to a building's overall energy balance.

Ventilation systems on rooftop. Source: leungchopan/Adobe Stock Photos.

Similarly, advanced lighting systems that respond to natural daylight conditions can significantly reduce electricity consumption. These are not just supplementary features but core elements that define the building's energy profile and contribute to certifications.

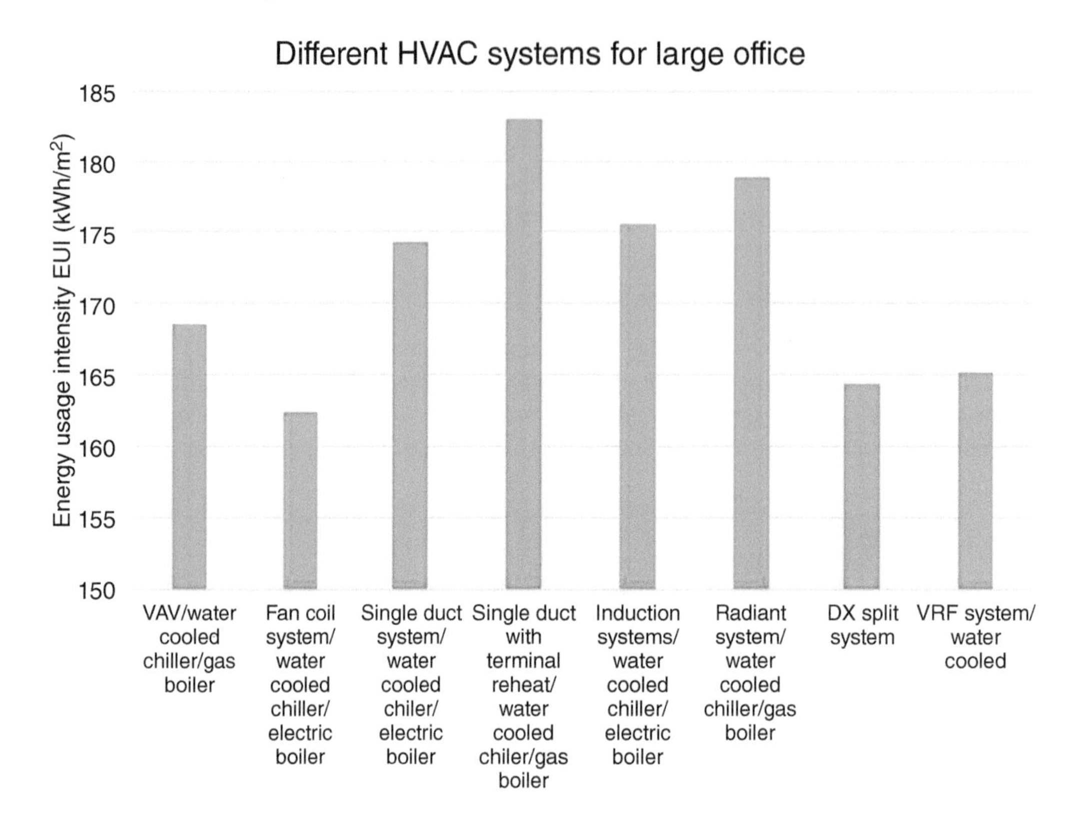

Comparing energy efficiency of typical HVAC systems.

To maximize the efficiency of HVAC and lighting systems, simulations become invaluable. These virtual environments allow architects and engineers to test various configurations, optimizing for both performance and energy efficiency. For instance, simulations can reveal how different HVAC layouts interact with other energy systems in the building, or how daylight-responsive lighting could vary its intensity according to the time of day. These "practice runs" enable fine-tuning of systems before construction, reducing both costs and environmental impact.

This leads us to the integration of onsite renewable energy sources, such as solar panels and wind turbines. Even though there are more limited applications of wind on the building level, it is now the cheapest form of energy in the world and significant tax credits are available for both solar and battery storage on site for maximum resiliency. These are not merely add-ons or afterthoughts; they are central components of a building's energy strategy. By generating electricity onsite, buildings can significantly offset their operational carbon and, in some cases, even produce more energy than they consume. Optimal angling of the solar is vital for maximum output but with the efficiency of new panel designs this concern should not govern. Sometimes the weather of a location can affect performance, like if a location is sunny in the morning and cloudy in the afternoon. In this case, one would want the panels facing southeast in the Northern Hemisphere, rather than due south. Active systems require simulation to size and orient properly within the urban context. This shift toward onsite renewable energy transforms buildings from being mere consumers of energy to becoming active participants in the broader energy ecosystem.

2.4.6 New Design Technologies are Shifting the Paradigm in Building Design

Generative design algorithms have revolutionized architectural planning by automating the generation of design options based on predefined objectives like emissions reduction or energy efficiency. This methodology allows for the rapid exploration of design possibilities, significantly broadening the scope of solutions that can be considered. Software platforms like Autodesk's Revit and Rhino, often augmented by plugins like Dynamo and Grasshopper, have become popular choices for architects delving into generative design.

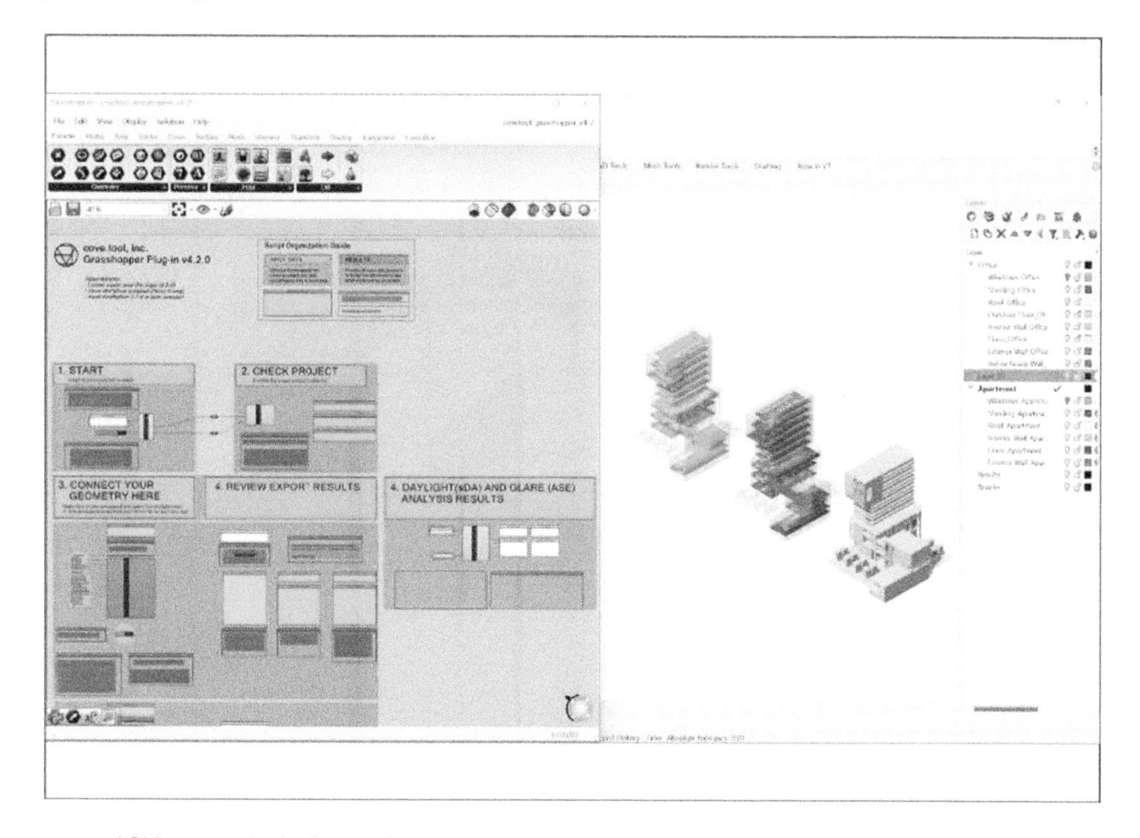

cove.tool Rhinoceros plugin. Source: Cove Tool, Inc.

The real power of generative design lies in its ability to discover innovative solutions that might not be immediately apparent through traditional design methods. By using computational power to navigate complex design spaces, these algorithms bring a new level of rigor and creativity to the planning stage. The result is a set of options that are not only diverse but also optimized for specific sustainability metrics by using simulations from cove.tool, Honeybee/Ladybug, or ClimateStudio to evaluate the generated designs.

BIM takes the design process several steps further by offering a multidimensional, interactive model of a building. This holistic approach enables architects to see how different elements—from materials to room layouts—interact with each other. Popular BIM platforms like Autodesk Revit and Bentley Systems OpenBuildings Designer provide comprehensive tools for this kind of modeling, including features for parametric design.

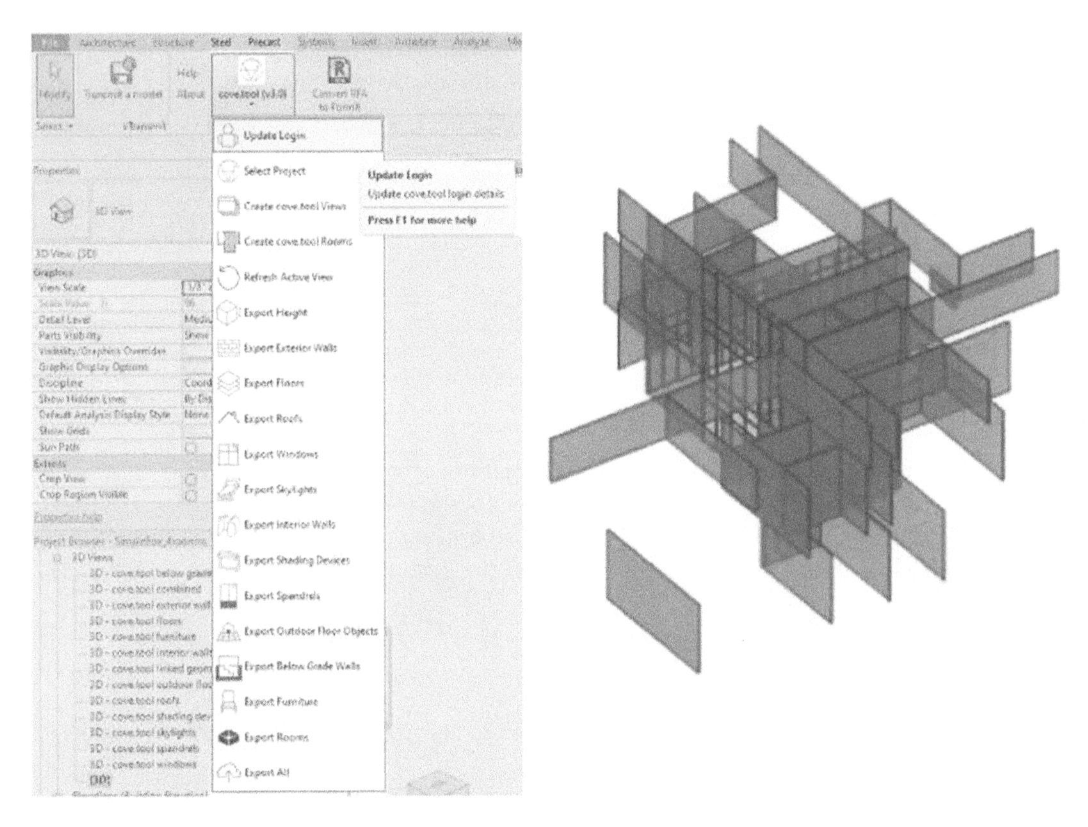

cove.tool Revit plugin. Source: Cove Tool, Inc.

The seamless integration between generative design software and BIM platforms has been a game-changer for sustainable architecture. Some sustainability platforms such as cove.tool can even integrate with BIM systems to offer real-time sustainability metrics. This symbiosis allows for a dynamic, responsive design process. As a result, architects can make data-driven decisions that align the project with its sustainability targets throughout its lifecycle, rather than just at the planning stage.

AI is no longer just a trending topic; it is on track to become a foundational technology in the AEC sector. Advanced AI algorithms have the capability to sift through vast datasets, offering optimized solutions for various building systems, including HVAC and lighting. This goes beyond simple automation; machine learning algorithms can adapt and improve building performance in real time.

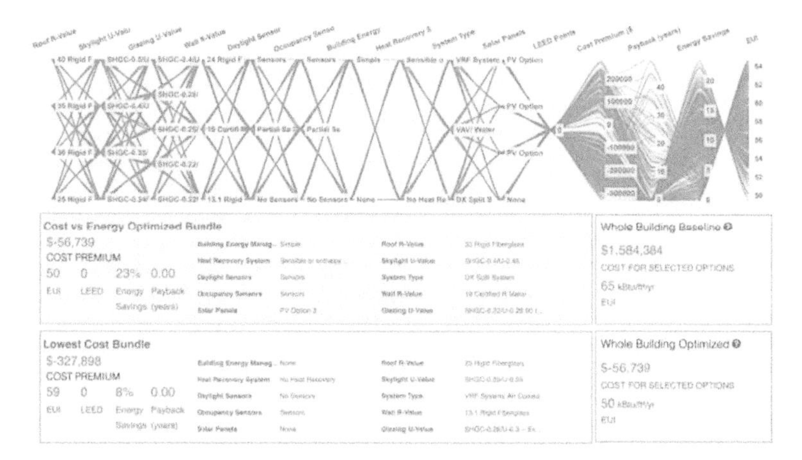

Each vertical line represents one decision. The blue line represents every possible combination and these can be filtered based on potential LEED points, EUI, Energy Savings, Payback, and First Cost Premiums. For a project with a target payback less than 10 years and EUI target of 60 or below, there were possible combinations costing under an additional US$100,000 to $250,000.

Several platforms are bringing AI's potential to the forefront of sustainable building design. Software solutions offer AI-driven tools that can significantly lower a building's operational carbon footprint, while technologies like Google's DeepMind have demonstrated how AI can optimize energy usage in complex environments like data centers. These AI solutions offer architects and builders the ability to make informed, real-time decisions, significantly elevating the standard for what is considered sustainable in building design. By the time this book is published, AI will be several versions ahead, so stay informed.

By incorporating these advanced technologies—generative design algorithms, BIM, and AI—architects and builders have an increasingly robust toolkit for creating buildings that are not just visually stunning but also exemplars of sustainability. These technologies enable each building project to be a dynamic, data-driven initiative, setting a new standard for sustainable design and offering an inspiring look into the future of the AEC industry.

2.4.7 Critiquing Designs Against Emissions Goals: The Reality Check

Before breaking ground, every sustainable building design should undergo a rigorous critique against established emissions goals. This critical evaluation serves as a reality check, ensuring that the design choices made are not only aesthetically pleasing but also aligned with sustainability objectives. Sometimes this scrutiny may lead to unconventional solutions, such as incorporating elements that improve natural ventilation or light penetration. These design adjustments are born from the need to meet or exceed emissions targets and are integral to the project's ultimate success.

Contractors visiting a site. Source: Seventyfour/Adobe Stock Photos.

No building exists in isolation, and achieving ambitious sustainability goals often necessitates collaboration beyond the design team. Early identification of solutions requiring external partnerships can be a game-changer. Whether it is working with local utilities for water recycling or collaborating with tech firms for energy management systems, these partnerships can open doors to innovative solutions that benefit not just the building but also the community at large.

Even the most robust sustainability plans can unravel without proper execution. Therefore, a close collaboration with contractors is not a mere recommendation but a necessity. From choosing construction methods to the selection of equipment, this partnership ensures that the project's sustainability goals are maintained throughout the construction phase. A well-coordinated team can make the difference between a project that meets its sustainability goals and one that falls short.

Despite all efforts, some level of emissions is often unavoidable. Preemptive planning for these scenarios is crucial. Whether it is through purchasing renewable energy credits, investing in community-based sustainability projects, or setting aside funds for future improvements, having a plan to offset emissions is an important part of any sustainable building strategy. This allows the project to maintain its integrity with regard to its sustainability ambitions. The important thing is not to back down from making sure the carbon is reduced as much as possible and avoid fake solutions like carbon credits that are difficult to verify and open the owner to legal exposure.

2.4.8 Empowering the Team: The Role of Knowledge Sharing

Knowledge is power, especially when it comes to sustainability. Creating comprehensive training guides for the team can elevate the project's sustainability goals from a conceptual stage to concrete action. These guides can cover a range of topics, from material recycling protocols to energy efficiency techniques. Such documentation ensures that everyone involved—from architects to contractors—is aligned and competent in implementing the project's sustainability plans.

The work is not done once the building is complete. Continuous monitoring is key to the project's long-term sustainability and increasing knowledge for how to improve design outcomes. Advanced building management systems (BMS) can provide real-time data on various metrics, from energy consumption to water usage. This ongoing monitoring not only confirms the effectiveness of implemented solutions but also allows for timely adjustments, ensuring that the building remains a living example of sustainability long after its completion.

By integrating these concluding steps—rigorous design critique, external collaborations, contractor partnerships, emissions offset planning, team training, and continuous monitoring—a sustainable building design can transition from a lofty ideal to a concrete reality. This multipronged approach ensures that decarbonization is not just a box to be checked but a guiding philosophy that permeates every aspect of the project from inception to completion and beyond.

2.5 Step 4: Create an Implementation Plan

The transition from a well-intentioned strategy to tangible results is a complex journey that requires meticulous planning and execution. Creating an implementation plan is the bridge that connects your sustainability vision to reality. This step is the guide for your project's decarbonization aspirations.

Understanding the parameters that make your building low or high carbon is critical. Source: EwaStudio/Adobe Stock Photos.

Let us dive into a detailed guide on how to develop a comprehensive plan that addresses every facet of implementation—from timeline construction and financial planning to risk mitigation and stakeholder approval. As you navigate through this guide, remember that an effective implementation plan is not just a checklist; it is ethos and a mind set. Add these sections to your plan and have all parties review so that things get done.

The cornerstone of any effective sustainability initiative is a comprehensive project plan that serves multiple purposes. Think of this plan as the DNA of your project—it carries the genetic code that will determine the project's characteristics and its future behaviors. In the most basic sense, a project plan is a blueprint, but it is also much more. This document is an actionable guide that informs each member of your team about the specific steps they need to take to materialize the project's sustainability objectives.

Beyond merely listing goals and wishes, a well-crafted plan delves into the nitty-gritty details. It specifies the technologies that will be employed, whether it is energy-efficient HVAC systems, solar panels, or cutting-edge insulation materials. Technology is only part of the equation. A robust project plan also covers human resource allocation, pinpointing who will be responsible for each task and what expertise is required to fulfill these roles. The project plan is where the abstract meets the concrete; it is where high-level sustainability targets are translated into actionable tasks, complete with deadlines and deliverables.

How much carbon do you think this building uses? It is impossible to determine the carbon impact of buildings by simply looking at them. Source: sandsun/Adobe Stock Photos.

Furthermore, the plan serves as a holistic guide that integrates various aspects of the project—from architectural design and engineering solutions to financial planning and stakeholder engagement. It is the canvas where all these elements come together in a harmonious composition, ensuring alignment across different departments and specialties.

One of the most critical roles of a comprehensive project plan is that it serves as a single source of truth. When questions arise—and they will—the project plan is the first point of reference. Whether it is a contractor wondering about material specifications or a financial analyst looking for budget allocation, the answers lie within the pages of this well-mapped and updated guide.

Timelines are not mere decoration; they are the skeletal structure that keeps your project upright and moving forward. A precise timeline factors in every little detail, from procurement of sustainable materials to the installation of energy-efficient systems. By allocating time to each task and subtask, you ensure that sustainability remains a priority, even when faced with the tightest of deadlines.

Money talks, especially when it comes to realizing your sustainability vision. It is imperative to prepare detailed cost estimates for each element of your project, be it a green roof, energy-efficient windows, or training programs. This enables you to secure the necessary funding and grants, turning financial constraints into opportunities rather than roadblocks.

Risk is the unwelcome guest at every project party, but ignoring it will not make it go away. Identify potential challenges early in the process, whether they relate to material sourcing, local regulations, or natural elements like wind and sun exposure. By planning for these contingencies and developing appropriate mitigation strategies, you ensure that risk does not derail your sustainability objectives.

In the journey toward sustainability, every team member is a crucial player. Clearly define roles, such as a 'Chief Sustainability Officer,' to oversee the implementation of green initiatives. Role allocation not only streamlines the process but also brings accountability, making sure that sustainability is not a casualty in the hustle and bustle of project management.

Sometimes the bird's-eye view is best seen from outside the nest. Bringing in an external sustainability consultant can offer fresh perspectives and specialized expertise. This can be particularly beneficial when you are aiming to achieve significant gains, like energy use reduction or waste management, that require a nuanced understanding of sustainability metrics. Not everyone can be an expert at everything.

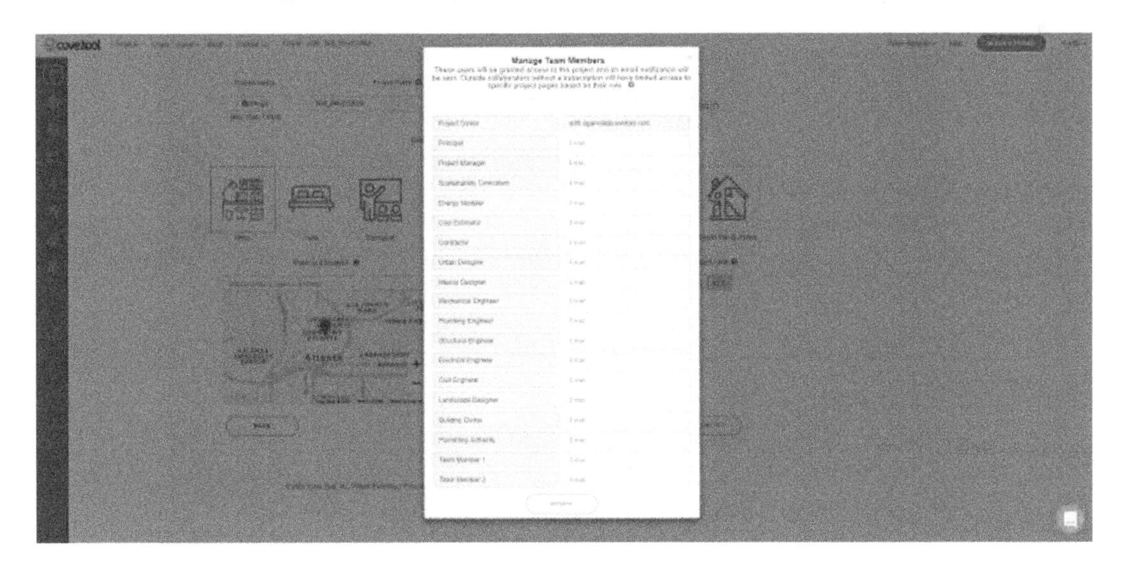

Bringing in different team members for their specific skill set is critical for collaboration. Source: Cove Tool, Inc.

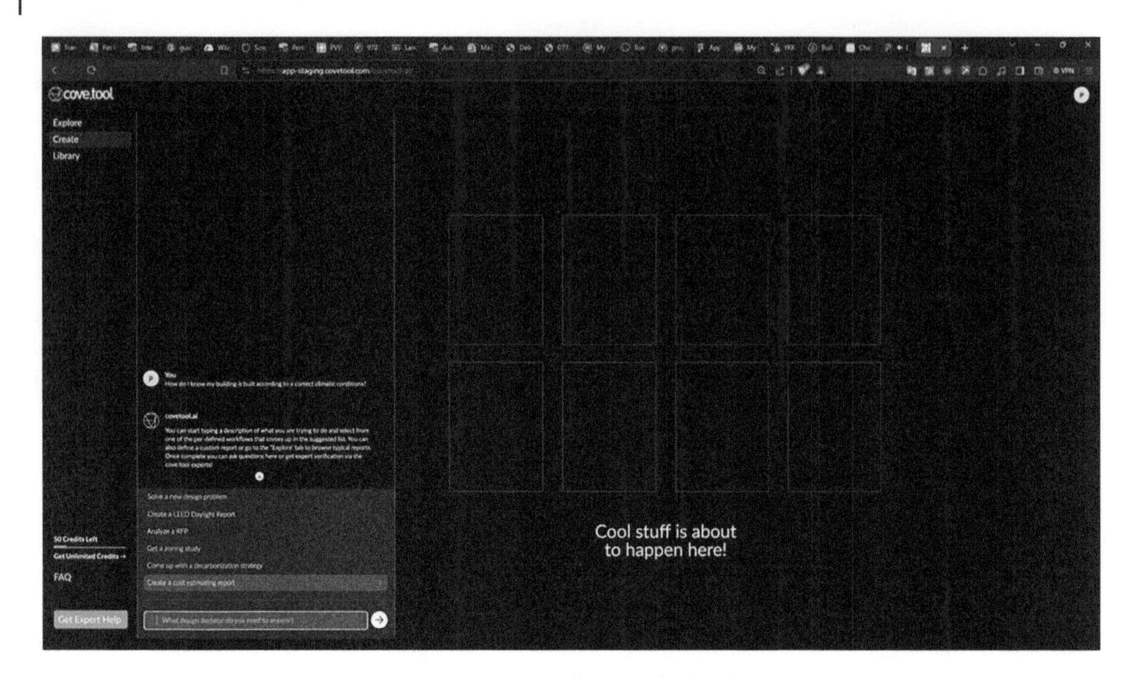

Source: Cove Tool, Inc.

What is a master plan without the masters to execute it? Investing in training equips your team with the specialized skills they need to implement complex, energy-efficient systems or materials. From workshops to on-site demonstrations, effective training turns your plan into a living, breathing, sustainable structure.

Contractors are more than just executors of the design team's vision; they are the artisans who bring the building to life. It is essential to recognize that the success of any sustainability project is heavily dependent on the craftsmanship and expertise of the contractors who turn construction drawings into built environments. As such, developing a robust contractor engagement plan is not just an administrative step; it is a strategic imperative for achieving your sustainability goals.

The engagement plan serves as a manifesto that articulates your expectations regarding sustainable construction practices. This goes beyond the usual terms and conditions to incorporate key elements like waste reduction strategies, responsible sourcing of materials, and adherence to energy-efficient construction techniques. However, a truly effective engagement plan goes one step further—it educates and aligns contractors with the broader sustainability ethos of the project.

Incorporating contractors into the sustainability framework transforms them from observers to active participants in the project's green journey. It is a way of empowering them to make informed choices, whether it is opting for recycled materials over conventional ones or employing construction methods that reduce carbon emissions. Moreover, an engaged contractor is more likely to offer their own insights and solutions, contributing to a culture of continuous improvement and innovation in sustainability practices.

This integrative approach can pay dividends. When contractors are aligned with the decarbonization goals, the chances of encountering delays or complications related to sustainability compliance are significantly reduced. Furthermore, contractors who are educated and committed to sustainable building can become long-term partners, providing valuable expertise for future projects and thereby raising the sustainability bar for the entire industry.

Last but certainly not least, get that crucial nod from your stakeholders. Present your comprehensive implementation plan, complete with financial estimates and timelines, to gain their approval. This final stamp of commitment ensures that everyone, from building owners to local authorities, is on the same page, propelling your project toward its goals.

An implementation plan is not merely a document that gathers dust on a shelf; it is a living, evolving roadmap guiding each decision and action taken during your project's lifecycle. A well-executed plan, rooted in the principles laid out in this guide, can serve as a cornerstone for your project's long-term sustainability. It ensures that the project not only meets but thrives within its environmental, financial, and social contexts. The journey toward sustainable buildings and cities is fraught with challenges, but with a robust implementation plan, you are not just surviving these challenges—you are mastering them. And in mastering them, you are contributing to a more sustainable and resilient world for future generations.

2.6 Step 5: Validating and Verifying the Finished Design

Sustainability in construction and building design is often laden with good intentions, but the true measure of success is in the outcomes. To ensure that your sustainability objectives translate into tangible results, you need to adopt a rigorous process of validation and verification. Step 5 outlines the key facets of this process, offering insights into how to objectively assess your project's performance in relation to its sustainability goals. This step ensures that you are not just doing good but doing it well, and provides the data to prove it.

The starting line is just as important as the finish line when it comes to validating sustainability outcomes. Before embarking on your green journey, establish a clear emissions baseline. This baseline serves as your yardstick for measuring the project's performance, enabling you to quantify the impact of your sustainability initiatives over time.

Smart city. Financial district and skyscraper buildings. Bangkok downtown area at night, Thailand. Source: tampatra/Adobe Stock Photos.

In the age of big data and smart buildings, digital tools are your best friends for tracking emissions and other sustainability metrics. Advanced BMS can provide real-time insights into energy usage, allowing you to compare current performance against the established baseline. These tools bring an element of precision to the otherwise complex task of emissions tracking.

Sustainability is not a one-dimensional endeavor; it permeates every aspect of a project. Continuous monitoring should extend beyond energy usage to include material sourcing, waste management, and even transportation-related emissions. A multipronged approach provides a more holistic view of your project's sustainability performance, enabling adjustments and fine-tuning as you go.

In the realm of sustainability, external validation can be a game-changer. Commissioning independent audits adds an extra layer of credibility to your sustainability efforts, offering unbiased verification that your project is meeting its green targets. This objective lens can often uncover areas for improvement that may have been overlooked otherwise.

Validation and verification are more than just bureaucratic steps; they are the analytical tools that turn your sustainability vision into an accountable, data-driven reality. Implementing these practices ensures that you are not only meeting your objectives but also setting new benchmarks for the industry. In a world increasingly mindful of its ecological footprint, the importance of proving the success of sustainability initiatives cannot be overstated. The final step of verification provides the framework for this essential process, helping you to not only reach but to substantiate your sustainability milestones.

2.6.1 Verify Material Properties

Before sketching the first line of a project, consulting materials databases for carbon emissions data is a critical step. These databases offer a wealth of information that can guide your selection process, allowing you to make informed choices. By understanding the emissions associated with various materials, you can steer your project toward options that align with your sustainability objectives.

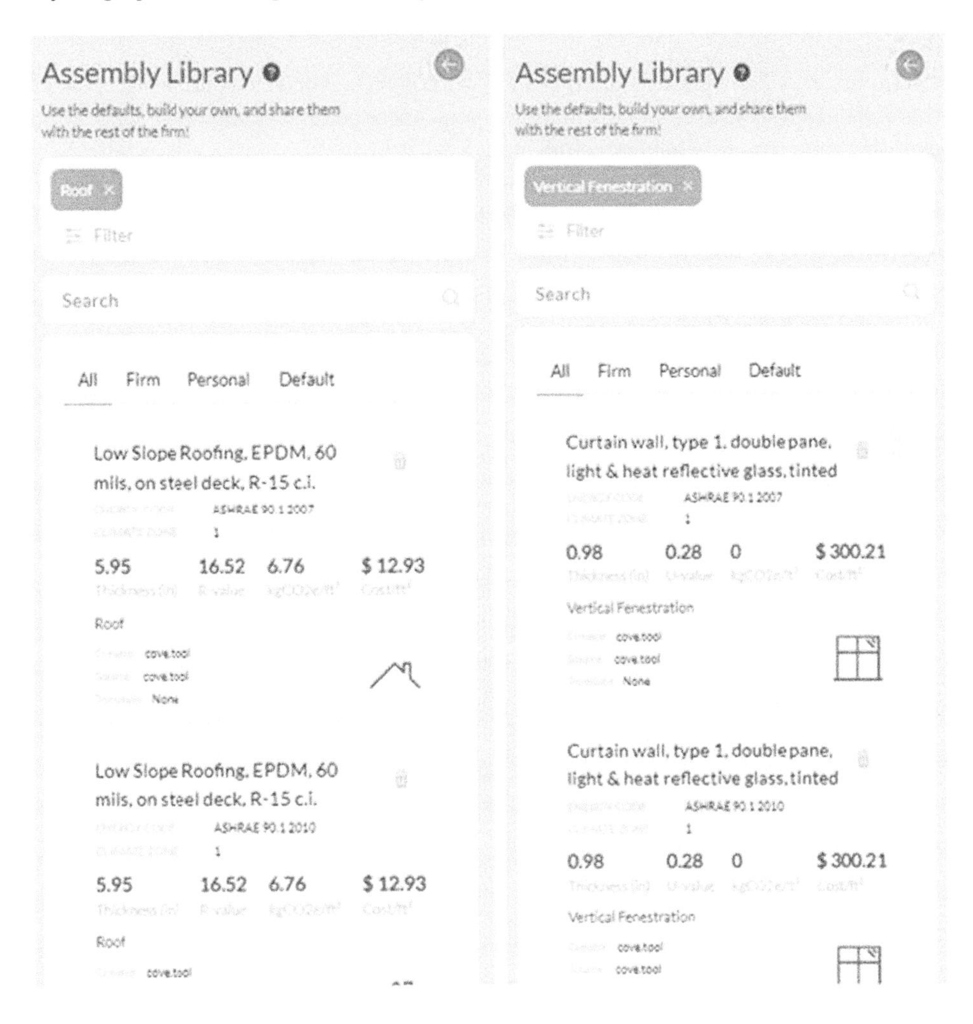

cove.tool's Assembly Library. Source: Cove Tool, Inc.

Operational carbon often gets the spotlight, but embodied carbon is an equally crucial player in a building's overall carbon footprint. Life cycle assessments (LCAs) can quantify this often-overlooked aspect, offering valuable insights into the carbon costs from sourcing to disposal. LCAs enable you to validate the efficacy of your material choices, whether it is sustainable timber, low-carbon concrete, or recycled metals.

Armed with data, the next logical step is to prioritize your material choices. This involves focusing on materials with lower carbon footprints and those sourced sustainably. Prioritization is essential for significantly reducing your project's environmental impact and achieving sustainability certifications.

Recycled and biobased materials offer unique advantages but also come with their own sets of considerations. For instance, some materials may have a higher upfront cost but offer benefits like longer life cycles and lower maintenance costs. Understanding these trade-offs is essential for a nuanced approach to material selection.

When manufacturers claim their products are 'low-carbon,' it is crucial to conduct independent evaluations. This due diligence ensures that the materials you choose genuinely align with your sustainability objectives, rather than simply being marketing hype.

Environmental Product Declarations (EPDs) offer a layer of transparency that is invaluable in sustainable construction. These declarations provide verified, third-party information on a material's environmental impact, and should be a standard requirement for any serious sustainable building project.

ENVIRONMENTAL IMPACTS

Declared Product:
Mix PN4888 • Quivas Plant
Description: 4,000 Non-Air Entrained
Compressive strength: 4000 PSI at 28 days

Declared Unit: 1 m^3 of concrete

Global Warming Potential (kg CO_2-eq)	457
Ozone Depletion Potential (kg CFC-11-eq)	1.19E-5
Acidification Potential (kg SO_2-eq)	1.36
Eutrophication Potential (kg N-eq)	0.55
Photochemical Ozone Creation Potential (kg O_3-eq)	28.0
Abiotic Depletion, non-fossil (kg Sb-eq)	8.00E-6
Abiotic Depletion, fossil (MJ)	503
Total Waste Disposed (kg)	3.76
Consumption of Freshwater (m^3)	0.63

Product Components: natural aggregate (ASTM C33), Portland cement (ASTM C150), admixture (ASTM C494), batch water (ASTM C1602)

Additional detail and impacts are reported on page three of this EPD

Embracing the concept of the "circular economy" involves looking beyond immediate construction needs. It involves selecting materials that are not only low in embodied carbon but also durable and reusable. This holistic perspective ensures that materials can have a second life, thereby minimizing their long-term environmental impact.

A comprehensive approach to sustainability involves scrutinizing a material's entire life cycle, from raw material extraction to transportation. LCAs offer this breadth of view, revealing hidden carbon costs and enabling more informed material choices.

Transportation is a significant but often overlooked contributor to a material's carbon footprint. Favoring local materials can help cut these emissions significantly. Additionally, sourcing locally can have economic benefits, supporting regional industries and communities.

Material choices often involve balancing various factors, such as weight and insulation capabilities. For example, heavier materials might offer better thermal insulation but could increase the project's overall carbon footprint due to transportation emissions. These complex considerations require meticulous calculations to strike the right balance.

Choosing non-toxic materials serves a dual purpose. On one hand, they contribute to a healthier indoor environment; on the other hand, they often have lower manufacturing emissions. Thus, selecting nontoxic materials aligns well with both health and sustainability goals. Items made with oil are typically bad for carbon and health.

Choosing materials for a sustainable construction project is not a linear or straightforward process. It requires a multifaceted approach that considers everything from carbon emissions data and lifecycle assessments to trade-offs and local sourcing. Each decision point is an opportunity to align more closely with sustainability objectives, reduce carbon footprint, and contribute to a project that stands as a testament to environmental stewardship. The key lies in rigorous verification, transparent data, and a holistic approach to material selection. When these elements come together, they form the backbone of a project that is not just built to last but built to sustain.

2.6.2 Sustainable Building Design Verification Checklist

With so many things to track, having a checklist is key. This example checklist is structured to align with the various phases of the building design process, providing up to 10 critical verification points for each stage. From the conceptual phase to post-construction and occupancy, each section offers actionable items that serve as both a guide and a means of accountability. When you use checklists, be sure that the people on the team understand the 'why' for best results.

The aim is not merely to meet regulatory benchmarks but to exceed them, to not just minimize harm but to contribute positively. Whether you are a seasoned architect or a client keen on sustainability, this checklist equips you with the framework to verify that your project's sustainability goals are not only set but also met and, ideally, surpassed. While certainly more or less items could be added, we recommend that design teams break them down by phase, assign responsibility, and provide regular updates on progress toward the goals.

2.6.3 Conceptual Phase

- **Preliminary Site Analysis:** Verify sunlight patterns and orientations for maximum natural light.
- **Initial Carbon Footprint Estimate:** Utilize databases to get a rough estimate of potential material carbon footprints and look for opportunities to reuse a structure of the existing building for example.
- **Stakeholder Alignment:** Ensure preliminary design aligns with sustainability goals of all stakeholders.
- **Local Material Availability:** Check for availability of local materials to reduce transportation emissions.
- **Regulatory Requirements:** Validate that the conceptual design is in line with local and national sustainability regulations.
- **Passive Design Elements:** Identify opportunities for passive solar heating or natural ventilation.
- **Energy Targets:** Set preliminary energy performance targets based on building type and location.
- **Water Efficiency:** Consider opportunities for rainwater harvesting or greywater reuse.

- **Public Transportation Access:** Verify accessibility to public transportation to reduce transportation-related carbon footprint.
- **Renewable Energy Sources:** Identify potential for on-site renewable energy generation.

2.6.4 Schematic Design Phase

- **Computational Modeling:** Validate solar orientation and shading via simulations.
- **Thermal Modeling:** Confirm thermal mass elements' efficacy.
- **Material Shortlisting:** Shortlist sustainable materials for construction.
- **Waste Management Plan:** Develop an initial plan for construction waste management.
- **Embodied Carbon Calculation:** Utilize tools for a more accurate estimate of embodied carbon based on design.
- **EUI Targets:** Set specific energy use intensity (EUI) targets.
- **HVAC Preliminary Selection:** Choose potential high-efficiency HVAC systems.
- **Natural Light Simulation:** Validate natural light levels in different spaces.
- **Acoustic Comfort:** Consider acoustic factors related to material choices and space design.
- **Finalize Window-to-Wall Ratio:** Validate energy modeling for thermal performance.

2.6.5 Design Development Phase

- **High-Performance Glazing:** Verify thermal performance through tests.
- **Cost Estimates:** Begin initial cost estimates for sustainable elements to ensure budget alignment.
- **Thermal Bridging:** Utilize thermal imaging to ensure design details eliminate thermal bridging.
- **LCA for Material Choices:** Conduct Life Cycle Assessments for shortlisted materials.
- **Water Efficiency Devices:** Specify water-efficient fixtures and fittings.
- **Indoor Air Quality:** Validate material choices for VOC emissions and indoor air quality.
- **Daylight Metrics:** Confirm Spatial Daylight Autonomy (sDA) and Annual Sunlight Exposure (ASE) metrics.
- **Accessibility:** Ensure design meets sustainable mobility and accessibility standards.
- **Resilience:** Verify that design elements contribute to climate and disaster resilience.
- **EPDs Requirement:** Confirm that Environmental Product Declarations (EPDs) will be available for major materials.
- **Energy Modeling:** Run a comprehensive detailed energy model to verify design against energy targets.

2.6.6 Construction Documents Phase

- **Airflow Modeling:** Confirm natural ventilation strategies.
- **Material Specifications:** Ensure all specified materials meet sustainability criteria.
- **Energy Code Compliance:** Validate that the design meets or exceeds local energy codes.
- **Final Waste Management Plan:** Update and finalize the construction waste management plan.
- **Detailed Cost Analysis:** Finalize cost estimates for all sustainable design elements.
- **Contractor Guidelines:** Include sustainability requirements in contractor guidelines.
- **Quality Checks:** Outline quality assurance procedures for sustainability metrics.

- **Documentation:** Prepare for collecting evidence for potential certification.
- **Contingency Plans:** Develop plans for potential deviations from sustainability targets.

2.6.7 Bidding and Negotiation Phase

- **Vendor Performance Testing:** Require energy performance data for HVAC and other systems.
- **Sustainability Criteria:** Ensure bidders can meet sustainability goals.
- **Local Sourcing:** Confirm commitments to source materials locally where possible.
- **Certifications:** Check contractor certifications in sustainable building practices.
- **Cost Validation:** Confirm that sustainability features are within budget.
- **Contract Clauses:** Include clauses for sustainability targets and penalties.
- **Material Alternatives:** Keep options for sustainable material alternatives.
- **Transportation Plan:** Validate the low-emission logistics plan for material transportation.
- **Lifecycle Costing:** Confirm lifecycle costs of sustainable technologies.
- **Stakeholder Review:** Ensure all stakeholders agree on final material and technology choices.

2.6.8 Construction Phase

- **On-Site Energy Audits:** Validate waste heat recovery and other energy-saving measures.
- **Material Receipt Checks:** Verify sustainable materials upon receipt.
- **Quality Assurance:** Conduct regular checks to ensure construction aligns with sustainable design.
- **Waste Management Monitoring:** Ensure waste is managed as per the plan.
- **Documentation:** Keep meticulous records for potential green building certification.
- **Water Testing:** Confirm the efficiency of water-saving devices.
- **Air Quality Tests:** Conduct tests to ensure indoor air quality standards are met.
- **Mock-Ups:** Use mock-ups to verify elements like natural lighting and thermal comfort.
- **Commissioning:** Preliminary commissioning of HVAC and other systems for performance.
- **Stakeholder Updates:** Keep stakeholders updated on sustainability metrics.

2.6.9 Post-Construction and Occupancy Phase

- **Post-Occupancy Evaluation:** Conduct comprehensive tests to validate energy and water performance.
- **Building Handover:** Ensure all sustainability features are fully functional at the time of handover.
- **Final Documentation:** Complete all documentation needed for green building certification.
- **User Training:** Educate occupants on sustainable features and best practices.
- **Monitoring Systems:** Verify the functioning of systems that monitor energy and water use.
- **Feedback Loops:** Establish channels for occupants to report issues related to sustainability features.
- **Renewable Energy Check:** Validate the performance of renewable energy systems.
- **Review Against Targets:** Compare actual performance against sustainability targets.
- **Lessons Learned:** Document insights gained for future projects.
- **Continuous Improvement:** Establish plans for ongoing performance monitoring and improvement.

The Sustainable Building Design Verification Checklist is more than a catalog of tasks; it is a blueprint for embedding sustainability into the very DNA of your project. By providing a structured approach to verification, the checklist serves as your navigational compass, ensuring that you stay true to your project's sustainability objectives at every phase of design and construction. It should be a dynamic tool, adapted to new advancements in technology and sustainable practices, allowing for both rigorous verification and innovative thinking. In a world grappling with the urgency of climate change, bringing tactical rigor to your project is a necessity. It is surprising how often teams make great goals but forget to verify with clear tracking. Use your checklist as both a guide and a gauge, measuring the project's true impact and steering it toward a future that is not just built but sustainably crafted.

2.6.10 Leveraging Technology

Technology plays an indispensable role in ensuring that design aspirations translate into tangible outcomes. While sustainability goals can be set with the best intentions, the final challenge lies in the verification of these objectives throughout the design and construction process. It is here that technology emerges as a linchpin, offering a suite of tools for real-time tracking, digital collaboration, and data-driven decision-making.

In the initial design and planning phases, using digital collaboration tools like BIM can vastly streamline communications among architects, engineers, and contractors. These digital platforms not only speed up decision-making but also minimize paper waste and reduce the carbon footprint associated with physical meetings. A digital-first approach ensures that all team members are on the same page, aligning with the project's overall sustainability goals.

During the material selection and procurement phase, digital tracking tools can offer an unparalleled level of transparency in the environmental impact of every material choice. These digital ledgers can inform real-time adjustments in the material selection process, ensuring that the most sustainable options are prioritized. It allows teams to verify the sustainability credentials of materials and make data-driven decisions.

As you transition to the construction phase, embracing digital supply chain management becomes invaluable for minimizing waste and emissions. By optimizing material orders and tracking them through a digital platform, you can significantly reduce both material waste and the carbon emissions associated with transportation. This is a crucial step in ensuring that the project stays true to its sustainability objectives.

In the design development and pre-construction stages, employing digital twins—virtual replicas of your building—can provide invaluable insights. These digital simulations allow you to test everything from energy efficiency to airflow, enabling adjustments to the design before construction begins. This preemptive verification ensures the design's optimal performance and minimizes its future environmental impact.

Once the building is operational, sensor technology can offer real-time data on everything from energy usage to occupancy levels. This allows for immediate adjustments to building operations, like dimming lights in unoccupied areas or modifying HVAC settings based on occupancy, thereby optimizing energy efficiency.

For the operational phase, the integration of smart meters can provide granular data on energy consumption. This real-time information enables continual adjustments to energy management strategies.

Such an approach ensures that the building remains not just compliant with, but at the forefront of, sustainability goals.

By systematically incorporating these technology-driven approaches at appropriate phases of your project, you can ensure that each design decision aligns with your sustainability goals, thereby making verification an integral part of your workflow.

2.6.11 Rethinking the Role of Certification Programs in Decarbonization and Sustainable Design Verification

The design and construction industry faces a critical juncture on certifications. The conversation has shifted from merely meeting established sustainability metrics to pushing the envelope to achieve real decarbonization. While certification systems like LEED and BREEAM have long been seen as the gold standard in sustainable design, their role is evolving in an era where local and international energy codes are setting increasingly ambitious targets, often surpassing the benchmarks set by these certification programs.

It is essential to recognize that while LEED, BREEAM, and similar frameworks offer a structured pathway to sustainability, they are not universally applicable to all projects. Assessing the unique needs of your project against the criteria set by these certifications can help pinpoint where efforts will yield the most significant impact. This exercise can also reveal whether such certifications align with your decarbonization goals or if they fall short.

The requirements for achieving different certification levels are often layered and complex. Using digital modeling can simulate various scenarios for compliance, from water-saving technologies to renewable energy solutions. This preparatory stage can serve as a litmus test for whether pursuing a particular certification level will genuinely contribute to your project's sustainability and decarbonization objectives.

Given the rising stringency of local and international building codes, these regulations now serve as an additional—and often more rigorous—layer of sustainability standards. Complying with such codes not only fulfills legal requirements but also provides a robust framework for sustainability that often exceeds traditional certification metrics.

The process of acquiring a sustainability certification isn't trivial. A strategic plan detailing timeline, required documentation, and responsibilities can significantly streamline the process. This roadmap serves as a continuous reminder of the commitment to achieving verifiable sustainability goals.

An early-stage assessment can identify the sustainability and decarbonization measures already in place and where the project falls short. This gap analysis can serve as a blueprint for targeted efforts throughout the design and construction phases.

Documentation serves as the backbone of the certification process. A well-organized submission, complete with evidence of implemented sustainability measures, can significantly ease the certification process, ensuring that your project's decarbonization efforts are verifiable and transparent.

Rather than treating certification as a final stamp of approval, it should be woven into every aspect of the design process. Regular meetings that integrate sustainability discussions ensure that the project stays aligned with its decarbonization and certification goals.

The frenetic pace of the construction phase often poses a risk to well-laid sustainability plans. Continuous tracking of certification requirements during this stage can safeguard against this, ensuring that the project remains on the path to sustainability and decarbonization.

The home stretch of the certification process is often the most labor-intensive, requiring meticulous documentation. A dedicated team for this final push can be invaluable in ensuring that all decarbonization and sustainability efforts are verifiable and result in successful certification.

Earning a certification is not the end of the road but rather the beginning of a commitment to ongoing sustainability and decarbonization efforts. Utilizing smart technologies to continually monitor performance can help ensure that your building remains a leader in sustainability, rather than becoming complacent.

In an era where decarbonization is not just an ideal but an imperative, traditional certification program must be re-evaluated. While they continue to offer valuable frameworks for sustainability, the industry is increasingly finding that these are starting points, not finish lines, in the race to mitigate climate change.

2.6.12 Navigating Challenges and Adaptations in Sustainable Design

In the quest for sustainable design and decarbonization, challenges are not just inevitable; they are integral to the journey. From financial constraints to data accuracy, these challenges test the robustness of your sustainability goals and the adaptability of your strategies. However, they also offer valuable opportunities for innovation and optimization. This section aims to delve into common challenges you may encounter and provides adaptive strategies to navigate them effectively. With a proactive approach and the right tools, these challenges can be transformed into stepping stones toward achieving your sustainability and decarbonization objectives.

One of the most daunting aspects of sustainable design is the initial cost. Cutting-edge technologies and sustainable materials often come with a higher price tag. However, it is critical to weigh these upfront costs against long-term lifecycle savings, such as reduced energy consumption and maintenance costs. Lifecycle cost analysis can provide a comprehensive view, allowing for informed decisions that account for both immediate expenses and future benefits.

Resource limitations can impede the immediate implementation of all planned sustainability features. However, a phased approach can still achieve impactful results. Begin with measures that provide immediate benefits and are easier on the budget, like energy-efficient lighting. This can create a foundation upon which more comprehensive and cost-intensive measures can be implemented in later phases.

Monetary constraints need not always hamper sustainability goals. Various public and private incentives, grants, and subsidies can serve as crucial financial catalysts. Understanding and leveraging these opportunities can bridge the budget gap and enable the implementation of ambitious sustainability measures.

The credibility of a project's sustainability claims rests on the accuracy of its data. Independent audits can provide a necessary layer of verification, validating emission metrics and other sustainability indicators. This not only adds rigor to your claims but also enhances the project's credibility and public trust.

Making informed decisions requires a nuanced understanding of a project's sustainability performance. Utilizing multiple data sources, such as in-house analytics, external audits, and industry benchmarks, can provide a more complete picture. This comprehensive approach allows for targeted improvements and ensures that decisions are grounded in robust data.

While digital models offer invaluable insights into a building's potential performance, they are not infallible. Physical tests can serve as a reality check, validating the theoretical benefits projected by these models. This dual approach strengthens the reliability of the project's sustainability claims and ensures they can withstand real-world conditions.

For sustainability measures to succeed, they must enjoy broad stakeholder buy-in. Interactive demonstrations, perhaps even virtual reality experiences, can help immerse stakeholders in the project's sustainability vision, fostering understanding and support. This emotional engagement can be a powerful motivator, turning stakeholders into advocates for the project's sustainability goals.

A successfully implemented sustainability project can serve as a beacon for others in the industry. Documenting and sharing these successes, perhaps through white papers or educational programs, can catalyze broader sustainability efforts, creating a ripple effect that extends far beyond the individual project.

The complexities of sustainable design often require specialized expertise. Partnering with sustainability consultants can bring invaluable insights into the project, from navigating regulatory requirements to optimizing material selection. This collective intelligence can be a game-changer in overcoming the myriad challenges that sustainability efforts often entail.

The long-term success of a building's sustainability features rests significantly with its facilities managers. Involving them early in the design process ensures that the sustainability measures are not only innovative but also practical and maintainable. This early engagement creates a seamless transition from the design and construction phases to the building's long-term operation, ensuring sustainability remains a focus throughout its lifecycle.

2.7 Conclusion

In a world increasingly attuned to the existential crisis of climate change, the steps outlined in this chapter serve as a comprehensive blueprint for building decarbonization. It is a call to action for design teams, contractors, and industry stakeholders. We began with the essential task of setting specific decarbonization and project goals, underlining that a well-defined objective is the cornerstone for any successful endeavor. From there, we delved into the often-overlooked importance of scrutinizing existing design and construction processes. The intent? To pinpoint inefficiencies and pave the way for targeted solutions that are both innovative and environmentally responsible.

The subsequent steps—developing targeted solutions and creating an implementation plan—act as the engine that propels a project from ideation to fruition. These steps speak to the power of foresight and planning. Tactical application always trumps good intentions. However, the final and perhaps most critical step is validation and verification. This is not merely a box to check off; it is the connection that ensures the project's sustainability goals are not only met but also quantifiably measured. The need for empirical data to validate the project's success underscores the narrative that sustainable design is as much a science as it is an art.

The chapter serves as a stark reminder that building decarbonization is not a lofty ideal but an attainable reality. It challenges the age-old wisdom that sustainable design is incompatible with economic viability. In fact, the two can and should be symbiotic. As we grapple with the urgency of climate change, these steps lay the foundation for a future where buildings are not just places we inhabit but are, in essence, living, breathing entities committed to the well-being of our planet.

2.8 Case Study: Georgia Tech Campus Center

It is increasingly important for institutions to do their part in mitigating the effects of climate change. Many educational and research institutions are upgrading their buildings to become carbon positive or zero emissions. Leading the charge, they set the example for future generations and prepare their students for the future that will be impacted by climate change. In the end, they have a moral responsibility to do so. As places of learning, they have a unique opportunity to educate about the importance of sustainability and the need to take action on climate change. They can inspire their students to make sustainable choices in their own lives.

Campus Center at Georgia Institute of Technology. Source: Cove Tool, Inc.

Spanning an impressive 300,000 square feet, the Campus Center at Georgia Tech stands as a transformative expansion from the heart of the campus, Tech Green, extending its reach westward. This ambitious project, evolving from the original Student Center constructed in the 1970s, marks a significant transition. Originally designed to serve 7000 individuals, the renovated and expanded Campus Center now caters to the dynamic needs of over 23,000 students and 7000 faculty and staff.

The project is a trifecta of architectural development: the revitalization of the Wenn Student Center, the addition of a modern Exhibition Hall, and the construction of a pavilion accompanied by a café. These structures are seamlessly connected by a newly developed network of outdoor pathways and plazas, fostering an environment of connectivity and community engagement.

Wenn Student Center, Georgia Institute of Technology.

At the core of this project is an unwavering commitment to redefining the standards of building performance. The collaborative design-build team, comprising Workshop, Cooper Carry, Newcomb and Boyd, and Gilbane, was instrumental in pushing the boundaries of architectural excellence. The team set ambitious goals, challenging conventional approaches and elevating the standards of campus infrastructure.

To realize these lofty objectives, the team employed an array of cutting-edge tools and methodologies. cove.tool, BIM 360, and Bluebeam were pivotal in ensuring an integrated approach to project design and management. This strategic application of technology manifested in three distinct phases: the pursuit phase, where these tools were critical in securing the project; the construction drawing phase, where they played a key role in minimizing design revisions; and finally, the construction phase, where they were instrumental in optimizing project costs.

This holistic approach to the Campus Center's development not only reflects Georgia Tech's innovative spirit but also sets a new benchmark in campus center design, showcasing the potential for harmonizing ambitious architectural goals with practical, cost-effective solutions. The project is as a testament to the power of collaborative innovation, sustainability, and the transformative impact of thoughtful, integrated design in the realm of higher education.

2.8.1 Pursuit Phase

Georgia Tech orchestrated an innovative ideas competition to select a design-build team for their ambitious Campus Center project. In this competition, three teams, comprising architects, contractors, landscape architects, engineers, and consultants, presented their visions for realizing the Campus Center within the predefined budget constraints.

The eventual winning team, an integrated design-build consortium featuring Workshop, Cooper Carry, Newcomb and Boyd, and Gilbane, distinguished themselves through their strategic use of cove.tool. During the critical pursuit phase, this tool enabled early-stage design analysis, facilitated real-time collaboration across disciplines, and fostered the development of a unique, cost-effective approach aimed at achieving the lowest carbon footprint. Combined with expertise of the team, it was a winning formula.

A key aspect of their proposal involved exploring four diverse massing strategies, leveraging automated energy and daylight performance modeling to demonstrate the efficacy of each approach during the interview process. This innovative use of technology was instrumental in showcasing their design capabilities and alignment with Georgia Tech's sustainability goals.

Central to Georgia Tech's vision was the commitment to identifying the most carbon-efficient solution. The team rose to this challenge, rigorously evaluating various massing options, including the potential renovation of existing buildings versus new constructions. An initial consideration involved the demolition of the existing structure, allowing for reorientation to mitigate the challenges posed by large east and west façades, unsuitable for Atlanta's warm climate. This strategy would enable the creation of high-performance façades, a significant improvement over the existing low-performance envelope.

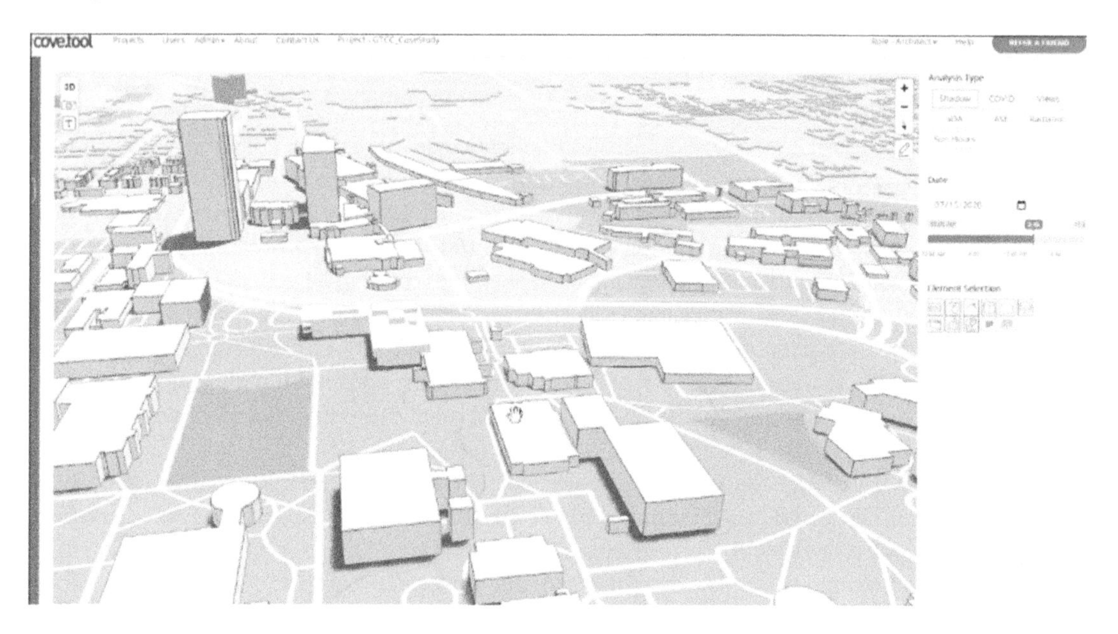

Shadow studies within cove.tool's analysis tool. Source: Cove Tool, Inc.

Radiation studies to understand the hot spots on façade.

View analysis to determine the areas within the floor plate that would have a view, impacting the wellness within the space.

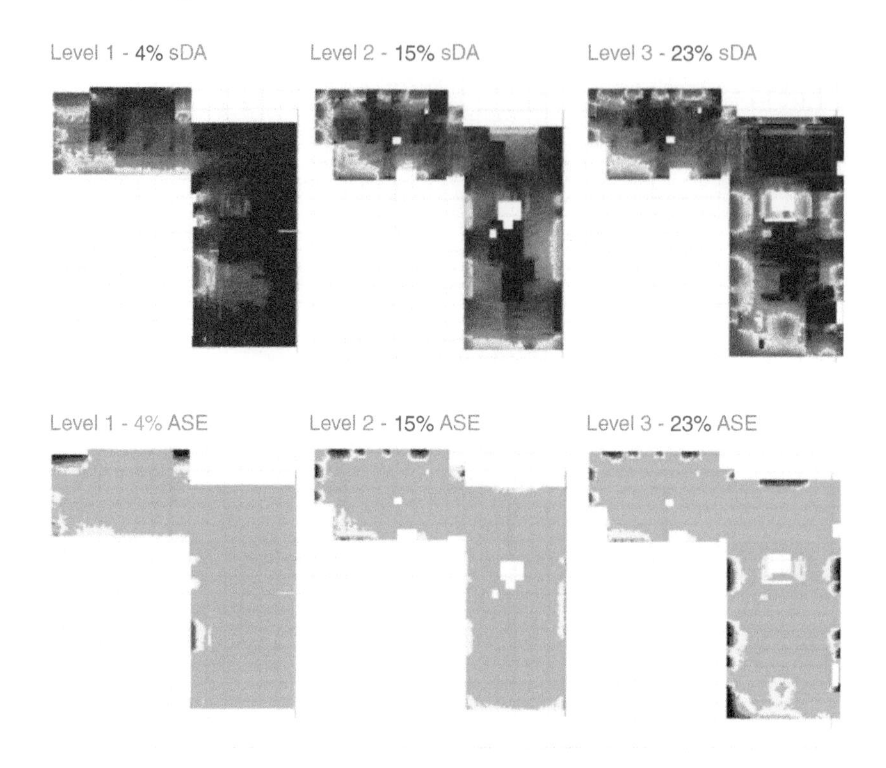

Daylight and glare studies in cove.tool's analysis tool.

cove.tool water use study.

However, a comprehensive analysis of both operational and embodied carbon revealed a striking insight: preserving the existing structure significantly reduced the overall carbon impact. This approach not only offered operational carbon savings but, more crucially, reduced the carbon footprint associated with new construction. This made renovation not just a superior choice in terms of carbon efficiency but also more cost-effective, presenting a win-win scenario from both environmental and financial perspectives.

This strategic recommendation, underpinned by their data-driven approach and technological prowess in the early design process, was a pivotal factor in their winning the competition. Their ability to utilize data and technology to present in-depth design performance insights and early recommendations aligned perfectly with Georgia Tech's sustainability criteria.

Moreover, the team's cost consciousness was evident throughout the pursuit. The automation and integrative collaboration allowed for efficient analysis within a tight budget and timeline. Each idea iteration in the pursuit phase required less than 60 minutes for analysis and documentation, enabling the team to effectively communicate preliminary performance data to the Georgia Tech selection committee. This approach not only showcased their commitment to sustainability but also demonstrated their ability to deliver innovative solutions within stringent budgetary and time constraints.

2.8.2 Design Development – Construction Phase

In the design-build project of the Campus Center at Georgia Tech, cost considerations played a pivotal role in the decision-making process throughout the design phase. The integration of Gilbane's preconstruction team with their proprietary cost estimating technology into the project's framework was crucial. This collaboration, coupled with the utilization of cove.tool's optimization feature, provided the team with a robust understanding and control over the project's budget.

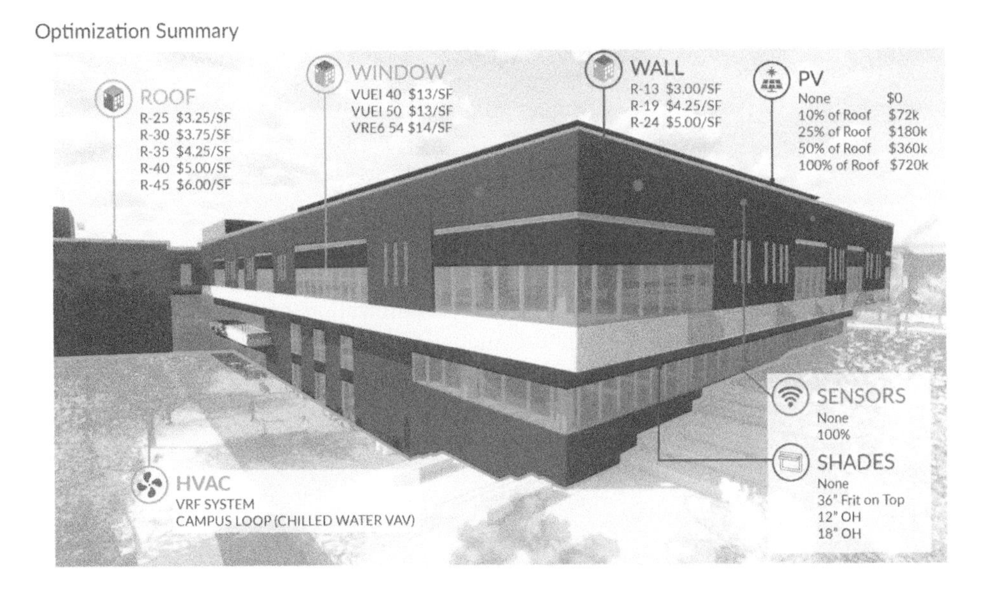

Design-build project of the Campus Center at the Georgia Institute of Technology, cost considerations.

The team used dual optimization for both initial (first cost) and ongoing (lifecycle cost) expenditures. The diagram mentioned highlights the complexity of these decisions. Critical choices had to be made regarding various building components, including window types, wall and roof insulation, HVAC systems, photovoltaic panels, shading strategies, and sensors. Each of these elements came with multiple options, necessitating thorough evaluation.

The challenge was compounded by the sheer number of potential combinations — a staggering 3600 — each presenting a unique profile in terms of energy usage and associated costs. Technology was instrumental in navigating this complex landscape. It enabled the team to methodically analyze each combination, thereby identifying the most cost-effective route to achieve their performance targets.

This approach exemplifies a meticulous, data-driven process, blending technological sophistication with practical budget management. It highlights the project's commitment to not only meeting performance standards but also ensuring financial viability, a critical aspect of sustainable and responsible design in the modern architectural landscape.

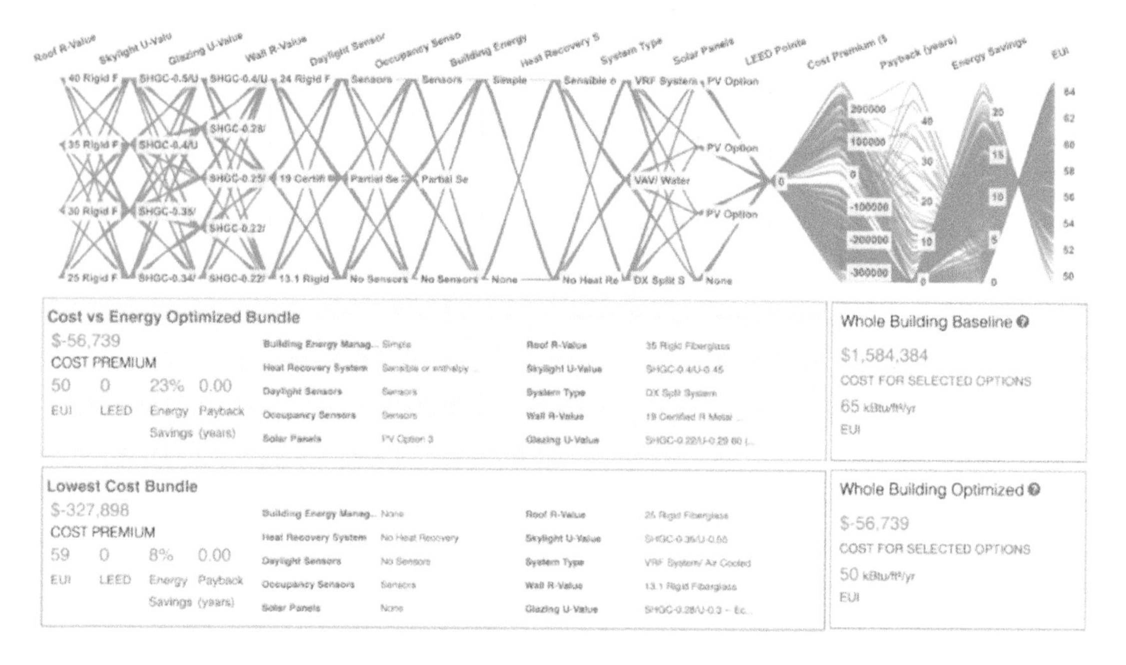

In the Parallel Coordinates Graph for Cost Vs Energy Optimization, each vertical line represents one decision. The blue line represents every possible combination and these can be filtered based on potential LEED points, EUI, Energy Savings, Payback, and First Cost Premiums. For a project with a target payback of less than 10 years and an EUI target of 60 or below, there were possible combinations costing under an additional US$100,000 to $250,000.

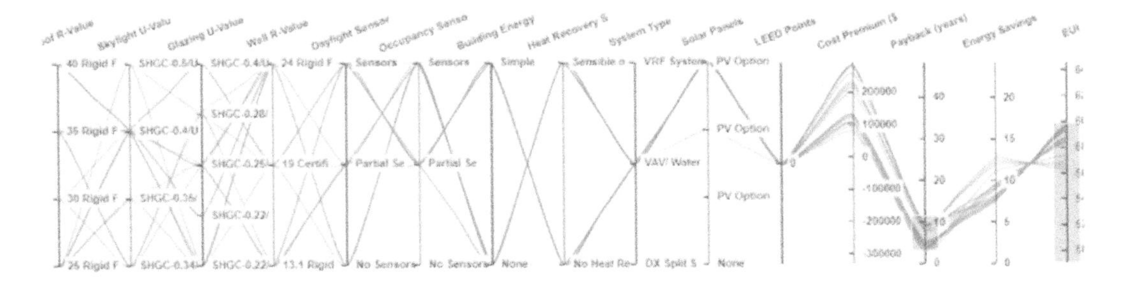

The design team used the sliders to find the absolute lowest cost option.

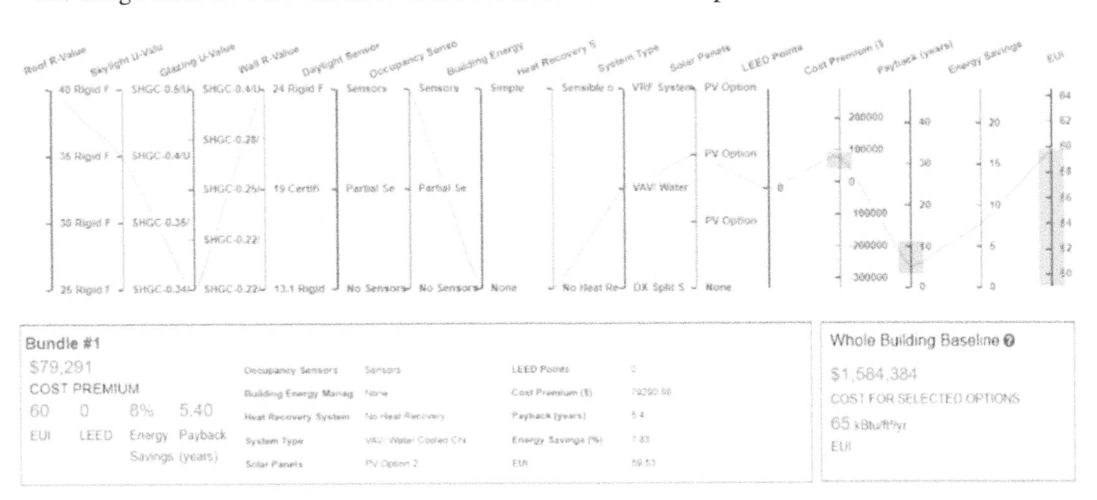

2.8.3 The Importance of Collaboration

The Campus Center project at Georgia Tech, characterized by its diverse team of professionals each bringing a unique expertise, presented both an asset and a challenge in project management. Ensuring alignment and real-time collaboration among team members was crucial for the project's success. This collaborative effort underscored the importance of harnessing a variety of perspectives and skill sets, transforming potential challenges into a cohesive, innovative force driving the project forward. Ultimately, the success of this project serves as a testament to the power of teamwork and the synergistic potential of combining diverse expertise in achieving architectural excellence and sustainable design.

3

What is Carbon Positive Construction?

3.1 Considering Variables Beyond Cost

Imagine a world where buildings become symbols of sustainability, each a fortress against climate change. A stretch? Perhaps. But the reality is that the construction industry has the potential to be a hero in the fight against global warming; it just needs a new set of plans.

Construction is a giant among industries, but it is a giant with a heavy footprint. In the United States alone, the sector is responsible for about 11% of the nation's total greenhouse gas emissions. Meanwhile, China and India, surging ahead in urban development, contribute significantly higher percentages to their national tallies but do not let the propagandist fool you that the global north takes the lions share of historical carbon emissions.

Where does all this construction carbon come from? Concrete, the industry's staple, is a notorious emitter. Excavators and cranes guzzle fossil fuels. Transporting materials over long distances adds another layer of emissions. Construction makes a massive difference in the selection of low carbon building materials since contractors often swap out materials to maintain project cost and time.

3.1.1 A Path to Transformation: Seven Pillars

The good news is that the industry does not need to reinvent the wheel; it needs to re-engineer it. There are seven major avenues that contractors can explore to reduce emissions and boost productivity:

1) **Reshape Regulation and Raise Transparency:** Streamline permits insist on data transparency and focus on outcome-based building codes.
2) **Rewire the Contractual Framework:** Shift from adversarial to collaborative contracting models, prioritizing value and performance over low-cost bids.
3) **Rethink Design and Engineering:** Adopt value engineering and apply manufacturing principles to construction.
4) **Improve Procurement and Supply Chain:** Use technology to enhance planning and coordination, with a focus on reducing waste.
5) **Improve On-Site Execution:** Embrace rigorous planning and employ digital tools for better communication and performance tracking.

Build Like It's the End of the World: A Practical Guide to Decarbonize Architecture, Engineering, and Construction,
First Edition. Sandeep Ahuja and Patrick Chopson.

6) **Infuse New Technology and Materials:** Adopt building information modeling (BIM) and explore new, sustainable materials.
7) **Reskill the Workforce:** Invest in continuous training and reskilling to keep pace with technological advancements.

3.1.2 Why Contractors Should Care

Why focus on contractors? Because they are the conductors of this complex orchestra, wielding the baton that can bring all these elements into harmony. Their decisions on materials, methods, and manpower can make or break a project's carbon footprint. It is time to roll up those sleeves; we have a planet to save. In the chapters that follow, we will dig deeper into each of these seven pillars. Let us explore technologies, strategies, and best practices that can be the turning point for an industry—and a planet—in crisis.

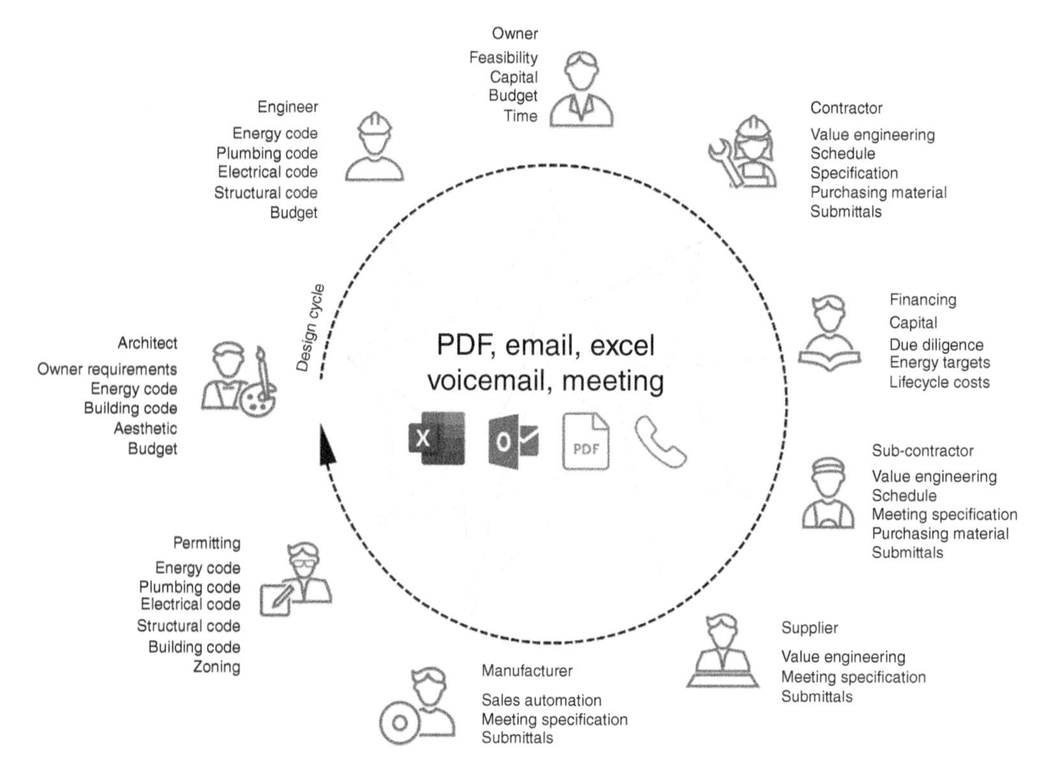

Incompatible tools and workflows from different participants leading to outcomes and costs are not transparent within the AEC industry.

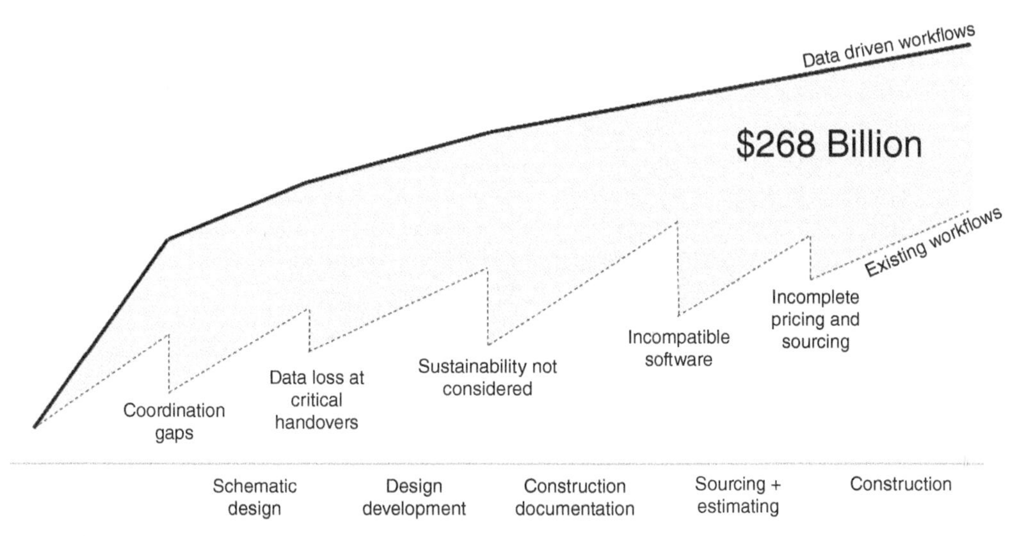

Each stage in the design and construction process destroys data from the previous step leading to outcomes and costs that are not transparent within the AEC industry.

3.2 Reshaping Regulation and Raising Transparency

Regulation and transparency are not merely buzzwords but the linchpins of a robust, ethical, and sustainable construction industry. The winds of change are sweeping through this sector, driven by technological evolution, heightened environmental scrutiny, and economic dynamism. However, this transformation also brings with it the darker elements of regulatory snags, opacity, and the ever-lingering specter of corruption. The subject is as delicate as it is urgent, demanding a recalibration of longstanding practices. Instead of lambasting the industry's pitfalls, let us pivot to an ethos of constructive solutions.

When we say "transparent policies," we are referring to a gamut of legislative actions. This ranges from policies that internalize the true environmental costs of carbon emissions to the formulation of green building codes and public disclosure requirements for construction emissions. The pages ahead will dissect how we can construct a new regulatory framework. This reimagined structure would not only smooth out the kinks in permitting and approvals but also inject greater transparency into cost and performance metrics. It aims to pivot from a cumbersome, one-size-fits-all approach to more agile, outcome-based building codes focused on zero carbon designs.

3.2.1 Reengineering the Permitting and Approvals Process

The labyrinthine permitting process stands as a formidable obstacle to the construction of zero carbon buildings. The permit is your ticket to the construction arena, confirming your compliance with existing standards. But today's process is an unwieldy behemoth, slow-moving and fraught with complexities.

A shift toward a cooperative and anticipatory regulatory approach could be the game-changer here. This could mean well-defined guidelines and timelines, pre-application consultations, and the digitization of the application and tracking process. Further streamlining can be achieved by employing technology to automate mundane tasks like data entry and document scrutiny. Such moves would make the pursuit of zero carbon building projects less daunting and more appealing for developers.

Moreover, embracing performance-based regulations can revolutionize the construction landscape. This strategy sets specific performance goals—be it energy efficiency or safety—while granting constructors the flexibility to innovate within those parameters. This is a win-win, ensuring compliance without stifling creativity.

3.2.2 Instilling Transparency in Cost and Performance Metrics

The absence of transparency is often the Achilles's heel of ambitious zero carbon projects. This opacity acts like a thick fog, enveloping critical details such as project costs, timelines, and performance metrics. The result? A precarious environment where accountability becomes an elusive goal and stakeholders are left navigating in the dark. There are powerful interests that work hard against transparency but that should not discourage us from pushing for reform.

One crucial antidote to this problem is for regulatory bodies to enforce more detailed, or "granular," reporting requirements. Applied poorly this could be a bureaucratic chore. However, this kind of reporting serves as the DNA of a project, revealing its essential traits and tendencies. It includes not just the conventional financial outlines but extends to performance indicators such as energy utilization, carbon footprint reduction, and safety metrics.

The idea is not merely to turn on the lights but to put everything under a microscope. This enables stakeholders to scrutinize projects at a granular level, peeling away layers of ambiguity and exposing the raw data for what it truly represents.

Ethical considerations are another layer that deserves explicit attention. This includes not just potential conflicts of interest but extends to other ethical dimensions like labor practices and sourcing policies. Such a multifaceted approach to transparency ensures that the project is not only economically and environmentally viable but also ethically sound. In other words, it guarantees that the project is built on a foundation of integrity, not just concrete.

To streamline this complex web of information, regulatory bodies should introduce uniform reporting templates. Think of these as the "Rosetta Stones" for deciphering the complex language of construction projects. Such templates can standardize the types of information that need to be disclosed, making it easier for stakeholders to compare projects on an apples-to-apples basis.

Moreover, these templates can serve as dynamic tools for accountability. Stakeholders can demand periodic updates in the same standardized format, making it more straightforward to track any deviations or alterations. This approach not only keeps everyone on the same page but also acts as a deterrent against delays and cost overruns. After all, it is harder to sweep things under the rug when the floor itself is made of glass.

Transparency is not just a good-to-have feature; it is a cornerstone for the success and credibility of zero carbon construction projects. By implementing granular reporting, focusing on ethical considerations, and standardizing this information through uniform templates, we can build a culture of openness and accountability. This is not merely an operational adjustment; it is a paradigm shift that could redefine how we view accountability in the construction industry, moving us closer to a future where transparency is the norm, not the exception.

3.2.3 Moving Toward Outcome-based, More Standardized Building Codes

Building codes serve as the skeletal framework that shapes the construction industry, laying down the safety and environmental standards that projects must adhere to. According to a *United Nations Environment Program report*, as of 2021, the landscape looked promising with over 80 countries embracing green building code standards. However, the devil is in the details or, more precisely, in the inconsistent implementation practices that muddy the regulatory waters.

3.2.4 The Promise of Outcome-Based Codes

Emerging best practices are steering us toward a new frontier: outcome-based, standardized building codes. These innovative guidelines shift the focus from prescriptive "how-to" rules to performance-based "what-to-achieve" goals. The primary targets are energy efficiency and safety.

This is a paradigm shift that has the power to revolutionize the construction sector. Rather than getting bogged down in the quagmire of methods and procedures, outcome-based codes liberate builders to innovate within set parameters. Think of it as giving a painter a canvas and a theme but allowing them the freedom to choose their brushes, colors, and strokes. It is an environment where efficiency is not just a buzzword but a culture, ripe for innovation and breakthroughs.

To sweeten the pot, financial incentives can be layered into this new framework. These could range from grants and tax benefits to technical assistance for energy-efficient upgrades. Such incentives can act as catalysts, accelerating the adoption of sustainable practices and technologies. The idea here is not just to make compliance palatable but to make it irresistible.

Effective execution of outcome-based codes requires a nuanced vetting process for potential contractors. This involves performance-based prequalification metrics that go beyond the usual suspects of cost and timeline. Criteria such as timeliness, quality of work, and safety records should be rigorously evaluated. This ensures that the contractors are not just capable but are also aligned with the project's sustainability and performance goals.

3.2.5 Setting the Stage for a New Era

By focusing on end goals rather than means, we are not just complying with codes; we are aspiring to exceed them. When we create an environment where the construction industry does not just meet regulatory benchmarks but sets new ones, we push the boundaries of what is possible. This shifts us closer to a future where sustainability is not just an add-on but the very core of construction practices.

The construction industry stands at a critical juncture. It faces a myriad of paths, each with its own set of regulatory and ethical challenges and opportunities. As we move forward, let us be audacious in our choices. We must take the road less traveled—the one paved with innovation, transparency, and sustainability. It is a journey that promises not just to fulfill but to exceed expectations, setting a global example and moving us closer to a future where the construction industry is a beacon of sustainability.

Crane and building construction site against blue sky. Source: vladdeep/Adobe Stock Photos.

3.3 Rewiring the Contractual Framework

3.3.1 Moving Away from Hostile Contracting to a Collaborative System

The conventional contractual ecosystem in the AEC sector takes place against a backdrop of stakeholders ensnared in self-interest rather than fostering collaboration. Antagonism sends projects spiraling through needless loops, nudging deadlines further into the horizon. The resultant scene is an unsavory concoction of costly delays, bitter disputes, and a crescendo into project failure. The evidence is that 80% of projects are overbudget and delivered late.

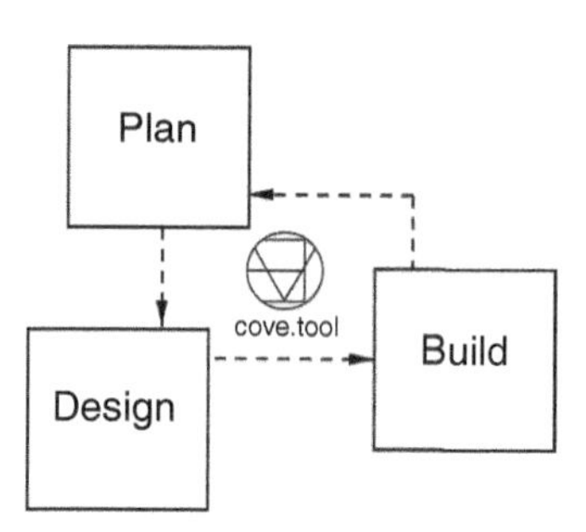

An interconnected data ecosystem for planning, design, and construction to keep costs and incentives aligned.

The call of the hour is a shift toward collaborative contracting, a model that seeds and nurtures relationships, aligning the interests of the myriad parties knitting the project fabric. A stark contrast to the conventional adversarial contracting models, this approach sows seeds of trust, transparency, and open communication among the stakeholders. The fruits are a bounty of enhanced project outcomes, especially in ventures demanding a high degree of innovation and synergy between architects, engineers, contractors, specialists, and building owners. The age-old contracting model in such scenarios digs a pit hard to climb out from. The essence of collaborative contracting is a sturdy scaffold of risk management, enabling early identification and mitigation of risks, thus diminishing the specter of project failure or cost overruns, a scenario often painted when new technologies and materials enter the fray.

The keystone of this model lies in the social tapestry of project management. This involves honing soft skills, mastering the art of negotiation, fostering regular dialogue, and sharing the helm of decision-making, all under the aegis of mutual benefits and a shared compass toward the project's objectives.

3.3.2 Revitalizing Tendering (Bidding) Processes: A Quest for Value and a Nod to Past Performance

The cadence toward zero carbon buildings is also orchestrating a new rhythm in how the AEC industry orchestrates tendering processes. Also known as bidding in some countries, tendering is a competitive procurement process that is used to select a main contractor or supplier for a construction project. The spotlight in the tendering process should focus on best value and past performance, ensuring that the baton only passes to qualified and capable contractors who resonate with the tenets of innovation, sustainability, and collaboration.

The reconfiguration of contractual frameworks nudges tendering processes toward a paradigm where best value and past performance are the cornerstones. This cog in the machinery is pivotal, especially in an arena where high-stake projects are the norm. The discerning lens should focus not merely on price, but delve deeper into a contractor's experience, expertise, and their track record of delivering sterling projects within the stipulated timeframe. Other facets to ponder include the contractor's safety dossier, their blueprint for sustainability and innovation, and testimonials on their ability to work in tandem with the client and other stakeholders. A transparent and unbiased selection panel, with a clear delineation of selection criteria and an equal platform for all bidders, lays the foundation for a fair and fruitful tendering process. It is simple—make carbon reduction be a part of the tendering process.

3.3.3 Anchoring a "Single Source of Truth" to Steer Projects

Embarking on the odyssey to disentangle complex contractual frameworks can indeed rattle the entrenched bureaucratic machinery. A seemingly simple goal can evolve into a maze when we consider that everyone stores their data in different ways. So, how do we create a unified understanding of the project?

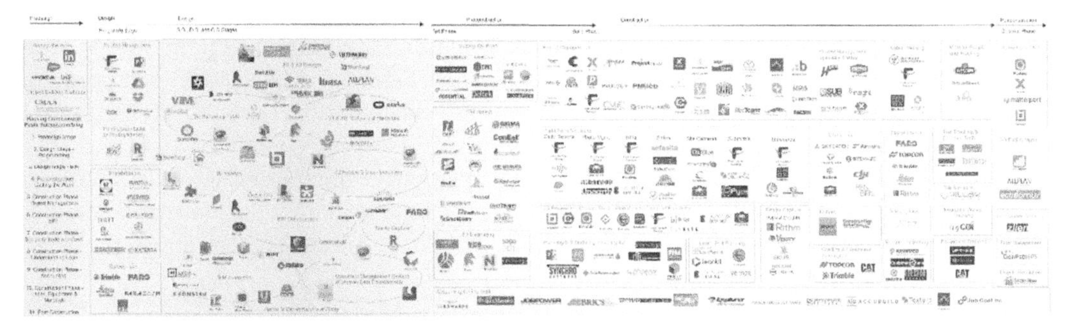

The AEC software market is highly fragmented without a single source of truth or interoperability. Solutions are designed for a specific task or user.

The concept of a "Single Source of Truth" (SSoT) emerges as a lighthouse amidst a stormy sea of disarray, embodying a centralized repository where all project-centric data finds a home—be it engineering blueprints, design documents, or contractual obligations. This nucleus serves as a common ground, ensuring every stakeholder is aligned, thereby shearing the chances of misunderstandings, errors, and delays that could hamstring the project's momentum.

The cultivation of an SSoT is not a fleeting endeavor but a sustained crusade, demanding a scrupulous regimen to ensure the data remains accurate, relevant, and timely. In this journey, deploying a cloud-based project management platform that stands accessible to all stakeholders presents itself as a viable conduit. Such a platform should not merely serve as a sentinel over sensitive data but should extend a user-friendly interface enabling effortless real-time access to complex data.

Let us delve deeper into the constituents and the journey of establishing an SSoT:

- **Assessment and Identification:** The voyage toward establishing an SSoT commences with a thorough assessment of existing systems and processes. This involves a meticulous examination of the current state of the project, an inventory of available data, and identification of data voids. Post-assessment, the project team is better positioned to chart the course toward establishing a centralized data repository.
- **Selection of a Suitable Platform:** The choice of platform is pivotal. A cloud-based project management platform is often favored for its accessibility, real-time data synchronization, and robust security measures to guard sensitive data. The platform should facilitate a seamless flow of information, with a user-friendly interface ensuring stakeholders can effortlessly navigate through complex data, fostering informed decision-making. It is likely that platforms will start to incorporate AI in the coming years to help manage complexity.
- **Data Aggregation and Organization:** The next stride involves aggregating and organizing data within the chosen platform. This step is crucial to ensure that data is not only centralized but also structured in a manner that is coherent and easily navigable.
- **Access Control and Security Protocols:** Establishing stringent access control and security protocols is imperative to protect sensitive data and ensure that only authorized personnel have access to specific data segments. This also encompasses setting up audit trails to monitor data access and modifications, bolstering accountability and traceability.
- **Continuous Upkeep and Review:** Establishing an SSoT is not a terminus but a continuous expedition. It necessitates regular reviews and updates to ensure the data remains accurate, relevant, and reflective of the project's evolving dynamics. This might require a dedicated team or individual tasked with managing and maintaining the centralized data system, ensuring its integrity and relevance over time.
- **Training and Awareness:** It is pivotal that all stakeholders are well-versed with the system, understanding the importance of the SSoT, and are trained on how to access and update the data accurately. This training should extend beyond the technical aspects, encompassing the cultural shift toward a collaborative and transparent modus operandi.
- **Feedback Loops:** Establishing feedback loops for continuous improvement is key. Gathering feedback from users and analyzing data usage patterns can provide invaluable insights into areas of improvement, ensuring the SSoT remains a robust, reliable, and indispensable asset in steering projects toward success.

In knitting together the myriad threads of project data into a cohesive tapestry, the SSoT transcends from a concept to a cardinal asset, illuminating the path in the complex journey of project execution in the AEC industry. Through this, we not only streamline operations but also foster a culture of transparency and collaborative engagement, which is quintessential in driving projects to their best.

3.3.4 Incentivizing Project Outcome Level Performance

The AEC industry has traditionally been tethered to fixed price or time and material contracts, which seldom incentivize the delivery of project outcomes or performance. The focus often narrows down to clocking the project on time and within budget, with a blind eye toward the final outcome or the project's performance in the long run. This tunnel vision eclipses the incentive to craft buildings that resonate with or transcend energy efficiency standards. So, what is the roadmap to unshackle this status quo?

- **Performance-based Contracts:** This avant-garde approach ties the contractor's payments to the building's performance, a narrative echoed in other sectors like transportation and healthcare. Clear, measurable, and achievable performance indicators form the bedrock of such contracts. For instance, a target to trim the building's energy appetite by 20% compared to a baseline year could be a viable indicator, fine-tuned to the building's function and the local climate.

Architect at construction site for achieving the building outcomes. Source: NDABCREATIVITY/Adobe Stock Photos.

- **Incentive-based Contracts:** Here, the contractor is incentivised with a bonus if the building's performance outshines the agreed targets, fueling motivation to push the envelope in project delivery.
- **Risk-sharing Contracts:** This model shares the performance risk between the contractor and the owner, creating a safety net for the owner while nudging the contractor toward energy-efficient outcomes.

All three types of contracts are viable approaches to achieve a reorientation toward the focus on the outcome. The key is to have performance indicators that are relevant to the building's use and the local climate.

3.3.5 Introduction of Appropriate Alternative Contracting Models

With the winds of change, it is imperative to revisit traditional contracting models in the AEC domain, which often venerate the lowest cost and risk-sharing, thereby stifling the shift toward zero carbon buildings. The need is to transcend this rut and embrace models that prioritize collaboration, transparency, and performance-based outcomes.

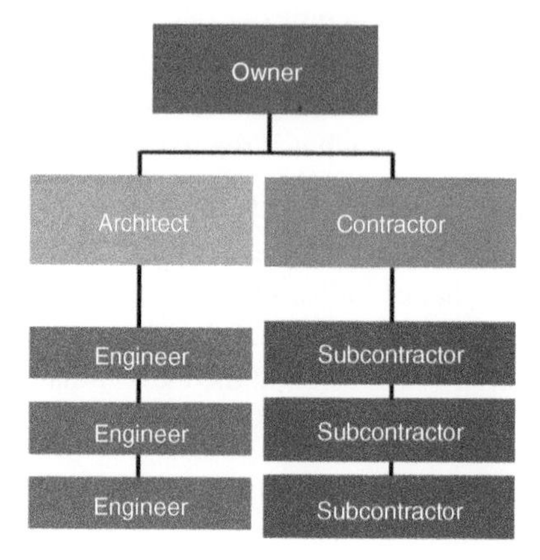

Traditional construction contracting model. Adapted from https://www.aiacontracts.org/contract-doc-pages/27166-integrated-project-delivery-ipd-family.

To overcome this limitation, the following alternatives could be introduced:

- **Integrated Project Delivery (IPD):** IPD is a symphony of collaboration, uniting the owner, designer, contractor, and key subcontractors from the get-go. This camaraderie births a shared vision and objectives, with risks and rewards of the project shared among the team, thus engendering a collective endeavor toward a common goal.

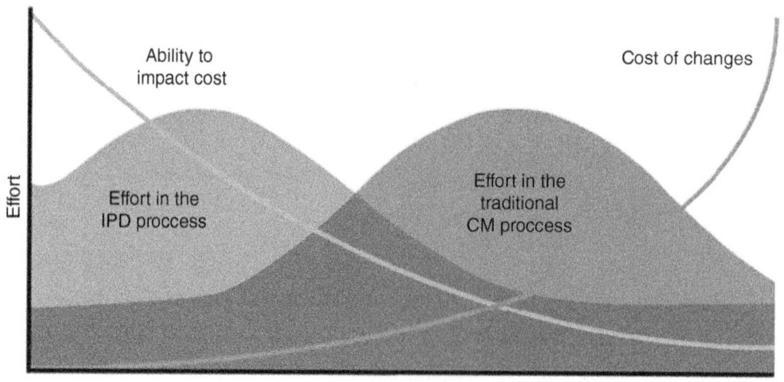

Integrated team

Owner

Architect Contractor

Engineer Subcontractor

Engineer Subcontractor

Engineer Subcontractor

Source: Adapted from AIA Contract Documents/https://www.aiacontracts.org/contract-doc-pages/27166-integrated-project-delivery-ipd-family.

- **Design-Build (DB):** The DB model is a contract between the owner and a single entity, the DB team, to design and construct the project. This model nurtures a collaborative nexus between design and construction teams, with the DB team helming the project within a fixed budget and schedule.

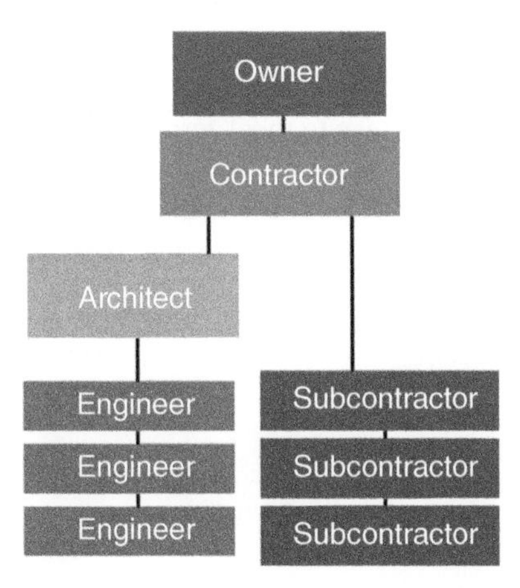

Source: Adapted from AIA Contract Documents/https://www.aiacontracts.org/contract-doc-pages/27166-integrated-project-delivery-ipd-family.

- **Performance-Based Contracting (PBC):** In PBC, the contract is a pledge to deliver a specific performance outcome, with incentives aligning the contractor toward delivering a high-performing building.

Transitioning toward alternative contracting models necessitates an embrace of a fresh culture of collaboration and transparency. This journey, albeit challenging given the entrenched traditional mindset, is a necessary stride toward fostering an industry that prioritizes sustainable goals over myopic cost-saving. The onus is on us to rewire the contractual framework, heralding a new dawn in the AEC industry where projects are not mere transactions but a voyage toward sustainable, innovative, and collaborative excellence.

3.4 Redefining Design and Engineering Approaches

3.4.1 Making Value Engineering a Habitual Aspect of the Design Process

The subsequent step after implementing a performance-based contractual framework involves incorporating value engineering (VE). The essence of this approach within the construction setup involves critically evaluating different facets of a project to find avenues to trim expenses without compromising on overall quality. It is akin to a surgical examination of various angles including, design, scheduling, materials, and locations, with an intent to optimize resource utilization. This methodology brings to light that value gains can stretch across numerous attributes, including monetary savings, increased productivity, better quality, time management, durability, environmental considerations, and even human behavior. The broader the attributes, the greater the benefits.

$$\text{Value} = \frac{\text{Function} + \text{Quality}}{\text{Cost}}$$

VE is far from ad hoc. The primary aim is to bolster the value of a project and its composite systems. The starting point involves identifying the fundamental functions of the product or system, followed by exploring alternative means to achieve those very functions but at a lower cost or with improved performance. For instance, a building's energy efficiency can be drastically improved by employing state-of-the-art simulation tools, which empower designers to model the impact of varying design elements such as insulation, ventilation, and shading. By dissecting the energy performance of different design options, designers can choose the most potent strategy to meet zero carbon targets.

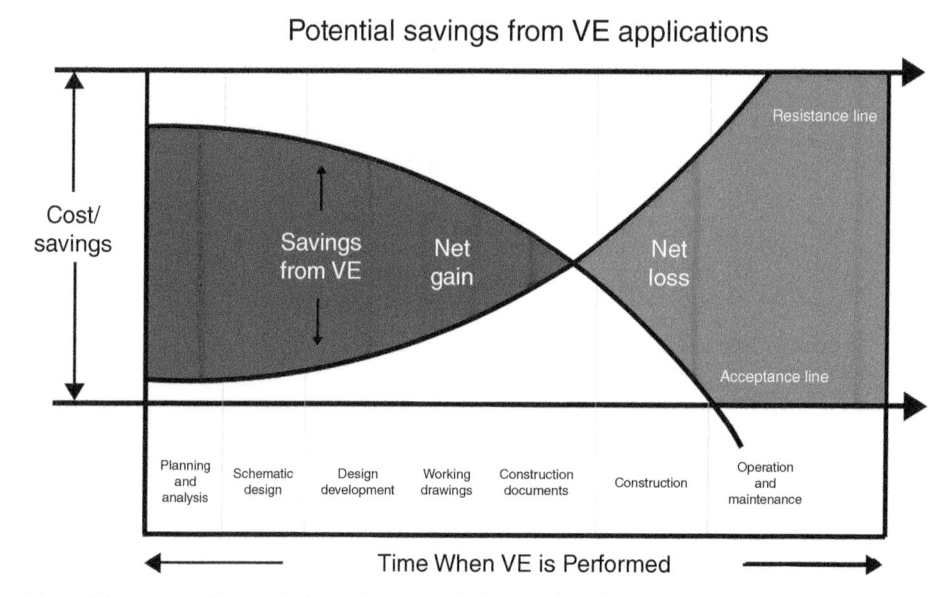

Source: Adapted from https://www.wbdg.org/resources/value-engineering/with permission of National Institute of Building Sciences.

Infusing VE into the AEC design process necessitates cultivating a culture of continuous improvement and a mindset that is open to challenging widely held assumptions and embracing innovative ideas. This will have to be buttressed by other aspects mentioned in this chapter.

3.4.2 Encouraging Off-site Manufacture

Off-site manufacturing is essentially about constructing building modules in a controlled factory setting, distant from the site of construction. These components are then transported to the site for assembly. This process significantly reduces the time and labor involved in on-site construction and brings to the table attributes such as waste reduction, improved quality control, and enhanced efficiency.

Prefabricated pod, built at a separate location, being placed on building site. Source: arska n/Adobe Stock.

Related to zero-carbon buildings, off-site manufacturing lessens the need for transporting materials and equipment, thus immediately reducing the carbon footprint. Furthermore, it is more precise and restricts the need for rework, further minimizing waste. Off-site manufacturing processes are given life by digital design and engineering, which create highly detailed 3D models of components and systems, paving the way for simulation and analysis before production and transport to the site. It majorly reduces chances of errors and waste during production. The concept of off-site manufacturing may initially be greeted with skepticism or resistance due to a perceived loss of control over the construction process. In actuality, there is a vast scope for control and, most importantly, prediction, if simulation techniques are employed.

Precast technology is characterized by the usage of prefabricated concrete panels, which are manufactured in a factory setting and then shipped to the construction site for assembly. This eliminates the need for on-site concrete pouring and curing, drastically reducing construction time, and minimizing disruption at the site. Panel assembly, however, involves the use of prefabricated wall, floor, and roof panels, which are assembled at the site in a manner akin to fitting together puzzle pieces. This method can be easily customized to meet specific project requirements.

Steel structure architecture construction. Source: jjfarq/Adobe Stock Photos.

The benefits of precast technology and panel assembly are plentiful. Precast is almost always faster and more cost-effective and reduces the risk of weather-related delays, material waste and on-site errors. More importantly, precast technology and panel assembly are more sustainable than traditional construction methods due to less raw materials that generate less waste on-site. The factory production process ensures high quality and less likelihood to require maintenance or replacement. Because the components are produced off-site, they can be customized to meet specific requirements, including size, shape, and color.

3.4.3 Encouraging Standardization and Parameter Specification

In the sphere of evolving technologies, standardization is resurfacing with a renewed significance. Once perceived through the lens of mass housing construction in the 20th century, its reputation was poor. However, the idea as a potent method for efficiency remains intact, only now it needs a modern adaptation. Standardization delineates the establishment and adherence to uniform specifications, methodologies, processes, and guidelines across products and systems. Although it may seem a bland subject, its indispensable nature shines in industries striving for consistent and efficient outcomes. Until AI allows for every project to be unique and cost-effective, embracing standardization and parameter specification is paramount. Unlike the rigid uniformity of the 20th century, contemporary standardization cultivates relevant categories, setting the stage for a more nuanced approach:

- **Building Codes and Regulations**: Our governing bodies orchestrate building codes and regulations, laying down the minimum standards in energy efficiency, insulation, ventilation, and other facets of architectural design and execution. This institutional framework serves as the cornerstone for standardization in the construction realm. Not only with current codes but the Inflation Reduction Act in the US and the CBAM tariffs for Europe also encourage energy efficiency standards.

US Energy Codes 2023

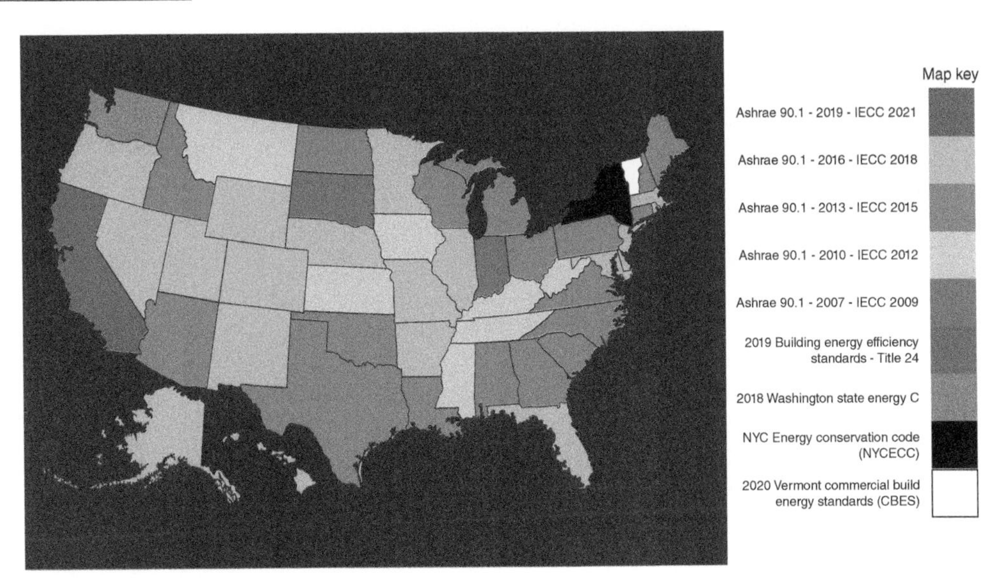

States that adopted energy codes.

- **Green Building Certifications**: Organizations such as the Leadership in Energy and Environmental Design (LEED) and the Passive House Institute provide certifications for buildings that meet specific standards for energy efficiency, sustainability, and environmental performance. Through a defined set of guidelines and requirements, they foster a culture of standardization and parameter specification among builders and developers.

Daylight analysis model in cove.tool. Source: Cove Tool, Inc.

- Standardized Building Materials: The use of standardized building materials that can be easily replicated, transported, installed, and mass-produced is highly beneficial. Engineered wood products, cross-laminated timber (CLT), or laminated veneer lumber (LVL) are sustainably harvested and lightweight and can replace traditional concrete and steel. All mentioned characteristics render a desirable set of parameters and can be used as standards.

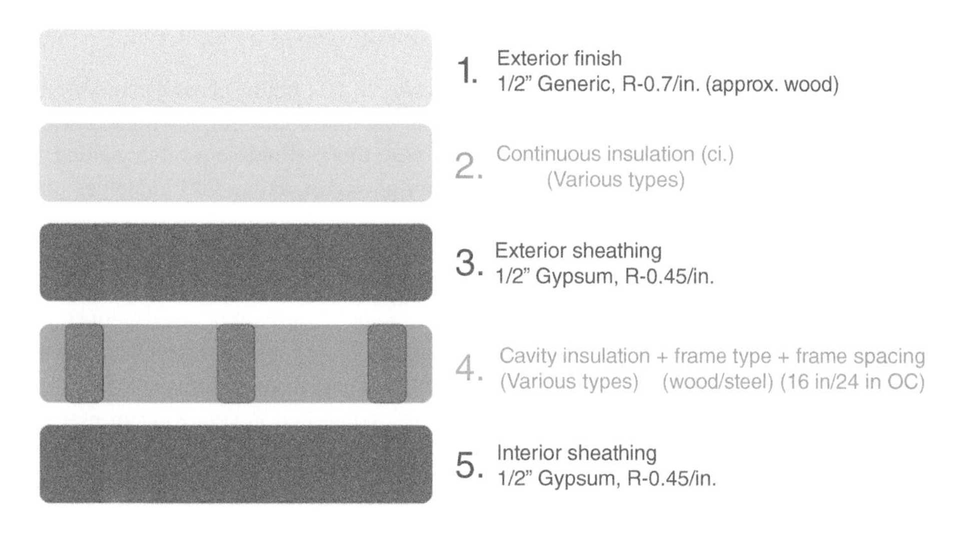

Assumed wall assembly parameters as outlined in cove.tool.

- **BIM**: BIM is a great tool for standardization and parameter specification that allows to test different design options and materials, identify potential issues and conflicts, and ensure that buildings are constructed to a consistent standard.

Building information modeling (BIM) showing an apartment complex floorplan.

3.4.4 Emphasizing a Move Toward Thinking About Construction as a Production System

The traditional narrative of construction often romanticizes it as a realm of craftsmanship and trade mastery. The story was artisans, be it carpenters, electricians, or plumbers, employ their refined skills to construct buildings one meticulous piece at a time. In the past, this methodology often yielded structures aesthetic allure and distinctive character – or so the story goes. Yet, as we find ourselves under pressure to achieve decarbonization, the quaint charm of traditional construction seems to falter against the pragmatic demands of consistency, efficiency, and waste mitigation.

Amidst the debate between tradition and innovation, a call for a shift resounds—envisioning construction as a production system. This ideology is not about mere incremental adjustments but a radical rethinking, akin to donning an industrial lens to scrutinize and re-engineer construction processes. The analogy lies in the essence of a factory, a well-oiled machine that churns out products through the assembly of standardized components. Embracing a similar ethos, construction sites too can metamorphose into orchestrated production arenas, employing prefabricated components in a modular construction approach. While modular construction is not novel per se—its prowess has been demonstrated well in the creation of bridges, tunnels, and other colossal structures—the building industry has been relatively slow in harnessing its full potential.

Source: Seventyfour/Adobe Stock Photos.

A vital linchpin in this progressive narrative is the integration of data analytics. The lexicon of factories resonates with data-driven optimization, enabling a rhythmic flow of production with minimized wastage and maximized efficiency. This data-centric philosophy, when transplanted into the construction ecosystem, holds the promise of similar dividends. It facilitates careful tracking of material usage, performance monitoring of building systems, thus painting a comprehensive picture of the construction process in real-time. With this trove of data, the scope for iterative improvements burgeons, enabling more enlightened decision-making. With mass customization or AI, we may be able to overcome the challenges of previous attempts at factory assisted construction.

However, this evolution does not merely stop at processes; it extends to the human capital, heralding a new era of interdisciplinary synergy. The new narrative necessitates a robust collaboration between architects and contractors, forging a symbiotic relationship to design and implement efficient construction systems. This tighter integration not only paves the way for bridging the traditional and the modern but also seeds a culture of continuous learning and adaptation.

This transition toward viewing construction as a production system is not a mere shift; it is a profound metamorphosis aimed at aligning the construction sector with the sustainability and efficiency benchmarks of the 21st century. It beckons a future where the elegance of architectural creativity seamlessly melds with the precision and efficiency of industrial engineering, propelling the construction industry into a new era of sustainable innovation.

3.5 Improving Procurement and Supply-Chain Management

3.5.1 Unveiling Transparency and Cohesion Among Contractors and Suppliers

The construction industry stands at an unprecedented crossroads. On one hand, it is an industry that is a linchpin of economic growth, providing the very infrastructure upon which society stands. On the other, it is a significant contributor to carbon emissions, making the quest for sustainability not just a moral imperative but also a practical necessity. In this context, procurement and supply chain management take center stage. Far from being mere logistical considerations, they become the canvas upon which a contractor can paint their sustainability ambitions.

It all starts with setting the stage. Contractors, suppliers, and vendors must not merely be on the same page; they must be on the same line and word when it comes to sustainability objectives. To facilitate this, robust communication channels are indispensable. These avenues are not just for relaying instructions but for fostering a culture of transparency and continuous improvement.

Enter tracking software and online collaboration tools, which act like the sinews that connect the muscle groups of a project. They make sure that everyone is working in tandem, and that deviations—whether they be in terms of timelines, budgets, or quality—are quickly identified and corrected.

At the heart of this orchestrated effort lies the procurement plan—a document that should be as living and evolving as the project itself. It should outline not just what is needed, but when and how. As projects scale, the complexity grows exponentially. A centralized procurement team can be the answer to this complexity, offering strategic cost-saving opportunities through economies of scale and more nuanced supplier relationship management.

Digitization is not just a buzzword here; it is a force multiplier. ERP systems, e-procurement platforms, and electronic data interchange (EDI) transform procurement from a series of tasks to an integrated workflow, turning data into actionable insights. Yet, digitization is not the end of the road. The real game-changer lies in predictive capabilities, enabled by AI, which could be the cornerstone of future advances in this area.

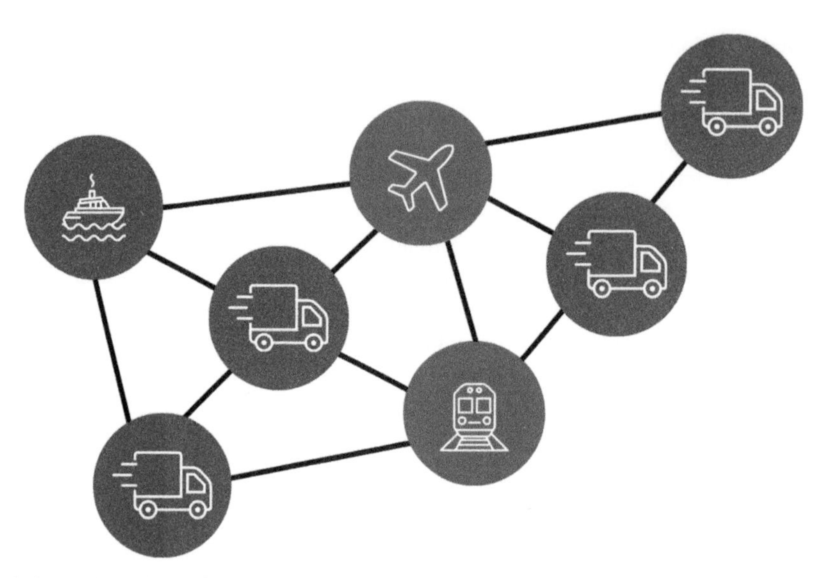

Connected logistics and supply chain concept with plane, trucks, train, and ship. Source: Adapted from Envato/ https://elements.envato.com/connected-logistics-and-supply-chain-concept-table-ZF3M95G.

3.5.2 Economies of Scale Through Properly Skilled Central Procurement Teams

The conventional approach to procurement—each project team or department working in its silo—is a relic of a bygone era. While this fragmented model might offer individual teams a semblance of control, it sacrifices the broader efficiency and coherence that come from a unified approach. The inception of centralized procurement teams is like the construction industry's own Enlightenment period; a moment of collective realization that a coordinated approach can yield dividends, both financially and environmentally.

Imagine a symphony where each musician plays their instrument independently, only loosely connected to the conductor's guidance. The result would be cacophony, not harmony. The same applies to procurement. A centralized procurement team acts as the conductor, orchestrating various elements—identifying suppliers, negotiating contracts, and making bulk purchases—to create a harmonious flow of materials and services.

Centralized procurement teams have the benefit of a helicopter view. They can identify the best suppliers not just in terms of price but also in terms of quality and sustainability. More importantly, these teams can negotiate better pricing through bulk purchases, which is particularly crucial when dealing with the often higher costs of green materials. These teams have the know-how to make informed decisions about product alternatives, taking into account the embodied carbon—a factor often overlooked but integral to a project's overall carbon footprint.

In a centralized model, relationship management with suppliers goes beyond transactional interactions. It fosters a partnership approach where both parties are invested in each other's success. In this mutually beneficial setup, suppliers are more likely to provide better terms and even collaborate on developing new, more sustainable materials or methods. A crucial advantage of having a centralized team is the ability to implement a feedback loop. Real-time data and analytics can inform future procurement strategies, ensuring that each project benefits from the lessons learned in previous ones. This continual learning process is vital for any contractor aiming to improve their sustainability credentials incrementally.

By bringing procurement under a centralized umbrella, contractors set themselves up for cost savings, operational efficiencies, and, most importantly, a concrete path to achieving their decarbonization goals. In an industry where margins are often razor-thin and the stakes for environmental impact are high, centralized procurement teams offer a tangible competitive advantage.

Multiple supply chain truck train and cargo ship working service shipping. Source: Yellow Boat/Adobe Stock.

3.5.3 Digitizing Procurement and Supply Chain Workflows

If centralized procurement is the Enlightenment of the construction industry, digitization is its Industrial Revolution—a seismic shift that redefines how we think about efficiency and scalability. The days of manual tracking and procurement activities conducted through isolated spreadsheets are not just inefficient; they are untenable for an industry pressed to achieve both cost savings and sustainability targets. Enterprise resource planning (ERP) systems, e-procurement platforms, and electronic data interchange (EDI) form the trinity of digital tools revolutionizing procurement. These tools transform isolated data points into a cohesive, actionable narrative. ERP systems offer a panoramic view of resources, streamlining the allocation and utilization of materials. E-procurement platforms bring a new level of transparency and accountability, enabling not just better tracking of spending but also ensuring compliance with sustainability benchmarks. EDI takes supplier relationships into the digital age, automating ordering, shipment tracking, and invoicing.

But these digital platforms do more than just collect and display data; they interpret it. Advanced analytics capabilities can identify patterns and trends, helping decision-makers to proactively address challenges and seize opportunities. Imagine being able to predict a spike in the cost of a particular material and adjust your sourcing strategy in real time. That is not just efficiency; it is a strategic advantage.

While the immediate benefits of digitization are compelling, the real promise lies in its predictive capabilities. Digital tools can forecast supply needs, identify potential bottlenecks, and even predict supplier performance based on historical data. This is not just an incremental improvement over previous methods; it is a qualitative leap forward.

The icing on the cake is the incorporation of AI into these digital platforms. AI algorithms can sift through massive data sets to make sense of complexities that would be humanly impossible to grasp. Whether it is optimizing routes for materials delivery or recommending alternative suppliers based on multivariable analysis, AI is set to play a pivotal role in the future of procurement and supply chain management.

The digitization of procurement and supply chain management is not an optional upgrade but a necessary evolution. It offers a pathway to not only streamline operations but also to leverage data for strategic decision-making, setting the stage for the next wave of innovations in this critical aspect of construction.

Source: N Lawrenson/peopleimages.com/Adobe Stock Photos.

3.5.4 Predictive Replenishment Informed by IoT

Traditional inventory management often resembles a high stakes guessing game. Despite best efforts, it is plagued by either of two extremes: overstocking, which ties up capital and leads to waste, or under-stocking, which interrupts workflow and may cause project delays. Both scenarios are suboptimal, and both directly clash with the overarching goal of sustainability.

Enter the Internet of Things (IoT), a constellation of sensors and smart devices that turns the physical world into a data-generating ecosystem. By equipping materials and inventory with IoT sensors, a real-time snapshot of stock levels becomes available at the click of a button. But IoT does more than just count; it communicates. These sensors can send data to centralized systems, providing a dynamic, continuously updated picture of inventory levels.

Armed with this real-time data, predictive algorithms go to work. They analyze not just current stock levels but also rate of usage, seasonal fluctuations, and even broader market trends. The result is a predictive replenishment system that knows exactly when new materials are needed. This is not merely automated reordering; it is a nuanced, data-driven approach that accounts for multiple variables to optimize inventory levels.

The economic benefits of predictive replenishment are palpable. By eliminating the costs associated with overstocking and the risks of understocking, contractors can achieve significant cost efficiencies. From an environmental standpoint, the implications are equally profound. Minimized waste translates into a smaller carbon footprint, aligning the supply chain with broader sustainability objectives. While IoT provides the data, the real analytical horsepower comes from AI. By incorporating machine learning algorithms, the system continuously improves its predictions based on new data. Over time, the predictive accuracy improves, creating a virtuous cycle of ever-increasing efficiency and sustainability.

In essence, predictive replenishment informed by IoT represents a paradigm shift in inventory management. It transforms a historically reactive process into a proactive, data-informed strategy. For contractors committed to sustainability, this is not just an operational improvement; it is a critical enabler of their decarbonization goals.

3.5.5 Digitally-Enabled Logistics Management

Logistics management has long been the "gotcha" of construction projects. The sector's logistical demands are unique; materials must not only arrive on time but often in a specific sequence and condition to meet project requirements. Traditional logistics management is fraught with inefficiencies: think of trucks idling at construction sites, waiting for their turn to unload, or the unnecessary movement of materials within the site, each adding to the project's carbon footprint.

In the era of digital transformation, "just-in-time" is not a lofty ideal; it is a manageable reality. Digitally-enabled logistics systems use real-time data to coordinate the arrival of materials precisely when they are needed. This timing minimizes idle times for transport vehicles and ensures that materials are used efficiently, reducing the likelihood of spoilage or waste.

Digital platforms collect and analyze data from multiple sources—traffic patterns, weather forecasts, and real-time project updates—to optimize logistics. This data-driven approach brings a new level of precision to logistics management. It is no longer about approximate delivery windows; it is about calculated arrival times based on a myriad of variables.

The environmental benefits are compelling. Precise, just-in-time delivery means fewer trucks idling, less congestion, and reduced emissions. In essence, digitally-enabled logistics management can make a construction project leaner, greener, and more cost-effective.

Importantly, digitally-enabled logistics management does not operate in a vacuum. It is most effective when integrated with other digital tools like ERP systems and e-procurement platforms. This integration creates a seamless, end-to-end digital workflow, from procurement to on-site delivery, each element informing and enhancing the others.

The future of digitally-enabled logistics management looks even brighter with the integration of machine learning algorithms. These algorithms can learn from each logistical operation, continuously refining and improving the system's predictive capabilities. Over time, what starts as a "smart" logistics system evolves into an "intelligent" one, capable of self-optimization.

Digitally-enabled logistics management is not just an incremental upgrade to traditional methods. The missing piece in the puzzle for contractors is a fundamental rethinking of how logistics can be optimized for both efficiency and sustainability.

3.5.6 The Future is Integrated

As the construction industry grapples with the dual challenges of economic pressures and environmental stewardship, an integrated approach to procurement and supply chain management emerges as a non-negotiable imperative. This is not merely about adopting new technologies or methodologies; it is about weaving them into a coherent, interconnected tapestry that spans from the initial planning stages to the final construction phases.

The benefits of such an integrated approach are manifold. Economically, the efficiencies gained through centralized procurement teams, digital workflows, predictive replenishment, and digitally-enabled logistics can translate into significant cost savings. These are not just numbers on a balance sheet; they are a crucial competitive advantage in an industry where margins are notoriously thin.

Environmentally, the impact is no less significant. The application of these innovations directly correlates with reduced waste, lower carbon emissions, and a more sustainable use of resources. In essence, they provide a roadmap for contractors committed to decarbonization, offering practical steps to translate sustainability ambitions into actionable strategies.

AI serves as the linchpin in this evolving ecosystem, offering the predictive capabilities that elevate these systems from being merely "smart" to truly "intelligent." As machine learning algorithms continue to evolve, they will further refine these systems, creating a virtuous cycle of continuous improvement.

For contractors aiming to decarbonize, the stakes could not be higher. Climate change is not a future problem; it is a current crisis, making the adoption of sustainable practices an urgent necessity. The integration of innovative procurement and supply chain management strategies is not just a pathway to sustainability; it is a clarion call to the industry. It demonstrates that not only is sustainable construction feasible but it is also efficient, cost-effective, and ultimately, the only way forward.

In conclusion, an integrated approach to procurement and supply chain management offers more than just operational efficiencies; it offers a new paradigm for the construction industry—one that aligns economic viability with environmental responsibility. For contractors pursuing decarbonization goals, this integrated approach is not just an option; it is the future.

3.6 Improving On-Site Execution

In the realm of construction, the phrase "time is money" takes on a palpable urgency. But in the race against time, the quality of on-site execution often gets compromised, leading to delays, cost overruns, and decarbonization goals being pushed to the back burner. As the construction industry makes strides toward a more sustainable future, the importance of meticulous on-site execution cannot be overstated. It is the crux that binds the planning and the end result, where the rubber meets the road. Improving on-site execution is not just about enhancing efficiency; it is about enabling a paradigm shift toward sustainable construction practices. To understand this better, let us dissect the critical components of on-site execution.

3.6.1 Rigorous Planning Process

Source: Kara/Adobe Stock.

In the high-stakes world of construction, where every delay can mean escalated costs and compromised sustainability goals, planning is not just a preliminary step; it is the bedrock upon which the entire project rests. Especially crucial is when the lens focuses on sustainability and decarbonization. Planning in this context is not just about logistics; it is about aligning every aspect of the project—from material selection to waste management—with the overarching sustainability objectives.

1) **Defining the Project Scope**: *The Blueprint* – The initial phase of defining the project scope serves as the blueprint for everything that follows. This is not just about listing deliverables; it is about setting the expectations for every stakeholder involved, ensuring everyone is aligned with the project's goals and objectives. It is the time when ambiguous ideas are molded into concrete plans, providing a foundation for all subsequent steps.
2) **Assess the Site**: *The Environmental Equation* – Once the project scope has been solidified, the next logical step is a comprehensive site assessment. This is not merely an exercise in measuring square footage or evaluating soil quality. It is a multidimensional analysis that takes into account environmental factors such as local climate, potential for renewable energy sources, and even the site's natural flora and fauna. The data gathered here will shape decisions on everything from building orientation to material selection.
3) **The Project Plan**: *A Collaborative Framework* – With a clear understanding of the project scope and site conditions, a detailed project plan can be developed. It requires the collaborative input of every team member, each bringing their expertise to the table. The plan serves as the project's operational manual, outlining timelines, budgets, and most importantly, the roles and responsibilities of each team member.
4) **Risk Management**: *The Safety Net* – The volatile nature of zero-carbon construction projects demands a robust risk management plan. This safety net identifies not only the obvious risks like

weather delays but also the more nuanced challenges such as regulatory hurdles specific to sustainable construction. A well-crafted risk management plan does not just identify these risks; it prepares the team for them, ensuring that contingencies are in place.

5) **Quality Control and Sustainability** – Quality control in this context goes beyond ensuring that a wall is straight or a foundation is solid. It includes verifying that materials are sustainably sourced, that the construction process minimizes energy usage, and that the project's carbon footprint is as low as possible. It is quality control with a green lens.
6) **Performance Monitoring**: *The Continuous Feedback Loop* – The final piece of the planning puzzle is performance monitoring. This is not a one-off task completed at the end of the project but a continuous process that feeds back into every stage of planning and execution. Through meticulous monitoring, teams can make real-time adjustments ensuring that the project stays on track not just in terms of timelines and budget but also in fulfilling its sustainability goals.

Rigorous planning serves as the navigational system for the construction project. It ensures that every step taken is a step toward not just project completion but also toward achieving the broader sustainability and decarbonization goals.

3.6.2 Reshaping Relationships and Interactions Between Owners and Contractors

The traditional owner-contractor relationship often mirrors a zero-sum game, where one party's gain appears to necessitate the other's loss. This adversarial dynamic can lead to a cascade of issues: cost overruns, delayed timelines, and most critically, compromised sustainability goals. This zero-sum mindset is not just outdated; it is detrimental to the industry's broader push toward sustainability. The need of the hour is a shift from an adversarial relationship to a partnership approach. In this evolved dynamic, both parties view each other as collaborators working toward a shared goal: a successful, sustainable project. This paradigm shift does not just benefit the project; it has the potential to redefine industry norms.

The cornerstone of this new relationship model is transparent communication. In the past, information was often guarded, leading to mistrust and conflict. In a partnership model, open dialogue is encouraged, and critical information—be it related to project risks, delays, or sustainability challenges—is shared freely. This transparency does not just mitigate issues; it prevents them from arising in the first place. Improving communication often requires flattening the traditional hierarchical structures that have long defined the construction industry. When information flows freely in a networked structure, decisions are made more collaboratively, and the focus remains firmly on the project's success, including its sustainability targets.

The benefits of a more egalitarian relationship are not merely economic—although reducing delays and avoiding conflicts certainly have financial advantages. There are also psychological benefits: teams function better when operating in an environment of mutual respect and shared objectives. This positive atmosphere can significantly improve on-site execution, leading to better quality and faster completion times. Improving the owner-contractor relationship has implications beyond the immediate project. It sets a precedent, offering a template that can be replicated across the industry. In a sector that is often resistant to change, successful examples of collaborative relationships can act as catalysts, accelerating the industry-wide move toward more sustainable practices.

Reshaping the owner-contractor relationship is not just a strategy for improving individual projects; it is a critical step toward modernizing the entire construction industry. By replacing adversarial dynamics with collaborative partnerships, we pave the way for an industry that is not just more efficient and cost-effective, but also more aligned with the urgent sustainability goals of our time.

Source: Sundry Photography/Adobe Stock Photos.

3.6.3 Agreement and Use of KPIs at Regular Performance Meetings

Agreement on key performance indicators (KPIs) is a good way to *track and evaluate performance* management systems. In the complex landscape of construction projects, relying solely on intuition or anecdotal evidence is a recipe for inefficiency, if not outright failure. The introduction of KPIs serves as a compass, offering data-driven insights that guide the project toward its objectives, including sustainability goals. While KPIs serve as crucial metrics, their real value is unlocked only when aligned with the project's broader goals. Whether the focus is on sustainability, efficiency, or cost control, the chosen KPIs should serve as a roadmap leading the project toward these objectives.

Holding regular performance meetings to review these KPIs is essential, but the frequency is a delicate balancing act. Too frequent, and the team could get bogged down in meetings, impeding actual work. Too infrequent, and critical issues could go unnoticed until they have escalated into major problems. The key is to find a rhythm that allows for real-time adjustments without becoming a hindrance. One size does not fit all when it comes to KPIs. Different departments or teams within the project will have different roles, responsibilities, and, therefore, different performance metrics. Tailoring KPIs to these specific roles ensures that each team is aligned with the project's broader objectives, creating a harmonious, effective work environment.

While KPIs offer quantitative insights, they should not be viewed in isolation. The regular review process should provide an opportunity for contextualizing these metrics, understanding the stories they tell and the real-world challenges or opportunities they point to. This nuanced approach to KPIs provides a fuller picture of the project's status. Performance meetings should serve as platforms for not just accountability but also celebration. Successes, even minor ones, should be recognized and celebrated, serving as motivation for the team. Conversely, shortcomings should be viewed as learning opportunities, providing valuable lessons that can be applied moving forward.

The strategic use of KPIs and regular performance meetings transforms them from mere administrative tasks into powerful tools for project optimization. They serve as the project's pulse, offering real-time insights that enable proactive decision-making, ensuring that the project remains aligned with its goals, both operational and sustainable.

3.6.4 Improvement in Mobilization for New Projects

In the lifecycle of a construction project, mobilization is often the unsung hero. It serves as the catalyst that turns plans into action, propelling the project from the drawing board into the real world. And yet, it is an aspect that is often given short shrift, leading to a cascade of downstream issues, from delays to compromised sustainability goals. At its core, effective mobilization is about seamless collaboration. But true collaboration goes beyond mere coordination; it is about understanding and leveraging the diverse skills, strengths, and even the weaknesses of every team member. This creates a synergistic effect, where the sum is greater than its individual parts, driving the project forward more efficiently and sustainably.

Technology offers unparalleled advantages in aiding project mobilization, from advanced software solutions to automation. However, it is crucial to remember that technology is a tool, not a panacea. Its effective utilization hinges on the team's ability to integrate it meaningfully into the project's broader objectives, including sustainability goals.

Mobilization is a human endeavor and not solely a logistical exercise that requires inspiration and leadership. A well-mobilized team is a motivated team, driven not just by deadlines but by a shared commitment to the project's success and its sustainability goals. Balancing assertive leadership with a collaborative spirit creates a dynamic team capable of overcoming challenges and innovating solutions. Effective mobilization is not a one-and-done event as it must be an ongoing process that evolves with each project. This continuous improvement is not just about fine-tuning logistics, but fostering a culture that questions established practices, challenges assumptions, and is open to exploring new methodologies and technologies. Effective mobilization draws upon a diverse set of skills and perspectives, from the analytical acumen of engineers to the creative insights of designers. This collective intelligence not only enriches the project but also makes it more resilient, adaptable, and aligned with sustainability objectives.

3.6.5 Careful Planning and Coordination of Different Disciplines On-site

The growing trend toward multidisciplinarity in construction is a double-edged sword. On one hand, the presence of diverse experts can enrich the project, offering a multifaceted approach to problem-solving and execution. On the other hand, coordinating these various disciplines can be akin to herding cats. Project managers must not only have a deep understanding of each discipline involved but also the skills to harmonize them. This involves establishing clear lines of communication, delineating responsibilities, and ensuring that everyone is aligned with the project's overall goals, including its sustainability objectives.

One of the biggest challenges in multidisciplinary settings is the potential for communication breakdowns. Different disciplines often have their own jargon, methodologies, and priorities. Bridging this gap requires creating a common language and framework that everyone can understand. This might involve visual aids for complex concepts or even cross-disciplinary training sessions. Coordinating multiple disciplines on-site also involves meticulous resource allocation. Whether it is human resources, machinery, or time, each discipline has unique needs. The project manager must piece together this jigsaw puzzle in a way that maximizes efficiency without compromising the project's quality or sustainability goals.

Conflicts are almost inevitable in multidisciplinary settings, especially when resources are limited, or deadlines are tight. Effective conflict resolution is not just about finding a compromise but aligning the conflicting parties with the project's broader objectives. This often involves reminding everyone of the big picture: the project's sustainability goals and its potential impact on the community and the environment. Despite its challenges, the multidisciplinary approach offers a payoff that is well worth the effort: a profitable and more sustainable outcome.

3.6.6 Implementation of Holistic Project Management Approach

The conventional approach to project management often treats each project as a standalone entity, isolated from the organization's broader ecosystem. This siloed perspective can lead to inefficiencies and missed opportunities, especially when it comes to sustainability. A holistic approach challenges this paradigm, proposing an interconnected model where projects are seamlessly integrated into the organization's overall operations. A holistic approach transcends the narrow focus on technical or financial metrics, embracing a triple bottom line that considers social, environmental, and economic impacts. Our processes must create sustainable, community-enriching, and economically viable buildings.

Traditional project management operates on three primary constraints: scope, time, and cost. A holistic approach introduces a fourth dimension: carbon. This added layer does not complicate the project but enriches it. The essence of a holistic approach lies in its integrative nature. We cannot just tack on sustainability as an afterthought since it weaves it into every aspect of the project, from planning to execution. This integration is made possible through multidisciplinary collaboration, advanced technologies, and a forward-thinking leadership approach. When a holistic approach is successfully implemented, it transforms the narrative. The focus shifts from completing projects to leaving legacies. These are structures that serve communities, respect the environment, and contribute positively to the economy for years to come.

Improving on-site execution is a multifaceted endeavor that demands a shift in both mindset and methodology. From meticulous planning to multidisciplinary coordination, each aspect plays a critical role in shaping the project's success. But the most transformative impact comes from adopting a holistic approach, one that aligns every facet of the project with the broader goals of sustainability and long-term impact. As we move forward in an era that demands both innovation and responsibility, mastering the art of on-site execution isn't just a skill; it is a moral imperative.

3.7 Infusing Digital Technology, New Materials, and Advanced Automation

In an age where technological prowess is often synonymous with operational efficiency, the construction industry finds itself at a pivotal juncture. With decarbonization, the sector is increasingly turning to digital technology, advanced materials, and automation. This trifecta not only promises to redefine the way we build but also how we conceive of buildings in the first place.

3.7.1 Making 3D BIM Universal Within the Company

The shift from 2D CAD to 3D BIM is not a mere upgrade; it is a transformation. While 2D drawings offer flat representations, susceptible to misinterpretation, 3D BIM presents an interactive model—almost a digital twin—of the building. This nuance is not just beneficial but crucial for complex projects.

However, universal adoption of 3D BIM is not a one-size-fits-all solution. Smaller firms with less complex projects might find the cost and time for BIM implementation disproportionate to its benefits. For larger entities the equation is different. Here, complexity breeds mistakes, and mistakes are expensive. In such scenarios, BIM serves as a cost-saving, error-minimizing tool. The transition to 3D BIM is as much about technology as it is about people. It necessitates training programs to ensure seamless integration with existing workflows. The goal is to augment human expertise with digital precision, not to replace it. A recent *survey* indicates a promising trend: over 44% of industry respondents have already integrated some form of digital technology, and the three-year projection stands at an impressive 70%.

3.7.2 Use of Digital Collaboration Tools, Drones, and Unmanned Aerial Vehicles

Digital tools are no longer just a fad; they are the new bedrock upon which modern industries are built. Forget the archaic constraints of time and space; these tools usher us into a world of remote control and operations, leaving traditional project management systems in the dust. Consider digital collaboration tools—software applications like Slack, Microsoft Teams, Zoom, and Google Drive—that have become indispensable in today's work environment. They *empower teams to collaborate* through instant messaging, video conferencing, and file sharing, transcending geographical barriers. While these tools are already revolutionary, imagine the possibilities when AI joins the party. We could soon have AI agents orchestrating construction logistics, making the future of collaboration even more intriguing.

Then there are drones and unmanned aerial vehicles (UAVs) changing the game in fields like construction. These are not just remote-controlled toys, but robust platforms equipped with cameras and sensors, capable of surveying vast terrains and inspecting terra cotta facades for signs of decay. The result? A reduction in manual labor and enhanced safety, particularly in environments too hazardous for human intervention.

However, it is not all smooth flying. Drones come with their own sets of challenges, mainly regulatory hurdles. In the United States, for instance, the Federal Aviation Administration (FAA) is the governing body laying down the law for drone operations. Rules like maintaining a visual line of sight and *flying below 400 feet* can cause problems if you need to inspect the top of a building. Failure to comply can land you in a legal quagmire, making it crucial for operators to be well-versed in local regulations.

The sophistication of digital tools is accelerating at a breakneck pace. What started as isolated tools are rapidly evolving into comprehensive platforms, making it easier to navigate complex processes. But here is the kicker: the real innovation lies in the increase of dimensionality. Beyond the confines of 3D space, new dimensions in digital tools are opening up *unprecedented operational capabilities*. This is not just evolution; it is a full-blown revolution in how we perceive and interact with the digital realm.

3.7.3 Use of 6D BIM Platform

We are familiar with 3D BIM as a digital twin of a building, capturing its physical and functional characteristics in three dimensions. But that is merely the tip of the iceberg. 6D BIM takes it up a notch by folding in elements like cost and time, effectively becoming a comprehensive playbook for a project. While 3D BIM sketches the "what," 6D BIM elaborates on the "how much" and "when."

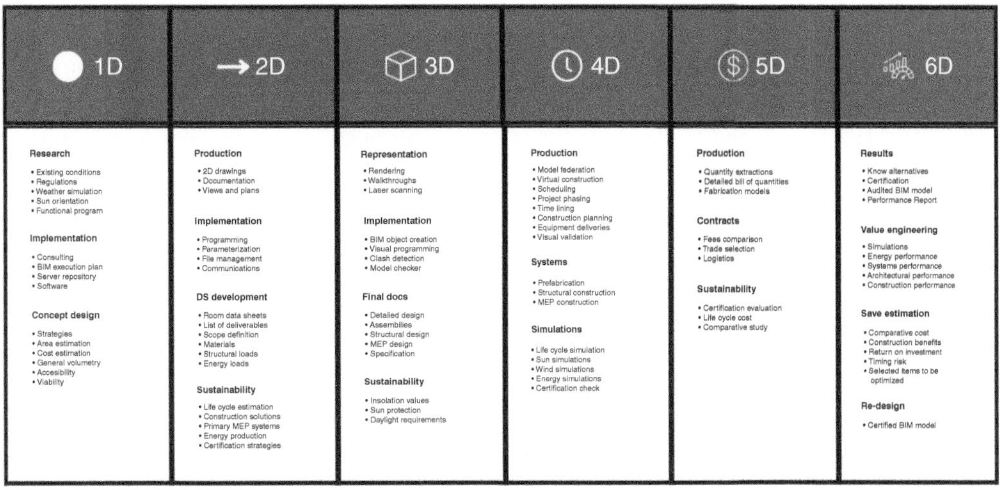

1D	2D	3D	4D	5D	6D
Research • Existing conditions • Regulations • Weather simulation • Sun orientation • Functional program **Implementation** • Consulting • BIM execution plan • Server repository • Software **Concept design** • Strategies • Area estimation • Cost estimation • General volumetry • Accesibility • Viability	**Production** • 2D drawings • Documentation • Views and plans **Implementation** • Programming • Parameterization • File management • Communications **DS development** • Room data sheets • List of deliverables • Scope definition • Materials • Structural loads • Energy loads **Sustainability** • Life cycle estimation • Construction solutions • Primary MEP systems • Energy production • Certification strategies	**Representation** • Rendering • Walkthroughs • Laser scanning **Implementation** • BIM object creation • Visual programming • Clash detection • Model checker **Final docs** • Detailed design • Assemblies • Structural design • MEP design • Specification **Sustainability** • Insolation values • Sun protection • Daylight requirements	**Production** • Model federation • Virtual construction • Scheduling • Project phasing • Time lining • Construction planning • Equipment deliveries • Visual validation **Systems** • Prefabrication • Structural construction • MEP construction **Simulations** • Life cycle simulation • Sun simulations • Wind simulations • Energy simulations • Certification check	**Production** • Quantity extractions • Detailed bill of quantities • Fabrication models **Contracts** • Fees comparison • Trade selection • Logistics **Sustainability** • Certification evaluation • Life cycle cost • Comparative study	**Results** • Know alternatives • Certification • Audited BIM model • Performance Report **Value engineering** • Simulations • Energy performance • Systems performance • Architectural performance • Construction performance **Save estimation** • Comparative cost • Construction benefits • Return on investment • Timing risk • Selected items to be optimized **Re-design** • Certified BIM model

The basics of a project: lean planning and preconstruction. Source: Adapted from BIM Community/https://www.bimcommunity.com/news/load/490/why-don-t-we-start-at-the-beginning.

Here are the five dimensions of 5D BIM:

1) **3D Modeling**: It is the first dimension of 5D BIM which involves creating a virtual model of the building project. This model includes all the details of the building, such as the floor plan, elevations, and sections.
2) **Time**: The second dimension of 5D BIM is time. This dimension involves creating a detailed schedule of the project, including timelines for each phase of the project.
3) **Cost**: The third dimension of 5D BIM is cost, which involves a detailed cost breakdown of the project, including labor, materials, and equipment.
4) **Sustainability**: This is the fourth dimension of 5D BIM. It involves creating a sustainable design for the building project, which includes energy-efficient systems and materials.
5) **Facility Management**: The fifth and final dimension of 5D BIM is facility management. Its content involves a detailed plan for the maintenance and operation of the building project after its completion.

The additional operations in the realm of sustainability and facility management is, of course, emblematic for our objective. The use of 5D BIM in the zero-carbon industry covers major categories of interest:

- **Material Selection**: 5D BIM can analyze the environmental impact of different building materials and help select those that have a lower carbon footprint.
- **Energy Modeling**: 5D BIM is used to create energy models of the building to optimize its energy performance.
- **Cost Optimization**: 5D BIM optimizes the cost of the building by analyzing the cost of different design options and selecting those that are both cost-effective and environmentally friendly.

- **Waste Reduction**: 5D BIM proposes ways to reduce waste during construction by optimizing the use of materials and reducing the amount of waste generated.
- **Life Cycle Analysis**: 5D BIM analyzes the environmental impact of the building over its entire lifecycle, from construction to demolition.

The sole possibility of creating a *5D operational model* capable of the former is based on the extensive amount of data sets that enter the system. The system operations are run by advanced algorithms that precedes advanced data analytics. This process is enabled through a combination of technology with these qualities:

- **Data Collection and Storage**: Advanced data analytics require large amounts of data, which are collected from various sources, sensors, and digital devices, all stored in a secure and scalable way.
- **Data Integration**: Data integration is the process of combining data from different sources into a single, unified, and comprehensive view.
- **Data Processing**: Advanced data analytics requires processing large amounts of data quickly, which is run by technologies such as distributed computing, parallel processing, and in-memory computing.
- **Machine Learning and Artificial Intelligence**: Machine learning and AI are key technologies that enable advanced data analytics, as they analyze large amounts of data, identify patterns, and make predictions.
- **Data Governance and Security**: Advanced data analytics requires a strong data governance and security framework to ensure that data is used ethically and responsibly, and that it is protected from unauthorized access and misuse.

3.7.4 Development of New Lightweight Materials and Construction Methodologies

The construction industry is undergoing a transformation, fueled by the imperatives of sustainability and carbon footprint reduction. At the heart of this revolution is the development of lightweight materials and innovative construction methodologies. But to fully appreciate the impact of these advancements, we must consider the complexities and tradeoffs involved, especially those surrounding weight and its correlation with a building's carbon footprint.

The weight of a building is not merely an engineering concern; it has substantial ramifications for the environment. Heavy structures require more foundational support, which often means more concrete and steel—two materials notorious for their high carbon emissions during production. The carbon footprint thus accumulates not just from the building itself but from the extensive support systems it necessitates. It is akin to the difference between an SUV and a compact car; the former not only consumes more fuel but also requires more resources for manufacturing and maintenance.

The use of lightweight materials like composites is a groundbreaking development. Composites combine two or more constituent materials, each with distinct physical or chemical properties, to create a substance with enhanced characteristics. Take fiber-reinforced polymer (FRP) composites, for instance, which meld fibers like carbon or glass with a polymer matrix. The resulting material is lightweight, yet impressively strong and resistant to corrosion. This makes FRP an attractive option for building structures from bridges to high-rises, offering considerable advantages over traditional materials like concrete and steel.

However, the composite story is not all roses, and we need to engage with manufacturers critically. Many composites are not easily recyclable, creating a future waste management challenge. Moreover, the manufacturing process itself can be carbon-intensive, offsetting some of the gains achieved through weight reduction. This is why it is vital to check the embodied carbon of every material used or substituted since plants and products are changing their footprint regularly.

Baku, Azerbaijan – December 2019: Heydar Aliyev Center architecture, the popular landmark for tourists and visitors was designed by architect Zaha Hadid. Source: uskarp2/Adobe Systems Incorporated.

CLT offers another compelling example. It is made by gluing layers of wood at right angles, resulting in a material that is both strong and lightweight. More importantly, it is a renewable resource, capturing carbon dioxide during its growth cycle. CLT is already being used in the construction of high-rise buildings, positioning wood as a serious contender in the future of sustainable construction. However, recently, concrete suppliers have engineered ultra-low carbon concrete mixes that rival the carbon content of CLT.

Vertical timber fins integrated within the glass facade. Source: creativenature.nl/Adobe Stock Photos.

In addition to materials, methodological innovations like modular construction and 3D printing are reducing the industry's environmental impact. Modular construction involves prefabricating components in a controlled factory environment, thereby minimizing waste and enhancing efficiency. Similarly, 3D printing enables the fabrication of complex geometries that would be impractical through conventional means. These methodologies not only save weight but also reduce the carbon footprint through more efficient use of materials and energy.

The quest for lightweight materials and innovative methodologies is a significant step forward, but it is fraught with complexities and tradeoffs. Weight reduction, while beneficial, is not a silver bullet as it must be part of a multifaceted strategy for sustainable construction. This includes a close collaboration with material manufacturers to assess the full lifecycle impacts of these new materials. As we architect the future, our blueprints must be as much about responsible material and method selection as they are about design and functionality.

3.7.5 Advanced Automated Equipment and Tools

In addition to drones, BIM, 3D printing techniques, and digital tools, let's explore some other advanced automated tools and equipment that are proven to harvest the *maximum efficiency*:

- **Robotic Total Stations**: They are advanced surveying tools that use robotic technology to automate the process of measuring and marking out construction sites. They are equipped with a remotely controlled motorized prism and are programmed to work autonomously, reducing the need for human intervention.
- **Autonomous Construction Equipment**: This is about excavators and bulldozers that are equipped with sensors and GPS technology. These tools can be programmed to perform specific tasks, such as digging trenches or moving materials, without human intervention.
- **Automated Guided Vehicles (AGVs)**: They are used to transport materials and products around a factory or warehouse. They are programmed to follow a specific path and can be used to transport heavy or bulky items.
- **CNC Machines**: Computer numerical control (CNC) machines are used to automate the manufacturing process. They are used to cut, drill, and shape materials with greater precision and accuracy than manual machines.
- **Automated Storage and Retrieval Systems (AS/RS)**: They are used to store and retrieve components in a distribution center, using robots or conveyors to move them to and from storage locations to sites.

All of them completely eradicated the need for physical human presence, which reduces danger of injury that comes with the manipulation of heavy equipment. Additionally, the precision and impact cannot be compared to anything involving purely human intervention.

The construction landscape is on the cusp of a technological metamorphosis, propelled by advancements in digitalization, materials science, and automation. Imagine 3D and 5D BIM as the orchestra conductors, harmonizing various instruments—cost, schedule, and sustainability analyses—into a symphony of integrated data. Lightweight champions like FRPs and CLT are rewriting the rules of architecture, much like the way carbon fiber changed the automotive industry. They are not just materials; they are enablers of a lighter, more durable built environment.

In the methodology arena, modular construction and 3D printing are the unsung heroes, reducing waste and environmental impact as effectively as a well-planned public transit system cuts down on urban emissions. We are also entering the age of construction robotics—autonomous vehicles, CNC machines, and even robotic storage systems are no longer the stuff of science fiction but an emerging reality. These technologies are the construction equivalent of smart home systems—optimizing tasks, enhancing safety, and making operations more efficient. Digital tools are the glue in this complex ecosystem, enabling remote collaboration and real-time decision-making as crucial as the role of air traffic control in coordinating flights in a bustling airport.

But here is the kicker: embracing these futuristic technologies is not a walk in the park. It demands significant capital outlay and a commitment to upskilling the workforce. Think of it as a venture capital investment for the planet; the initial stakes are high, but the returns—in terms of efficiency, waste reduction, and carbon footprint minimization—are more than justifiable. As these technologies gain traction, they are set to transform the construction industry into an automated, eco-conscious, and tech-driven powerhouse. This is not mere speculation; it is a forecast backed by the current trajectory of innovation. Companies willing to invest in these transformative technologies will not just be staying ahead of the curve; they will be defining it, gaining a competitive edge that is as structurally sound as the buildings they will construct.

3.8 Reskill the Workforce

3.8.1 Investment in Retooling an Aging Workforce

In our journey through advanced tools and technologies, we find a seismic shift shaking the traditional work paradigms. This disruption is not solely a function of an aging workforce; it is equally about the breakneck pace of technological change. Retraining the workforce is a cornerstone for any large enterprise. As workers age, equipping them with updated skills and knowledge becomes vital. It is not merely about current job roles, but about preparing them for future shifts in technology and market demands.

The role of education in the workforce. Source: FotoArtist/Adobe Stock Photos.

3.8.2 Continuous Reskilling and Training to Use Latest Equipment and Digital Tools

Remarkably, this evolution is so swift that even younger employees require these upskilling programs. Interestingly, many employees educate themselves out of personal interest and then pivot into roles involving advanced technology and sustainability. Investment in ongoing training not only benefits employee retention but also arms workers with skills vital for a rapidly changing job market. In an ever-shifting job landscape, staying ahead is crucial. Continuous training pinpoints employees' skill gaps and addresses them, ensuring that the workforce operates at peak efficiency. This also minimizes failures and mistakes along with the associated costs, creating a win-win in the long run.

It is important to recognize that everyone learns differently. Training sessions can be both in-house and outsourced, or even provided via online platforms. Leveraging learning management systems or online educational resources offers flexibility in training schedules.

3.8.3 Implementation of Apprenticeship Programs

Apprenticeship programs have been around for centuries to train and develop a skilled workforce. An apprenticeship program is a structured training program that combines on-the-job training with classroom instruction. The goal is to provide individuals with the skills and knowledge they need to become proficient in a particular trade or occupation. Implementing an apprenticeship program can be a significant investment for a company, but it can also yield the most benefits:

- **Developing a Skilled Workforce**: Apprenticeship programs are designed to provide individuals with the skills and knowledge they need to become proficient in a particular trade or occupation. This is an opportunity for companies to train the workforce to meet the specific criteria they need.
- **Increasing Employee Retention**: Programs can help increase employee retention by providing employees with a clear path for career development. When employees feel that they are valued and invested in, they are more likely to stay with the company.
- **Improving Productivity**: Apprenticeships provide skills and knowledge that workers need to perform their jobs efficiently, which can only benefit the quality of work in addition to reducing costs.
- **Promoting Diversity and Inclusion**: Programs promote diversity and inclusion by providing equal opportunities for all individuals, regardless of their background or experience.

Identifying the right people for these programs can be a subject for wider business strategy, because it requires careful planning and execution. And even though it can be a major investment for the company, it can yield long-term success for both sides.

3.8.4 Breaking Seasonality and Cyclicality for Increased Stability in Workforce

In the construction sector, two types of fluctuations challenge workforce stability: seasonality and cyclicality. Seasonality refers to short-term changes in demand, often dictated by the calendar—think construction upticks in summer and downticks in winter. Cyclicality is a longer-term phenomenon, influenced by economic factors like recessions or booms. Both culminate in a precarious employment situation: feast or famine, so to speak. Some of the top solutions to mitigate this are:

Diversifying Services

- **Why**: Specialization is a double-edged sword. While it allows a company to excel in one area, it also makes it vulnerable to market fluctuations within that niche.
- **How**: Companies can offer a range of services, such as branching from residential to commercial construction or even infrastructure projects. This diversification cushions the blow from downturns in any single area.

Geographical Expansion

- **Why**: Local dependency is risky; a downturn in local demand can cripple a business.
- **How**: By expanding into new regions—or even countries—companies can tap into different markets with different cycles, effectively hedging against local downturns.

Embrace Technology

- **Why**: Technology is a potent ally in boosting productivity and efficiency, which are critical during low-demand periods.
- **How**: The use of advanced tools and software can streamline operations, reduce waste, and minimize errors. Technologies like BIM can enhance collaboration and planning, making the workforce more resilient against seasonal and cyclical shifts.

Offer Training and Education

- **Why**: A skilled workforce is a flexible workforce.
- **How**: Companies can invest in continuous training programs to keep employees up-to-date with the latest industry trends and technologies. This adaptability makes them valuable assets who can pivot to different roles as needed, thereby mitigating the impact of demand fluctuations.

Strategic Partnerships

- **Why**: No company is an island, and strategic alliances can offer a lifeline during tough times.
- **How**: Partnering with firms in related sectors—like engineering and architecture—opens up new project opportunities and shared resources. These collaborations can counterbalance the risks associated with dependency on a single market.

3.9 Conclusion

Source: Nour Betar/Wirestock/Adobe Stock Photos.

In the sprawling world of the global construction industry, the stakes are paradoxical. On one hand, it is a titan of economic activity; on the other, an insatiable monster guzzling resources and belching out carbon like there is no tomorrow—literally. This chapter is not just a polite suggestion; it is a clarion call for a seismic shift, articulated through seven strategic levers that could change the game entirely.

Let us start by dismantling the ivory tower that is regulation and policy-making. Contrary to the bureaucratic haze it is often lost in, policy is a mindset, albeit a complex one. It is the first domino in a long chain; tip it right, and you can set off a cascade of sustainable practices. Think transparency, ethical underpinnings, and a circular economy that does not just spin, but evolves. This is not just about eco-friendly building codes but stitching accountability into the industry's DNA.

Now, let us talk money—specifically, the contracts that make the construction world go around. Forget the old incentives; the new age demands performance-based rewards. It is not charity; it is enlightened self-interest. Contractors and companies need assurance that virtue is more than its own reward. They are asking for a stake in the outcome, an alignment of risks and rewards, and concrete benefits to collaborative practice. Without these, we are just building castles in the air.

It's not just about rethinking; It is about retooling—literally. Introduce modularization and prefabrication into the workflow, sure, but let us also marry bricks with bytes. Digital transformation—through BIM, IoT, robotics, 3D printing—is not a gimmick. Yes, it is a challenge that demands new skills and disrupts old routines. But the payoff? A quantum leap in quality and productivity. Forget job displacement; we are talking about workforce empowerment.

The elephant in the room? The sheer inertia that makes adopting these changes seem Herculean. But here is the kicker: you are not alone. Our interviews and case studies show that the vanguard is already marching; the pioneers are already building oases in the desert. The first step is huge, but once you join the ranks, you are part of a collective thrust toward a cleaner, more sustainable reality.

So here is where we land: change is not a solo act; it is a symphony of efforts, an amalgamation of individual actions that could crescendo into a transformative force with planetary consequences. As public sentiment shifts, as demands for a sustainable future crescendo, the construction industry does not just have a chance to jump on the bandwagon; it has the potential to take charge and make even greater profits.

3.10 Insights from Natalie Terrill, Director of Sustainability at Beck Group

In her role as Director of Sustainability at Beck Group, Natalie Terrill is propelling the company toward new heights in sustainable building design and construction across the United States and Mexico. She spearheads discussions on sustainable business practices, carbon-neutral design, and environmental impact reporting, particularly in terms of GHG emissions.

Man in "Beck" construction vest looking at wall of work orders.

Natalie has been an integral part of the Beck team since 2018. Her influence extends across the built environment, where she bridges the gap between intricate technical concepts and stakeholders. Throughout her career, Natalie has achieved third-party certifications for over 15 million square feet of built space under multiple rating systems. Her expertise spans environmental management, encompassing land and water use, waste diversion, air quality, sustainable material sourcing, health and well-being, and energy efficiency.

The imperative to demonstrate cost savings serves as a linchpin in the drive for decarbonization. This encompasses a meticulous evaluation of equipment efficiency, electrified equipment availability, rental/purchase costs, and the practicality of charging compared to traditional fuel consumption. Questions about charging infrastructure, space, and power capacity must be addressed.

3.10.1 Metrics for Material Selection

Terrill and her team continually seek material transparency statements, including environmental product declarations (EPDs), health product declarations (HPDs), declare labels, and emissions test results. They prioritize low-carbon materials such as steel, concrete, and insulation due to their ubiquity in construction. Notably, steel and concrete, with their high embodied carbon, are focal points for improvement.

Construction equipment loading prefabricated building section.

3.10.2 Benefits of Integrated Project Delivery

Integrated project delivery fosters seamless communication and collaboration, particularly when teams are co-located. It accelerates decision-making and enhances cost predictability. In the realm of sustainability, it affords early participation in crucial discussions.

3.10.3 Incorporating Data into the Construction Process

Beyond conventional construction documents, specific requirements are embedded within bid documents. Structured plans, procedures, and workflows ensure the clear communication of these requirements.

Beginning construction site.

3.10.4 Coordinating Subcontractors for Decarbonization

Effective coordination with subcontractors is crucial. Initiatives include orientation sessions and kick-off meetings to clarify the expectations and requirements for decarbonization.

Construction workers meeting.

3.10.5 Emerging Trends and Opportunities

Natalie envisions a growing focus on creating healthier job sites, which entails improving job site trailers and electrifying construction processes, including vehicles and equipment. Shifting from diesel generators to alternative fuels and on-site solar solutions is another promising trend. Additionally, the drive for prefabricated components and greater material transparency is expected to continue emphasizing low carbon solutions.

3.10.6 Impact on the Carbon Dialogue

The Beck Group's journey in sustainability has been marked by continuous education and advocacy. Natalie emphasizes the importance of educating leaders, clients, consultants, and sub-consultants across the spectrum of the built environment. This holistic approach, extending from boardroom discussions to on-site practices, has been a cornerstone of Beck's sustainability efforts. The company's involvement in industry organizations, including the Large Firm Roundtable Sustainability Committee, the USGBC Chapter, ASHRAE, the Carbon Leadership Forum, and the AGC Climate Action Task Force, underscores its commitment to staying informed and amplifying its influence.

Beck Construction site at night.

3.10.7 Design-Build Teams and Building Decarbonization

Design-build teams, with their enhanced control over projects, are well-positioned to integrate decarbonization effectively. Early involvement of designers, sustainability teams, and preconstruction experts enables tailored solutions aligned with clients' visions and budgets. Beck's approach begins with a thorough understanding of climate conditions, addressing building orientation and massing for improved energy efficiency. Cost management revolves around identifying opportunities, such as downsizing mechanical systems through improved glazing. Collaboration with like-minded MEP 2040 and SE 2050 signatories further enhances decarbonization discussions. Rigorous material reviews and specifications play a pivotal role in shaping sustainable design solutions.

3.10.8 Contractors' Role in Decarbonization

Contractors play a significant role in decarbonization, particularly during construction. Their impact extends to waste management, dust and air pollution control, reducing idling, introducing electric vehicles and equipment, and exploring alternative fuels when electrification is not feasible. While construction adheres to construction documents, contractor involvement in design reviews amplifies their contribution to decarbonization efforts.

4

The Business Case for Carbon Positive Buildings

The building sector stands as an emblem of inefficiency; a silent behemoth that guzzles energy only to waste it through leaky windows, antiquated heating, ventilation, and air conditioning (HVAC) systems, and subpar insulation. This is not merely an architectural or economic hiccup; it is a colossal environmental misstep. However, the sector's inefficiency should not be viewed as a problem but as an unparalleled opportunity. Retrofitting existing structures and adopting sustainable construction methods could represent the golden ticket to a new economic boom. This is not mere speculation; it is supported by data indicating that green construction could be a driving force for economic growth, job creation, and innovation in multiple fields, from renewable energy to materials science.

However, the clock is ticking loudly, its hands moved by the gears of climate change. The longer we delay, the more we risk. While the potential for growth and innovation in the building sector is dazzling, the risks of inaction are staggering. Climate change, if left unchecked, has the power to unleash damages that defy quantification. And let us be clear: the burden of these consequences will not be evenly distributed. The most vulnerable populations will bear the brunt of environmental degradation, making our choices today not just an economic or environmental matter, but a deeply ethical one.

So here we stand at a crossroads. On one path, we see a future enriched by sustainable buildings that act as pillars of economic growth, technological innovation, and environmental responsibility. On the other path lies a grim tableau of missed opportunities and escalating climate catastrophes. The latter is not merely a possibility; it is a guarantee if we do not act now. The evidence is irrefutable and the choice is stark: embrace sustainability and unlock unprecedented opportunities or procrastinate and pay an unimaginable price. The decision seems obvious; yet the urgency to act cannot be overstated.

4.1 The Interconnection Between Climate Risk, Finance, Insurance, and Economic GDP

The intricate web that ties together climate risk, finance, insurance, and economic GDP is both dynamic and volatile. It is not simply a matter of if climate change will impact these sectors, but how and when. In the United States alone, from January to August of 2023, there were US$27 billion climate disasters. Each event, each dollar lost, reverberates through the economic landscape, affecting everything

Build Like It's the End of the World: A Practical Guide to Decarbonize Architecture, Engineering, and Construction,
First Edition. Sandeep Ahuja and Patrick Chopson.

from insurance premiums to the cost of construction materials. This is more than a footnote in a financial report; it is an alarm bell ringing across sectors. For instance, construction companies are already feeling the pinch. The cost of insurance is rising, as is the price of materials, and the need to invest in climate-adaptive technologies is becoming non-negotiable. In one telling statistic, climate risk was responsible for a 12% increase in homeowner insurance premiums from 2021 to 2022. In cities more exposed to climate disasters, like New Orleans and Miami, the annual insurance cost can skyrocket to US$5,000 per year.

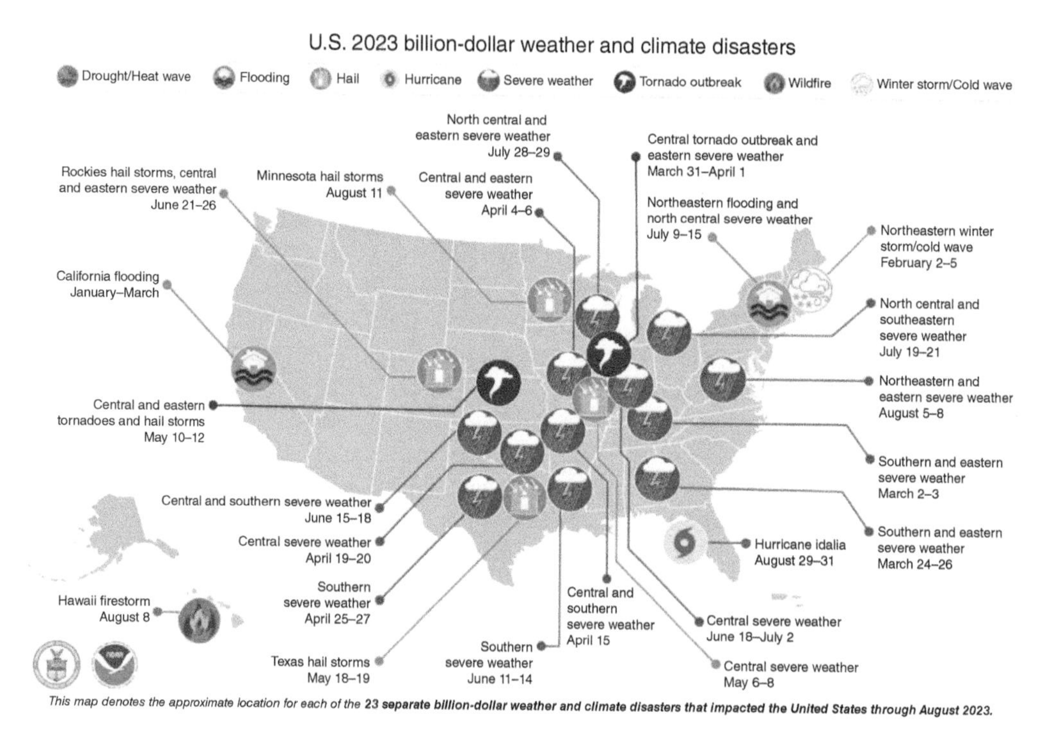

Source: NOAA/https://wpo.noaa.gov/dras/Public Domain.

Insurance, traditionally a safety net for businesses and homeowners alike, is becoming both more expensive and less available due to the rising frequency and severity of extreme weather events. This is particularly problematic for the construction industry. Firms may soon find it unaffordable or even impossible to secure coverage unless they adopt sustainable practices as a countermeasure. But the impact of climate risk is even more sprawling; it has the potential to negatively influence GDP through direct and indirect channels. These include direct economic losses and a cascade of secondary effects that impair supply chains, inflate costs, reduce productivity, hamper investments, and even dent tourism revenues.

	Temperature rise scenario, by mid-century			
	Well-below 2°C increase	2.0°C increase	2.5°C increase	3.2°C increase
	Paris target	*The likely range of global temperature gains*		*Severe case*
Simulating for economic loss impacts from rising temperatures in % GDP, relative to a world without climate change (0°C)				
World	−4.2%	−11.0%	−13.9%	−18.1%
OECD	−3.1%	−7.6%	−8.1%	−10.6%
North america	−3.1%	−6.9%	−7.4%	−9.5%
South america	−4.1%	−10.8%	−13.0%	−17.0%
Europe	−2.8%	−7.7%	−8.0%	−10.5%
Middle east & africa	−4.7%	−14.0%	−21.5%	−27.6%
Asia	−5.5%	−14.9%	−20.4%	−26.5%
Advanced asia	−3.3%	−9.5%	−11.7%	−15.4%
ASEAN	−4.2%	−17.0%	−29.0%	−37.4%
Oceanio	−4.3%	−11.2%	−12.3%	−16.3%

The projected GDP loss with different climate change models based on information from Swiss Re Institute. Source: Adapted from https://www.weforum.org/agenda/2021/06/impact-climate-change-global-gdp/with permission of World Economic Forum.

However, let us not lose sight of the flip side: the extraordinary economic benefits that come with transitioning to a more sustainable model. By shifting toward the use of sustainable materials and energy-efficient designs, we stand to gain in terms of increased investments and job creation in the construction industry. Renewable energy adoption not only slashes energy costs but also fortifies energy security, bringing with it a gamut of broader economic advantages. While transitional costs, such as those tied to insurance, cannot be ignored, the long-term economic outlook is overwhelmingly positive. Investments in zero-carbon construction are not just ethical imperatives; they are savvy economic strategies promising job creation, enhanced energy security, and a resilient economy for the decades to come.

4.1.1 The Financial Benefits of Adopting Low-carbon Building Investments

The financial case for net zero investments in construction is compelling, both in the short term and across a longer horizon. It is a misconception that sustainability is just a "feel-good" endeavor; in reality, it is a strategic financial move. The upfront cost savings from implementing energy-efficient systems and using sustainable materials are significant. Additionally, the long-term gains manifest in multiple ways: lower operating expenses due to reduced energy usage, greater asset value from green certifications, and a myriad of financial incentives. In fact, projects with a focus on sustainability often have such appealing financial returns that justifying and funding them becomes a straightforward exercise.

Green buildings are not just eco-friendly; they are also economic magnets. The certifications they earn often translate into higher rents and sale prices, enhancing their asset value. The financial sector is catching on, with many institutions offering preferential rates and terms specifically tailored for green construction projects. On the policy front, governments are stepping up their game, providing a range of incentives like tax credits, rebates, and grants to make low-carbon building investments more attractive. Beyond the balance sheet, the improved indoor environments of these green buildings can have a marked impact on occupant productivity and performance. This not only offsets some of the initial investment costs but adds another layer of long-term financial benefit. As carbon regulations become more stringent, buildings designed with low-carbon strategies will face fewer financial risks, providing an additional buffer against market volatility.

Combining all these factors—lower operating costs, increased asset value, a plethora of financial incentives, and productivity gains—creates a robust financial portfolio for low-carbon building strategies. Plus, let us not forget the reputational boost that comes with being a socially responsible organization. In a market increasingly driven by consumer perception and social values, this is not a negligible asset; it is a strategic advantage. Businesses that invest in sustainable construction not only stand to improve their bottom line but also enhance their brand equity. In summary, low-carbon building strategies are not merely an ethical or environmental choice; they represent a comprehensive financial strategy that promises both immediate and enduring returns.

4.2 The Role of Climate Risk in Finance, Insurance, and Economic GDP

4.2.1 The Impact on Businesses and Economies

Should climate change remain unchecked over the coming decades, we face the risk of significantly undercutting GDP growth. The cascading effects will wreak havoc on supply chains, slash corporate profits and consumer spending, undervalue assets such as real estate, and impair financial systems that grapple with accounting for climate risks.

View of the pollution in Shanghai. Source: Collab Media/Adobe Stock Photos.

As for businesses, climate risks manifest as losses from facility damage, supply chain disruption, litigation, healthcare costs, as well as reductions in productivity, sales, and brand value. Failure to adapt can mean a halt in business operations, bankruptcy, or premature closure, while early action can generate new opportunities. Transitioning toward sustainability is key to economic resilience and competitiveness in a world altered by climate change.

Impact of global supply chain disruption. Source: Yellow Boat/Adobe Stock Photos.

In the insurance sector, climate change intensifies the frequency and severity of insurance claims, jacks up coverage costs, and increases the chance of policyholders facing multiple catastrophic losses simultaneously due to a single extreme weather event. And if regions became uninsurable, it would imperil the business viability and financial stability of the industry. Governments and development banks would be grappling with debt burdens to provide aid, thereby crowding out other public investments imperative for long-term prosperity.

When it comes to economies, the costs of damage inflicted by climate change will take shape as lost GDP, tax revenues, jobs, productivity, and welfare. The brunt of the impacts falls most heavily, of course, on poor and developing nations with the smallest economies and emissions profiles, which will only serve to deepen global inequality. Based on research by the New Climate Economy annual report, there are suggestions that 2% of lifetime GDP could be jeopardized by unchecked climate change, with costs due to damage from extreme weather alone potentially reaching a jaw-dropping US$100 trillion this century.

4.2.2 The Costs of Inaction

Experts caution that failure to take decisive action on climate change could mean incurring colossal costs to the global economy, society, and environment. Without adopting significant measures, we court the risk of incurring massive economic losses. The World Economic Forum's (WEF) *Global Risks Report 2019* calculations show that failure to meet the aims of the Paris Agreement could equate to losing over US$40 trillion in economic output by 2030. The *WEF calls for an additional US$3.5 trillion* in total global

Imagery highlighting the impact from natural disasters, including hurricane, tornado, and more. Source: ungvar/Adobe Stock Photos.

spending by governments, businesses, and individuals on energy and land use systems, an essential step if we are to get to net zero by 2050. The costs of inactivity are massive and escalating. The risks break down along the following big categories.

- Economic losses from extreme weather events
- Disruption to global supply chains
- Instability in financial systems
- Risk of stranded assets
- Health, food, and water security

The financial toll exacted by extreme weather events is staggering. Between 2000 and 2019, these natural disasters have caused an estimated US$3.3 trillion in economic damages. These are not just isolated incidents but indicators of a growing trend. If global temperatures continue to rise unchecked, such catastrophes are projected to become both more frequent and more devastating. The implications are far-reaching, affecting everything from local economies to national budgets, and even international financial markets. The message is clear: failure to mitigate climate change is a gamble with increasingly unaffordable stakes.

Highlighting the impact of supply chain and logistics. Source: h368k742/Adobe Stock Photos.

Climate change poses a significant risk to the intricate web of global supply chains that underpin modern commerce. Events like port closures due to storms or flooding, infrastructure damage from natural disasters, and supply disruptions caused by extreme weather conditions could have a cascading effect. Businesses, industries, and jobs that depend on the timely delivery of goods and components from around the globe could be severely impacted. In a world increasingly reliant on just-in-time delivery systems, these disruptions could reverberate through the economy, causing widespread instability and job loss.

The WEF has sounded the alarm that climate change has the potential to undermine the financial systems that are the backbone of global economic prosperity. We are not just talking about isolated defaults or short-term market volatility; we are looking at the possibility of full-scale market meltdowns. In such a scenario, the ripple effects would extend far beyond Wall Street, endangering livelihoods, investments, and overall economic growth.

As we move toward tighter climate regulations and technological advancements, assets like fossil fuel reserves face the risk of becoming stranded—essentially worthless. This could trigger what is known as a "climate Minsky moment," a sudden market collapse triggered by the reckless accumulation of climate-related financial risk. The repercussions would be severe, affecting both individual and institutional investors, and could lead to a widespread economic downturn.

Source: Stuart Monk/Adobe Stock.

We cannot overlook the indirect financial implications associated with the health, food, and water security risks posed by climate change. Rising temperatures, the spread of diseases, and worsening air and water quality not only jeopardize human life but also have economic costs. These range from increased healthcare spending to losses in agricultural productivity and heightened food and water scarcity, which can lead to social unrest and further economic instability.

The financial ramifications of climate change are not confined to any single sector but are a complex interplay of numerous factors that pose serious risks to global economic stability. These are not hypotheticals or distant future scenarios; they are urgent issues demanding immediate and coordinated action.

4.2.3 The Importance of Climate Risk Management for the Finance Sector

The Network for Greening the Financial System (*NGFS*) has been at the forefront of understanding the monumental impacts of climate risks on the financial sector. Established in 2017 by an initial coalition of eight central banks, the NGFS has grown to include 81 member organizations, spanning central banks, supervisors, and international bodies. These members are not merely discussing climate risk in theoretical terms; they are formulating actionable steps to integrate climate considerations into the foundational practices of the financial sector. The urgency of their work stems from a multidimensional understanding of how climate risks can infiltrate and destabilize financial systems.

Abandoned unfinished building located along a highway in the forest wild valley of Rhodope Mountains. Source: YouraPechkin/Adobe Stock Photos.

Environmental water pollution. Source: TRAVELARIUM/Adobe Stock Photos.

The NGFS outlines several key areas that accentuate the importance of climate risk management for the financial sector:

- **Physical Risks**: Financial institutions are not immune to the physical impacts of climate change. These impacts can manifest in ways that seriously affect asset valuations, the creditworthiness of borrowers, and the scope of insurance claims.
- **Transition Risks**: The move toward a low-carbon economy is not without its hazards. Financial institutions may find themselves grappling with stranded assets, fluctuating asset valuations, and disruptive changes to business models in sectors that are late to adapt.
- **Underpricing of Risks**: A failure to adequately manage and disclose climate risks can result in the misallocation of capital. This can lead to a dangerous accumulation of risks within the financial system, making it susceptible to sudden shocks.
- **Financial Stability**: The systemic nature of climate risks means they have the power to threaten the overall stability of financial markets. Scenarios include widespread defaults, forced asset sales, and spillover effects that can cascade through various markets and institutions.
- **Regulatory Pressures**: As climate change gains prominence on the global agenda, financial regulators are stepping up their demands for climate risk management and transparent disclosures. Financial institutions that lag behind face not only regulatory sanctions but also reputational damage.
- **Fiduciary Duty**: Asset managers and other financial institutions are bound by a fiduciary duty to their clients and stakeholders. This extends to the proper identification, assessment, and management of material risks like those associated with climate change.
- **Business Opportunities**: On the flip side, effective climate risk management also opens the door to new avenues of green finance and sustainability-linked business. These are not just ethical imperatives but lucrative opportunities to support the transition to a low-carbon economy.

Given the scale and complexity of these challenges and opportunities, the NGFS has set an ambitious goal: to mobilize US$100 trillion for sustainable finance. Central banks have a pivotal role to play in reaching this milestone. Through the development of specific recommendations, partnerships, and tools, the NGFS aims to build the momentum for systemic action, integrating sustainability as a core tenet across the financial sector.

4.2.4 The Influence of Climate Risk on Insurance

Munich Re retains the title of one of the world's biggest reinsurance companies, zealously tracking disaster trends. It possesses a substantial amount of data on escalating costs and frequency of climate-related events. Their *findings* indicate that the economic toll inflicted by disasters has grown considerably over the past few decades. For evidence, consider the costs of hurricanes in the Atlantic which have seen a rise of over 300%, after adjustments for inflation, since the 1980s. The costs of floods worldwide have also followed suit. A large part of this surge is attributed to increasing exposure of lives and assets in vulnerable areas as well as rising disaster severity tied to climate change.

Home insurance prices are rising rapidly

Change in premiums from January 2022 to July 2023

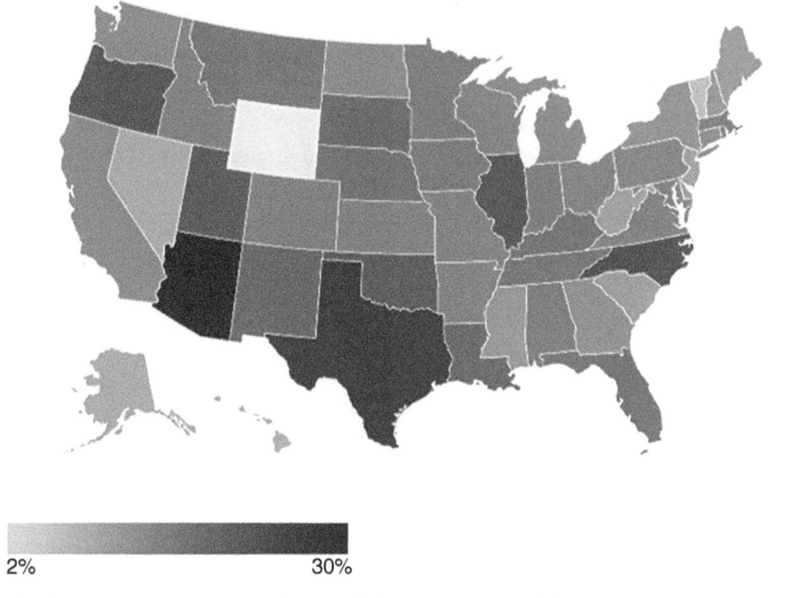

Since January 2022, 31 states have witnessed double-digit rate increases. Six states saw increases of 20–30%. Based on data from council on foreign relations study on climate change and U.S Property insurance.

The scientific consensus is that climate change is projected to lead to more frequent and more intense extreme weather events like heatwaves, heavy rain and snowfall, flooding along coasts and rivers, along with more frequent and severe wildfires. These types of extreme weather events tend to cause the largest economic losses and damage, and as the situation evolves, they anticipate the frequency and severity of such events to keep spiraling upwards in a noticeable manner.

Munich Re takes data on disasters, claims, and risks and puts it under a microscope to discern trends and project potential impacts. These cautionary notes are grounded in extensive data and research, and are not sheer conjecture. As a reinsurance firm, their goal is to delve deep into risks to ensure they can provide financial cover and support vulnerability reduction. Government figures on costs of damage often underestimate the toll, as they neglect many impacts, and only factor in immediate, quantifiable costs.

To get a real sense of risks and to allocate resources appropriately, partnerships between the public and private sectors is a must. No individual sector has the power or financial resources to rise to the occasion. Collaboration is key across governments, NGOs, scientists, businesses, finance leaders and more if we are to devise and implement the solutions we need.

Swiss Re acknowledges the main threats *looming over different time horizons*. This spells trouble for the ability of insurers and reinsurers to offer comprehensive coverage and stability to policyholders.

As a result, insurers must adapt their models, limits, and strategies to cover for climate risks if they hope to stay in the game over the long term.

Source: tampatra/Adobe Systems Incorporated.

Traditional insurance models were not designed to deal with the scale and type of risks generated by climate change. Regular increases in coverage limits, geographic spreading of risk, and averaging outcomes over long time periods will not fully account for risks such as more powerful hurricanes, rising sea levels, and more frequent wildfires.

Anticipation of more frequent so-called "tail risks" is part and parcel of climate change (extreme events causing catastrophic losses). They are referred to as "tail" because even though they have a low chance of occurrence, they represent dire threats in the tails of a probability distribution.

Standard insurance models are not engineered to absorb the costs of frequent tail risks, which endangers the financial stability of insurers. New models are required to better gauge and limit exposure to such. One challenge to ponder for their implementation is geographic spreading of risk, which provides less benefit as climate change impacts become more widespread. When damage becomes systemic rather than localized, the costs cannot be mitigated by lower risks elsewhere. As some areas could turn into no-insurance zones due to high risk, insurers will have to factor that into their insurance models. Another facet to take into account is a disorganized shift to a low-carbon economy which can pose risks due to a drop in demand for fossil fuel extraction/transportation insurance, which in turn affects revenues. As new renewable energy technologies come to the fore, they may usher in unfamiliar risks. Insurers need to mull over how to modify their underwriting models and stay competitive as system changes speed up.

4.2.5 The Effects of Climate Change on Economic GDP

Potential Reductions in GDP due to Climate Change

The International Monetary Fund (IMF) warns that if climate change goes unattended, global GDP could shrivel by up to 20% by 2100 due to damage wreaked by extreme weather events. However, the severity of outcomes largely depends on our ambitions for cutting emissions and adaptation strategies put to effect in the coming decades. Some of the sectors and areas that will likely be dealt the harshest blows may see GDP shrink by over 30–50%. This includes regions such as parts of sub-Saharan Africa, Southeast Asia, coastal communities, and those dependent on agriculture/tourism.

Less developed nations with fewer resources have to face consequences which they had little hand in creating. Wealthier nations have more means to shield themselves, at least for the time being. But no country will be spared from severe damage, and this poses threats to global stability.

Source: FiledIMAGE/Adobe Stock Photos.

According to predictions by the IMF and researchers, adapting to climate change will call for spending trillions worldwide on an annual basis. Massive investments need to go into areas like infrastructure which is resilient to climate change, biodiversity conservation, fortifying health systems, improving emergency management, managing migration, and adopting sustainable agriculture.

The Role of Climate-resilient Infrastructure and Investments in Maintaining Economic Growth

Infrastructure and investments which fail to account for climate change risks pose a potential threat to economic stability. If the damage inflicted by extreme weather events pulls ahead of the pace of repair/replacement costs over time, it could significantly hamper growth.

Source: onlyyouqj/Adobe Stock.

Climate-resilient infrastructure refers to designing, constructing, and maintaining infrastructure (like roads, bridges, water systems, energy grids, emergency management facilities, and housing) in such a way as to better withstand the probable impacts of climate change. As per experts, this means using more durable materials, fortifying redundancy, strategic location, and factoring in higher risk impacts into planning.

Global partnerships and financing mechanisms need to come into play to support this goal, in addition to investments and transitioning of jobs. The likes of green bonds, loans linked to sustainability, dedicated climate funds, international agreements, modifications to financial regulations, policy incentives, and vehicle retirement programs will necessitate coordinated action across borders. Developed nations should take up the mantle of providing substantial funding and support for developing nations to go down the sustainable route.

4.3 The Role of Policy and Regulation in Encouraging Low-Carbon Building Investments

4.3.1 The Implementation of Carbon Pricing to Promote Sustainable Investments

Carbon pricing is increasingly recognized as a potent policy tool for incentivizing sustainable investments. As emphasized by the Carbon Pricing Leadership Coalition, this mechanism is particularly effective when used in conjunction with other regulations to encourage the private sector to invest in low-emission technologies and solutions. In the context of the building sector, carbon pricing can serve as a catalyst for investments in energy-efficient upgrades, on-site renewable energy sources, and other sustainable construction materials. However, the barriers of high upfront costs and extended payback periods often deter building owners from making these crucial "green" investments. Herein lies the opportunity for policy intervention: by putting a price on carbon through mechanisms such as emissions trading systems or carbon taxes, governments can level the playing field.

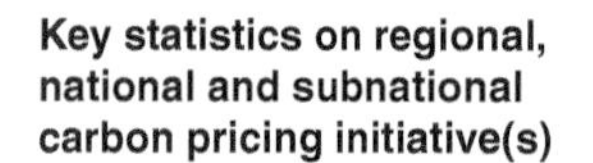

73 Carbon pricing initiatives implemented

39 National jurisdictions are covered by the initiatives selected

33 Subnational jurisdictions are covered by the initiatives selected

In 2023, these initiatives would cover 11.66 GtCO2e, representing 23% of global GHG emissions

Summary map of regional, national, and subnational carbon pricing initiatives

- ETS implemented or scheduled for implementation
- ETS or carbon tax under consideration
- ETS implemented or scheduled, ETS or carbon tax under c...
- Carbon tax implemented or scheduled for implementation
- ETS and carbon tax implemented or scheduled
- Carbon tax implemented or scheduled, ETS under consider...

Carbon pricing dashboard: Key statistics on regional, national and subnational carbon pricing initiative(s) based on data from Carbon Pricing Leadership coalition. Source: Adapted from https://carbonpricingdashboard.worldbank.org/.

The strategy is straightforward but powerful: make traditional high-emission systems, such as conventional heating systems, financially less appealing compared to their low-carbon counterparts. The aim is to tip the scales of financial viability, nudging building owners toward sustainable investments with a fiscal prod. However, for carbon pricing to truly be effective in reshaping investment patterns, the carbon price needs to be more than just symbolic; it needs to be substantial enough to make stakeholders reevaluate their investment strategies seriously.

But carbon pricing is not a silver bullet; it is part of a broader policy ecosystem. Revenues generated from carbon pricing can be funneled back into the system to offer incentives and subsidies, helping to mitigate the financial risks associated with transitioning to greener solutions. Beyond that, governments can set regulatory mandates specifying minimum levels of energy efficiency for both new and existing buildings, establishing a baseline that carbon pricing aims to exceed. Over time, these carbon prices will likely need to be incrementally increased to sustain meaningful change and to meet increasingly stringent climate goals.

Achieving the emissions reductions needed in the real estate sector requires a nuanced blend of financial incentives, funding mechanisms, and regulatory standards. Carbon pricing stands as a pivotal element in this policy cocktail, acting as both a deterrent against unsustainable practices and an inducement for greener alternatives.

4.3.2 The Adoption of Green Building Standards and Certifications

Green building standards and certification systems like LEED and BREEAM have emerged as pivotal tools in steering the building industry toward more sustainable practices. These programs offer a tiered system of certifications—typically Basic, Silver, Gold, and Platinum—that serve as a roadmap for sustainability. Earning these certifications is more than just a badge of honor; it entails meeting rigorous criteria that cover a spectrum of sustainability issues, such as energy efficiency, water conservation, and the use of sustainable materials. The process requires building owners to collect, document, and submit a range of performance metrics, which are then verified by an independent third party to ensure that the building meets the set sustainability standards.

Source: twinsterphoto/Adobe Stock Photos.

The landscape of green building certifications is diverse and increasingly specialized. While LEED and BREEAM offer a holistic approach focused on overall environmental impacts, others like the WELL Building Standard specialize in health and well-being, and the Passive House and PHIUS certifications target ultrahigh-performance buildings. Moreover, certifications like Energy Star and Green Globes are more narrowly focused on energy performance. Despite their differences, all these systems offer a structured framework for green building design, allowing for benchmarking and recognition of sustainability achievements.

Source: Processing Instruction teamjackson/Adobe Stock Photos.

Public and private entities have started to integrate these certifications into their operational and policy frameworks. For instance, some jurisdictions mandate minimum certification levels for new public buildings, while others use them as a prerequisite for accessing government funding and incentives. Financial institutions are also taking notice, offering favorable financing terms for certified green buildings, recognizing their intrinsic higher value and lower operational and systemic risks. On the flip side, critics argue that the focus on individual buildings may overshadow broader urban planning goals. Others question the rigor of these certifications, suggesting that they sometimes award points for measures with only marginal environmental benefits.

Green commuting is a low carbon strategy that can be broadly supported by the right urban infrastructure. Source: tirachard/Adobe Stock Photos.

Key Considerations in the Adoption of Green Building Standards

- **Benchmarking**: These standards offer a comparative framework against which the sustainability of traditional buildings can be assessed.

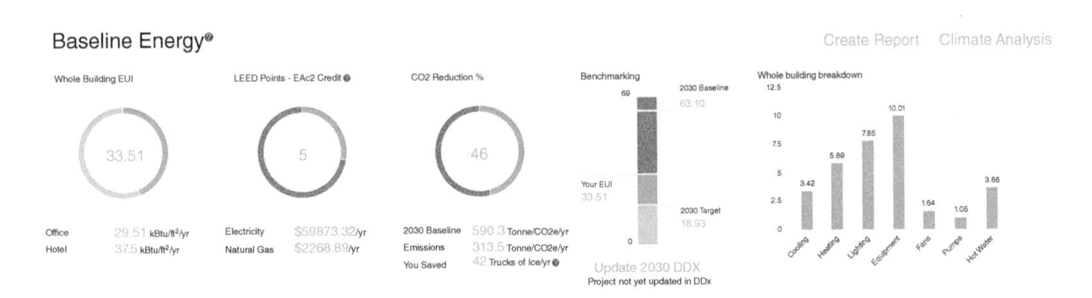

Baseline energy model from cove.tool.

- **Technological Adoption**: Certification criteria often encourage, or even mandate, the adoption of cutting-edge low-carbon technologies and materials.
- **Data-Driven Improvements**: The data collected during the certification process can inform ongoing efforts to improve building performance.

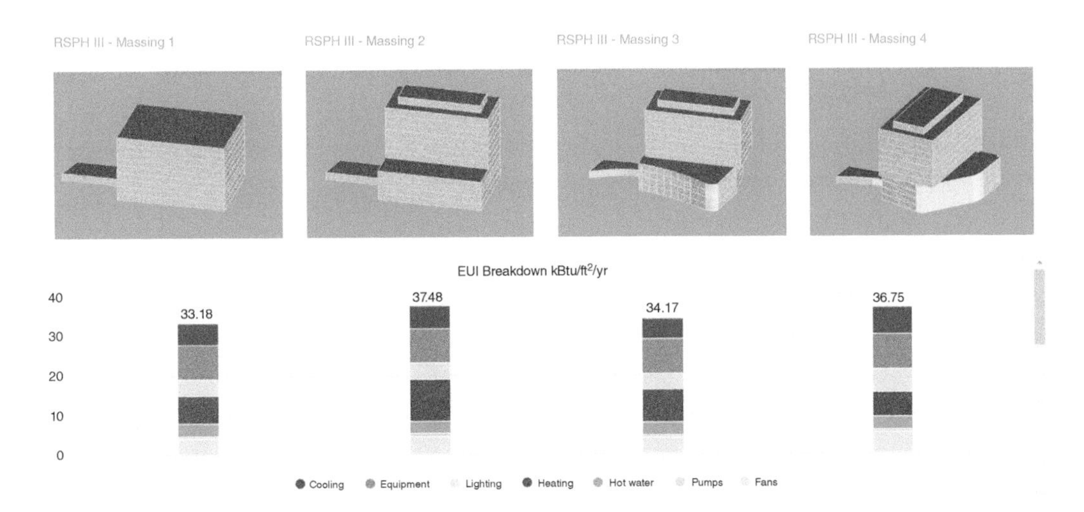

Comparing the energy use breakdown of 4 different building massing options.

- **Third-Party Verification**: This lends credibility to the building's sustainability claims, providing assurance to buyers, tenants, and investors.

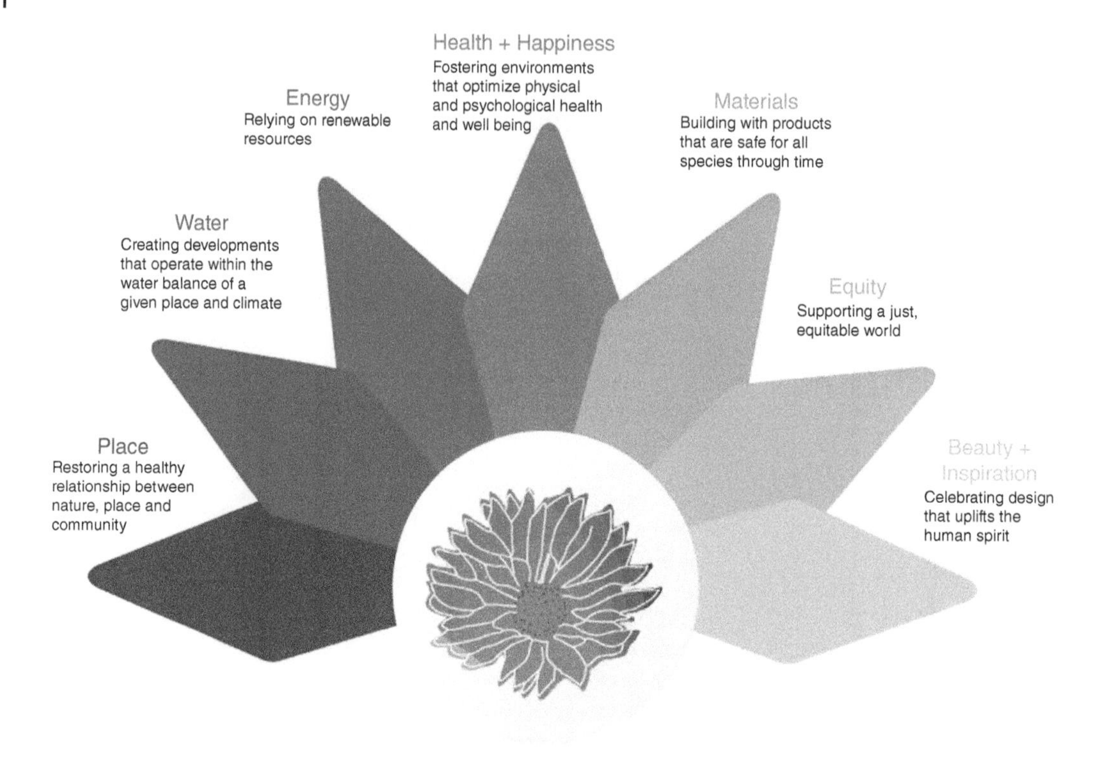

The various petals within the Living Building Challenge.

- **Influence on Building Codes**: Green building standards are increasingly serving as a reference point for developing more stringent building codes and regulations.

The energy code evolution map. The deeper the color of red, the more stringent the energy code.

- **Occupant Benefits**: Certified buildings often provide enhanced indoor air quality, natural lighting, and general well-being for occupants.

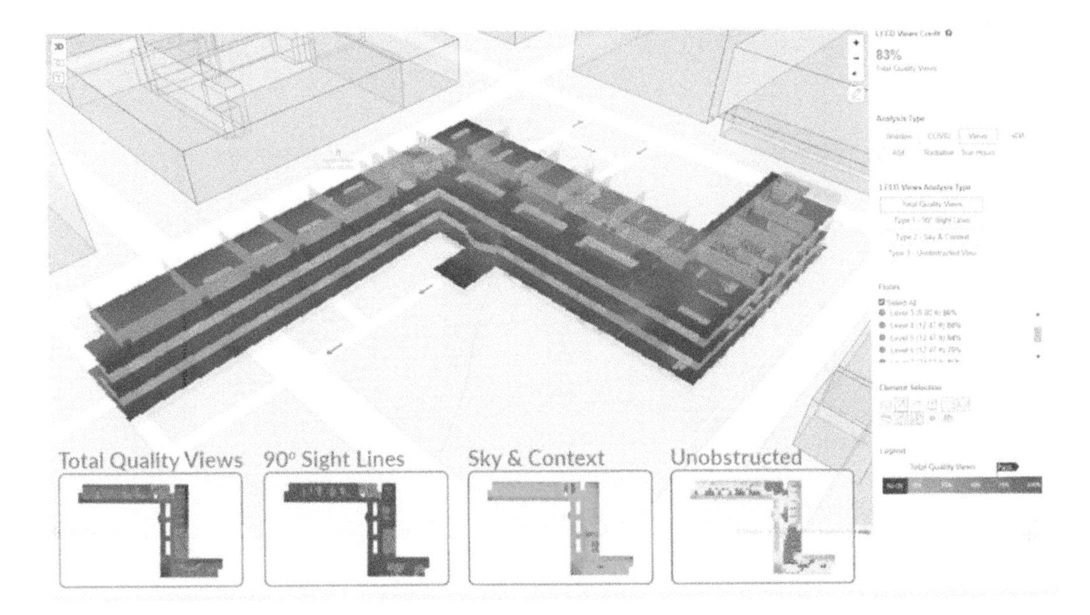

Understanding the various type of View Quality.

- **Market Demand**: With both individual and corporate buyers showing a preference for certified buildings, there is a market incentive for builders to seek these certifications.
- **Continuous Improvement**: Certification bodies regularly update their criteria, encouraging the industry to continually raise the bar on sustainability efforts.

When thoughtfully designed and implemented, green building standards and certifications can act as a lever in governmental efforts to encourage low-carbon building investments. They effectively use a "carrot and stick" approach, combining the allure of incentives with the weight of mandates, to drive sustainable building practices at scale.

4.3.3 The Creation of Financial Incentives for Sustainable Construction

Financial incentives are crucial policy tools for catalyzing the transition to low-carbon buildings. As the International Energy Agency notes, governments have lots of fiscal options to spur private investment in sustainability. Tax credits and deductions for energy efficiency upgrades and renewable energy systems help offset the higher upfront costs of going green. Low-interest loans, grants, and rebates also make these projects more financially feasible by overcoming limited access to capital. Also, performance-based incentives that rise with higher energy savings give building owners skin in the game to maximize efficiency.

Jurisdictions can target incentives across the building lifecycle—from design to construction to operations; for example, subsidies for architects who meet green building standards or rebates for contractors who install high-efficiency HVAC systems. The mix of financial carrots and regulatory sticks is essential to rapidly scale green building. But, as the IEA emphasizes, incentives must be easy to access, stable over time, and provide sufficient value to change decision-making. Robust public funding, streamlined administration, and effective promotion also increase success rates.

Building designed with daylight in mind can significantly enhance the experience while cutting on need of electric lighting. Source: AnnaStills/Adobe Stock Photos.

Financial incentives for sustainable construction may include:

- Targeted subsidies
- Green mortgages
- Green bonds
- Property tax reductions
- Bulk purchasing programs
- Streamlined permitting
- Access to financing

Government intervention can be pivotal in making sustainable construction economically viable. One way to accomplish this is through targeted subsidies aimed at specific low-carbon building technologies and materials. For instance, subsidies could be offered for the installation of high-efficiency HVAC systems, solar panels, quality insulation, or the use of sustainable timber. By directly lowering the upfront costs associated with these eco-friendly alternatives, governments can incentivize builders and homeowners to opt for sustainable solutions, thereby fostering a market shift toward greener construction practices.

Financial institutions have a role to play in driving sustainability, and one avenue is through offering green mortgages. These are specialized mortgage packages with preferential rates for buildings that meet certain green or sustainability certifications. The rationale here is twofold: first, sustainable buildings often have lower long-term operating costs; second, they typically have a higher market value. Both of these factors can mitigate financial risks for lenders, making green mortgages a win-win for both banks and builders focused on sustainability.

Another innovative financial instrument is the green bond, typically issued by governments or development banks. These bonds are earmarked exclusively for financing sustainable building projects and other green initiatives. By providing a secure and ethical investment opportunity, green bonds attract capital from investors interested in both financial returns and environmental stewardship, thereby funneling more resources into sustainable construction.

Fiscal policy can also serve as a lever for encouraging sustainable building practices. One example is offering property tax reductions for buildings that achieve a specified level of energy efficiency or attain certain green certifications. By linking tax incentives to sustainability metrics, governments can make the long-term ownership of green buildings more economically attractive, further encouraging the adoption of sustainable construction methods.

Economies of scale can make a significant impact in reducing the costs associated with sustainable building materials. Governments can facilitate this by setting up bulk purchasing programs for materials crucial to green construction, such as high-efficiency windows or quality insulation materials. By pooling demand, these programs can negotiate lower prices, making sustainable materials more financially accessible to individual builders and contractors.

Understanding the impact of solar panels and other renewables. Source: Lategan/peopleimages.com/Adobe Stock Photos.

Time is money, especially in the construction industry where delays can significantly inflate project costs. To incentivize sustainable building, jurisdictions can offer a streamlined and expedited permitting process for projects that meet certain green criteria. This not only reduces the soft costs associated with delays but also serves as an additional motivator for builders to adopt sustainable practices from the outset.

Capital is often the most significant barrier to adopting sustainable construction methods. Governments can help overcome this hurdle by facilitating access to below-market rate financing specifically targeted at sustainable building projects. By reducing the cost of borrowing, this form of financial support makes it more feasible for builders to invest in green technologies and materials, thereby accelerating the transition to sustainable construction.

When thoughtfully designed, financial incentives act as catalysts rather than long-term supports. They help demonstrate the viability of efficiency improvements that pay off over the building lifespan through energy savings. Incentives kickstart the transition, opening the floodgates for private investment. Ultimately, though, improving the underlying economics is key for sustainability to become the new normal. Still, well-targeted incentives remain invaluable tools for accelerating the flow of private capital into low-carbon building projects. They tip the scales toward investments that serve both financial and environmental returns.

4.4 The Opportunities for the Finance Sector in Supporting Low-Carbon Building Investments

4.4.1 Green Bonds and Loans for Sustainable Building Projects

The finance sector has increasingly recognized the transformative power of financial instruments like green bonds and green loans in steering investments toward sustainable buildings. Endorsed by the Climate Bonds Initiative, these instruments are deemed vital for directing large-scale investments toward environmentally responsible projects, especially in the construction and real estate sectors.

Green bonds serve as a unique category of debt securities explicitly designed to finance projects with environmental benefits. They are governed by a set of rules that stipulate the use of proceeds for eco-friendly endeavors, such as the construction of green buildings or the installation of renewable energy systems. Emerging rating criteria and market standards are enhancing transparency, integrity, and disclosure within the green bond market, ensuring that these bonds contribute to genuinely sustainable projects.

Connecting the urban landscaping into the building design can create both a great space, but also natural glare mitigation with tree canopies. Source: Artinun/Adobe Stock Photos.

While similar to green bonds, green loans are generally geared toward institutional investors. These loans offer favorable terms, often featuring reduced interest rates as an incentive for funding sustainable projects. Both green bonds and loans serve as vital financial mechanisms for sourcing additional funding for purposes like retrofitting existing buildings or constructing new, sustainable structures.

Key Dynamics and Advantages of Green Bonds and Loans

- **Reduced Financial Risks**: Both green bonds and loans often come with preferential terms, such as lower interest rates or extended repayment periods. These incentives help mitigate the financial risks associated with sustainable building projects, which might otherwise be perceived as novel or uncertain investments.

- **Market Development**: The consistent issuance of green bonds and loans contributes to the maturation of the broader green financing market. It also serves to normalize sustainable real estate projects as mainstream investment opportunities, thereby attracting more capital into the sector.
- **Project Pipeline**: A considerable backlog of potential sustainable building projects is in need of financing. Green bonds and loans offer an effective mechanism to match these projects with available capital, thereby accelerating the transition to sustainable construction practices.
- **Revenue Generation for Governments**: The funds raised through green bonds and loans can serve as a revenue stream for governments. This revenue can be channeled into programs designed to further incentivize low-carbon investments in the real estate sector, creating a virtuous cycle of sustainable development.
- **Transparency and Accountability**: The rise of comprehensive frameworks, standards, and disclosure requirements ensures that green bonds and loans genuinely contribute to environmental sustainability. These mechanisms enhance transparency and hold issuers accountable for delivering on their environmental promises.
- **Attracting ESG Investors**: The growing interest in environmental, social, and governance (ESG) investing means that an increasing number of both institutional and retail investors are seeking responsible investment opportunities. Green bonds and loans allow these impact investors to focus specifically on the real estate and construction sectors, thereby channeling more funds into sustainable building projects.

Green bonds and loans are not just financial instruments; they are catalysts for change. By providing favorable terms, ensuring transparency, and matching projects with capital, they hold the potential to significantly accelerate the adoption of sustainable practices in the building industry.

4.4.2 The Role of ESG Investing in Driving Low-carbon Building Investments

Endorsed by the *Principles for Responsible Investment*, ESG investing is emerging as a potent force in mobilizing capital toward the transition to low-carbon and sustainable buildings. ESG investors extend their evaluation criteria beyond traditional financial metrics, incorporating companies' performance in ESG aspects. This comprehensive approach to investment decision-making effectively incentivizes building owners to enhance the sustainability and resource efficiency of their properties, which are increasingly becoming investment prerequisites.

The industry has witnessed a consistent uptick in the proportion of assets under management (AUM) that incorporate ESG factors. This growing influx of capital is driving market demand for low-carbon building technologies and innovations. However, the impact of ESG investing on building sector decarbonization is not without its challenges. Issues like the absence of universally agreed-upon metrics for evaluating ESG performance and gaps in data disclosure from the real estate sector hinder the effective channeling of capital toward high-impact projects.

Despite these challenges, the financial industry is innovating with targeted solutions like green bonds, ESG-linked financing, and sustainability-linked real estate investment trusts (REITs). While still in its formative stages in the real estate sector, ESG investing has the potential to become a significant source of private capital for low-carbon building initiatives—provided that frameworks, standards, and transparency evolve to meet the growing demand.

Key Advantages and Dynamics of ESG-Linked Financing

- **Development of Metrics**: The growing prominence of ESG investing is catalyzing the development of standardized metrics and frameworks. These tools are crucial for accurately evaluating the sustainability performance of buildings, thereby enhancing the efficacy of ESG investments in the sector.
- **Market Shifts**: The influx of capital contingent on sustainability targets is putting building owners under increasing pressure. The need to improve energy efficiency and reduce resource intensity has never been more urgent, both for existing properties and future developments.
- **Transformative Impact**: As ESG considerations transition from being a niche focus to a central component of mainstream real estate investment strategies, the potential for transformative change in decarbonizing the building sector becomes increasingly plausible.
- **Incentivizes Innovation**: The growing demand for sustainable buildings is driving real estate companies to innovate. Whether it is the development of new sustainable technologies or the conceptualization of eco-friendly properties, the market is responding with solutions that align with investor expectations.
- **Data Transparency**: The push for enhanced transparency, data availability, and standardized reporting is making ESG financing more effective. These developments are crucial for directing capital toward the most impactful opportunities in sustainable building projects.

ESG investing is not merely a trend; it is a paradigm shift in how capital is allocated in the real estate sector. By driving demand for sustainable building practices, fostering innovation, and ensuring transparency, ESG-linked financing is poised to play a pivotal role in the transition toward a more sustainable and low-carbon building landscape.

4.4.3 Innovative Financing Models for Sustainable Construction

Traditional financing models utilizing bank loans and equity funding are at this point insufficient to meet the large capital requirements of sustainable construction projects. Innovative financing models and mobilization of the private sector capital at scale are needed to close the gap between current funding sources and the financing needs of sustainable infrastructure. That requires taking steps toward making these projects attractive for private investment, which is based on de-risking them.

From the *perspective of Global Infrastructure Hub (GI Hub)*, green bonds are, again, a key tool but need to be further expanded. While the green bond market has grown rapidly in recent years, green bonds still only account for a small fraction of total bond issuance. GI Hub is working with public and private partners to grow and standardize the green bond market to enable trillions in sustainable infrastructure investment. Blended finance can be particularly powerful when combined with policy reforms to crowd in private investors. However, it currently makes up only a small share of infrastructure investment. GI Hub aims to help governments design and implement optimal blended finance structures that can leverage private capital while ensuring socioeconomic returns.

Public-private partnerships are seen as an essential tool but one that requires careful design and implementation to achieve value for money. GI Hub assists governments in benchmarking and monitoring their PPP programs to implement best practices that maximize both economic efficiency and socio-environmental outcomes.

We have already introduced several models of alternative financing in the green sector, but let us recap and include some important points of how this can be achieved in regard to the cooperative effort of the public and private sector:

- **Sustainability-Linked Bonds**: They are a variant of green bonds where the coupon rate paid to investors depends on the issuer meeting predefined environmental targets.
- **Social Impact Bonds**: They allow private investors to fund social programs, with returns tied to program outcomes measured by governments or other intermediaries. This model could be applied to sustainable construction programs to engage impact investors.
- **Insurance Products**: They are like green mortgages, catastrophe bonds, and parametric insurance, they help manage environmental and climate risks. With the right insurance in place, projects seen as too risky may become investable.
- **Carbon Markets**: Like cap and trade systems or offset programs, they can generate revenue streams for sustainable projects by monetizing their reductions of greenhouse gas (GHG) emissions. The sale of carbon credits to regulated emitters can improve project economics.
- **Energy-as-a-service and Operations-as-a-service Business Models**: They can allow builders and developers to retain ownership of sustainable projects while contracting operations and maintenance to specialized firms. It attracts firms focused on performance and energy savings instead of upfront capital costs.

While these give many countries a green light, access to climate finance from global funds and donors still remains limited for most developing countries. Scaling up technical assistance, capacity building, and policy reforms is crucial to enable these countries to design bankable projects and tap into growing sustainable finance flows. The best way is to work with partners across sectors to develop, pilot, and scale up innovative financing models that can help close the global infrastructure investment gap in sustainable and responsible ways. With concerted action and enabling policy frameworks, these models collectively have the potential to mobilize the trillions needed for the sustainable infrastructure of tomorrow.

The path to a low-carbon building future is not a solo endeavor but one featuring both public policy and private finance. Government actions like carbon pricing create the framework, while the financial sector fills in the details with innovative tools like green bonds and ESG investing. This is not a search for a singular solution, but a dynamic interplay of multiple strategies—sticks, carrots, and nudges—that collectively move the needle.

For the building sector to be a meaningful player in achieving global climate goals, we must recognize that the toolkit is already at our disposal. It includes policy, regulation, and inventive financing models. The real challenge now is not a lack of resources; it is a test of our collective will to deploy these tools effectively to fast-track a low-carbon future.

4.5 The Competitive Advantage of Decarbonized Buildings

The business case for sustainable, low-carbon buildings has become increasingly clear and compelling in recent years. Beyond cost savings from efficiency gains, green buildings provide tangible advantages in attracting tenants, enhancing company reputation, and creating positive public relations. As both companies and consumers demonstrate growing preference for environmentally responsible options, buildings that minimize carbon footprints and environmental impacts have emerged as differentiators that generate real competitive edge.

This section will explore the mechanisms through which low-carbon buildings create value and competitive advantage for owners and occupiers alike. We will examine the rising market demand for green buildings, their ability to command rent premiums and higher occupancy rates, their role in advancing corporate social responsibility, and their capacity to garner positive public relations. The evidence indicates that sustainability has transitioned from a niche consideration to a core component of competitive strategy in the real estate sector.

4.5.1 The Role of Sustainability in Attracting and Retaining Tenants and Customers

Sustainability has become a key purchasing factor for both businesses and consumers, driving growing demand for green buildings and properties that minimize environmental impacts. *Nielsen* research finds that among consumers, two-thirds say it is extremely or very important that the products they buy are eco-friendly. Younger consumers in particular express a strong preference for purchasing from companies committed to environmentally responsible practices. These attitudes extend to preferences for living and working in green buildings. Consumers indicate a willingness to pay higher rents for properties that are sustainable and energy efficient.

Thinking through innovative financing solutions can help cover the demand for green buildings. Source: chokniti/ Adobe Stock Photos.

On the business side, Nielsen reports that more than 90% of companies view sustainability as important for their future success. Nearly half say rising customer demand for sustainable offerings has driven their company's sustainability efforts. As a result, businesses are increasingly valuing sustainability as a criterion in real estate decisions and willing to pay a premium. This demand stems from several factors:

- Green buildings offer operational cost savings through lower utility bills.
- They enhance employee attraction, satisfaction, and productivity through better indoor environmental quality.
- They provide alignment with the sustainability values and brand positioning, increasingly important to businesses and consumers alike.

For building owners, the ability to attract and retain tenants and customers precisely because of a property's environmental and sustainability credentials has become a source of competitiveness. Green leases and higher occupancy rates can help justify higher upfront investment costs for green building features. The visibility and marketing opportunities offered by sustainability performance certification also enhance a property's appeal.

Building owners often cite higher occupancy rates and shorter vacancy periods as key potential financial benefits of sustainability improvements and green building certification. Research from sources including a construction.com study indicates that green buildings exhibit measurable advantages in tenant retention and leasing velocity. For occupier businesses, green buildings offer important value propositions that can drive rental and relocation decisions. From cost savings to efficiency, green buildings provide well-documented benefits for employee productivity, health and satisfaction. This "indoor environmental quality" has been shown to reduce absenteeism and attrition while also improving performance. Green buildings also align with the sustainability priorities and brand values of many companies.

Credit: Casa imágenes/Adobe Stock Photos.

As a result, green buildings tend to have an easier time attracting and retaining high-quality tenants compared to conventional properties. Multiple studies including a study from the better buildings challenge in Los Angeles have found that rent premiums and higher occupancy rates are positively correlated with green building certification and efficiency. Successful green repositioning of existing properties have demonstrated measurable increases in both rental income and occupancy. For new developments, green building features and certification can reduce tenant turnover and marketing time during lease-up phases. LEED-certified buildings, in particular, have been found to achieve occupancy one to two years faster on average. The mechanisms through which sustainability leads to these outcomes include:

- Better alignment with tenants' sustainability goals and values proposition
- Enhanced visibility, reputation, and marketing opportunities
- Improved employee satisfaction and productivity within tenant spaces
- Lower operating costs that benefit tenants directly based on a publication from US Department of Energy focused on studying the post occupancy evaluation of green buildings
- More resilient properties capable of withstanding future risk

Source: yu/Adobe Stock.

The evidence indicates that sustainability-related investments in buildings can generate meaningful advantages in tenant attraction and retention that translate to higher occupancy rates and reduced vacancy periods. These financial returns, when combined with efficiency savings and other factors, strengthen the business case for low-carbon retrofits and development.

4.5.2 Enhancing Company Reputation and Brand Value Through Sustainability Efforts

The Importance of Corporate Social Responsibility in the Modern Business Environment

Corporate social responsibility (CSR) is an important part of the business imperative. Beyond philanthropy and compliance, CSR encompasses how companies manage the environmental and social impacts of their full operations and value chains. Its effective form requires systems to identify material issues, manage risks, engage stakeholders, and ensure accountability, while the CSR programs particularly focused on climate change and resource efficiency are especially important.

Research from Harvard Business Review finds that a strategic and authentic form of CSR can create long-term business value by mitigating risks, enhancing reputation, increasing employee engagement, driving operational efficiencies, and accessing capital, thereby building competitive advantage. However, the most important factor for stakeholders is authenticity, and that is why the *CSR initiatives* must seem integral to a company's business model, not merely for marketing purposes.

The Harvard Business Review analysts cite several factors driving *the rising importance of CSR*:

- Rising consumer and customer demand for sustainable products and business practices. As awareness of social and environmental issues grows, consumers and customers seek out companies that demonstrate good stewardship and responsibility.

- Need to attract and retain top talent. Young professionals especially expect their employers to have a positive impact on society and the environment. CSR strategies help companies recruit and retain the best talent.
- Risk mitigation and resilience. Social and environmental risks present growing threats to business operations. A proactive approach to managing these risks through CSR enhances corporate resilience.
- Reputational benefits and brand value. Consumers and other stakeholders reward companies seen as responsible citizens, while punishing those viewed as irresponsible. CSR thus directly enhances corporate reputation and brand value.
- Pressure from investors. Institutional and impact investors increasingly consider ESG factors when allocating capital. This "sustainable investing" trend pressures companies to demonstrate effective management of environmental and social risks.
- Regulatory and NGO scrutiny. Stricter rules, reporting requirements, and advocacy from NGOs compel companies to improve CSR performance and transparency.

In this context, sustainability-related efforts like low-carbon building decarbonization offer companies clear avenues for demonstrating CSR leadership along key dimensions that matter to stakeholders. From stakeholder engagement to efficiency improvements to supply chain emissions reductions, decarbonization strategies can meaningfully advance corporate responsibility goals while generating business value.

The Potential for Positive Media Coverage and Public Relations Benefits

Usually, sustainability and decarbonization efforts by companies tend to generate or have the potential to generate significant positive media coverage and improve public relations. This comes from both traditional news outlets and the increasing volume of niche environmental and sustainability media. Proactive sustainability actions that demonstrate corporate responsibility and stewardship often receive favorable coverage in mainstream business and technology publications. Articles highlighting such initiatives enhance a company's image, reputation, and visibility among key stakeholders.

These media outlets provide opportunities for targeted, positive publicity around specific sustainability programs. Announcements of innovative low-carbon initiatives, new partnership programs, and performance milestones often qualify as "newsworthy" for these publications, generating further beneficial attention. For companies, the positive coverage and public relations outcomes generate many benefits, such as:

- Heightened visibility and branding as a sustainability leader.
- Increased credibility and trust among stakeholders regarding sustainability commitments.
- Enhanced external reputation and improved corporate image in the eyes of consumers, investors, and potential employees.
- Indication of strategic foresight and capacity for innovation around global sustainability challenges.
- Demonstration of authenticity and "walk the walk, not just talk the talk" action on CSR objectives.
- Marketing and reputational benefits that create competitive differentiation among peers.
- Opportunities to attract new customers and business through alignment with sustainability values.

These achievements have genuine news value and interest for a variety of media outlets. The positive media coverage and public relations benefits generated meaningfully reinforce a company's broader sustainability communications strategy while contributing to credentials as a responsible corporate citizen.

Companies should consider putting additional effort to increase the return on positive media coverage and public relations benefits as following:

- Proactively pitching sustainability stories to relevant media outlets—both general business publications and specialized environmental outlets—to maximize coverage of initiatives.
- Effectively communicating the business case, financial impacts, and innovative aspects of sustainability programs helps frame initiatives as compelling stories of interest to a variety of audiences.
- Sustainability leaders and champions within companies—including building decarbonization experts—make ideal spokespeople for interacting with reporters and journalists. Their passion and insights enliven coverage.
- Major announcements and milestones often garner the most media attention—for example, when a company achieves a significant reduction in emissions, receives green building certification, or launches a new low-carbon technology. But steady progress and ambitious goals also merit coverage.
- Both traditional press coverage (articles and interviews) and social media promotion through company channels can help amplify the positive messaging associated with sustainability initiatives.
- Authenticity remains vital—coverage must reflect the genuine motivations and impacts of initiatives, not just spin for marketing purposes. Companies should be prepared to discuss both challenges and successes openly.
- Over time, a track record of successful sustainability initiatives and positive media coverage can help establish a company as a "go-to" source for commentary on related issues, further enhancing its sustainability credentials and reputation.

The confluence of market forces, stakeholder expectations, and reputational factors has made low-carbon buildings a source of competitive differentiation and business value creation. Companies occupy decarbonized spaces not just for efficiency gains but for the advantages they confer in attracting talent, demonstrating authentic CSR commitment and earning positive public perception. For their part, building owners invest in sustainability to meet tenant demand, reduce vacancy rates, and boost property prestige.

In essence, sustainability has been mainstreamed in the real estate sector, and as climate concerns grow, first-mover advantages will likely accrue to companies and buildings outpacing peers in decarbonization efforts. Still, competitive benefits rest on authenticity and transparency regarding motivations and performance. Greenwashing will fool no one in an era of rising eco-consciousness. Ultimately, low-carbon buildings offer a winning value proposition if their advantages are realized through a strategic integration of sustainability principles into core business models and practices. The market increasingly rewards such efforts.

4.6 The Business Case for Low-Carbon Buildings

Beyond the competitive advantages and asset value benefits of low-carbon buildings, opportunities also exist for direct energy cost savings and revenue generation. Energy efficiency upgrades, on-site renewables, and innovative financing models create pathways to cut energy bills, monetize efficiency gains, and access incentives. Building owners can capitalize on underutilized revenue streams while advancing sustainability.

This section explores the financial mechanisms through which buildings can capture value from improved energy performance. We examine energy performance contracts that guarantee savings to repay project costs, on-site generation that offsets purchases from the grid, and incentive programs that reduce initial outlays. In combination, these strategies demonstrate the capacity for low-carbon buildings to move beyond cost savings into direct revenue opportunities. The prospects are substantial for those seizing them.

4.6.1 The Financial Benefits of Low-carbon Building Investments

Lower Operational Costs Through Energy Efficiency

Energy efficiency improvements and renewable energy systems can significantly reduce operational costs for buildings through lower utility bills. The US Department of Energy estimates that more than 40% of total US energy consumption and corresponding energy costs stem from commercial and residential buildings. By *reducing this energy demand* through *efficiency upgrades* and on-site renewables, building owners and operators can achieve substantial savings over the lifespan of a property.

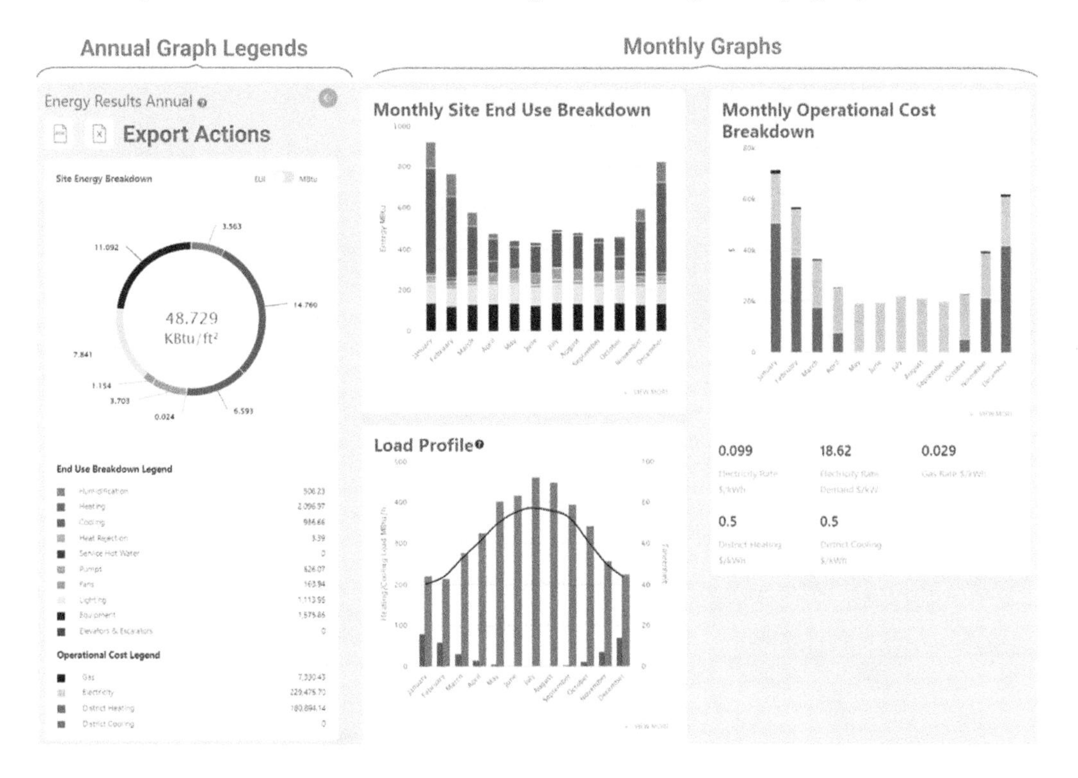

Source: Cove Tool, Inc.

Studies including a study from National Renewable Energy Laboratory have found that well-designed low-carbon buildings can achieve simple payback periods of 5–10 years on average for efficiency investment. Beyond that point, the building generates purely financial gains from further energy and cost reductions. A full bundle of energy efficient net zero measures tailored for a specific building can commonly reduce total energy usage by 20–50% or more, leading to proportional decreases in utility spending.

The business case for low-carbon buildings rests significantly on the fact that using less energy costs less money in the long run. Lower operational costs enable investors to earn higher returns and building owners to benefit from more stable expenses, greater resilience to utility price fluctuations, and higher asset value over time.

Green building councils and other experts argue that *sustainability credentials* have become an important differentiator for properties in attracting both investors and tenants. Those seeking lower-risk opportunities that are aligned with their environmental goals are increasingly drawn to properties with strong sustainability performance. For both groups, investors and tenants, this offers several advantages.

For investors, green building certification and higher efficiency means:

- Lower operational costs and higher cash flows due to efficiency improvements.
- Potential for higher rents, occupancy rates, and returns given proven tenant demand for sustainable spaces.
- Ability to access rapidly growing pools of "green capital" seeking investments with social and environmental benefits.
- Diversification of risk through exposure to assets resilient to future environmental challenges.
- Alignment with impact investment objectives for some investors.

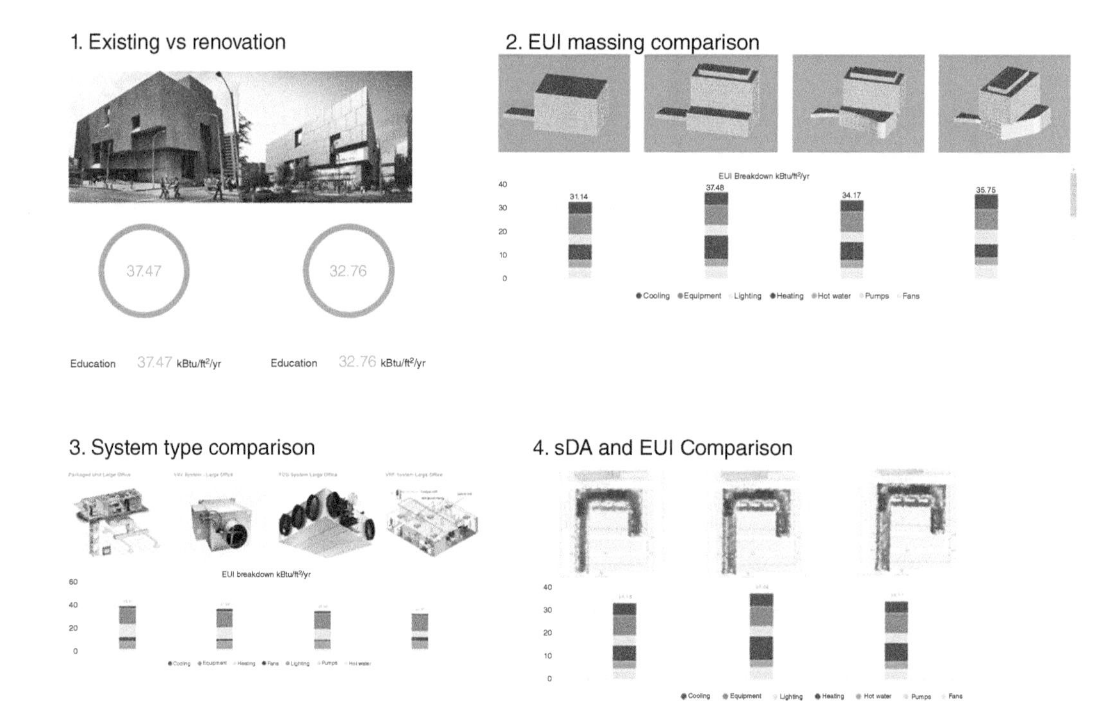

1. The energy use intensity difference in the building before and after the renovation 2. Comparing various massing schemes and their impact on energy use break down 3. Comparing the impact of typical HVAC systems on energy use breakdown 4. Comparing the impact of changing glazing percentage on daylight and energy use breakdown. Source: https://help.covetool.com/en/articles/3359626-how-to-do-a-massing-study-in-cove-tool/with permission of cove.tool.

As a result, research shows including from world green building council that green-certified properties command lower capitalization rates and higher valuations, indicating premiums of up to 16% for investors.

Empty office during renovation work. Source: denboma/Adobe Stock.

For tenants, the value propositions offered by sustainable buildings include:

- Operational cost savings passed on through lower rents or reduced expenses built into leases.
- Higher employee productivity, satisfaction, and lower turnover due to better indoor environments.
- Enhanced brand alignment with customers and clients that value sustainability.
- Opportunity to achieve sustainability goals within real estate footprints.

Additionally, the willingness of tenants to accept higher rents and commit to green leases further strengthens the financial case for investors. Sustainability surely has become an important differentiator in real estate, with green buildings able to command a "green premium" due to their ability to attract high-quality investors and tenants seeking alignment with environmental goals. This forms a central component of the business case for low-carbon building strategies, to which there are many more benefits that support the green case, such as:

- Leading sustainability certifications can verify that a building meets strict criteria for resource efficiency, emissions reductions, and indoor environmental quality.
- *Research* from the Cambridge University Land Society on the financial rewards of sustainability shows that buildings with green certification experience lower vacancy rates, higher rents, and reduced tenant turnover—indicating they are better able to attract and retain high-value occupants.

- Properties achieving deep efficiency retrofits or undergoing low-carbon redevelopments can effectively "reposition" themselves as premium sustainable assets, despite having older vintages. This competitive differentiation helps attract capital.
- *Evidence* from a study published by Harvard Business Review called 'Building the Green Way' suggests that energy efficiency improvements equal to or greater than 30% can translate to valuation premiums for properties of 5–15%. Deep retrofits may unlock even higher returns for early adopters.
- There are now entire funds focused exclusively on sustainable real estate, indicating strong investor appetite for green buildings. Major institutions are allocating significant capital to such assets.
- Beyond operational cost savings, sustainable buildings offer productivity benefits, well-being amenities, and brand association that many knowledgeable workers and companies highly value.
- Tenants committed to sustainability are often willing to accept higher rents and sign green leases that share financial benefits with owners—further evidence of the "green premium" that sustainable buildings can command.
- The ability to attract capital from investor pools focused on impact and sustainability, in addition to mainstream capital, enhances resilience and potential returns over time.

4.6.2 Cost Savings in Construction

The Efficiency of Modern Construction Technologies

McKinsey research finds that new technologies have the potential to significantly increase efficiency, productivity, and sustainability in the construction industry. Their analysis states that advanced technologies could save up to 20–30% of total construction costs while reducing emissions. A key factor driving these potential savings is the ability of technologies like AI, automation, advanced materials, and digital twins to reduce waste and inefficiency in construction processes. Currently, as much as 30% of materials on construction sites go to waste due to issues like damage, theft, inaccurate estimates, and poor planning. Technologies like AI and automation can reduce such waste by enabling more precise planning, material optimization, quality control, and on-site management.

Advanced materials offer opportunities to reduce costs through light-weighting, quicker assembly, and less rework. For example, mass timber construction techniques using cross-laminated timber panels are up to four times faster than traditional methods yet provide equal load-bearing strength. 3D printing of construction components likewise promises more precise manufacturing with less material usage. Similarly, with digital twins and virtual design models, which allow for testing of construction plans, sequencing, and optimizations in virtual environments before implementation—reducing physical prototyping, change orders, and execution issues. Extended reality (virtual and augmented reality) tools can further improve worker training, safety and efficiency during construction.

Smart applications of technologies like AI, automation, advanced materials, and digital tools will fundamentally uplevel the design and construction processes, unlocking up to 30% in cost savings while enhancing sustainability and resilience. These efficiency gains represent an opportunity for the construction industry to reinvent itself through technology and innovation, yielding business value for all actors involved in building projects. To summarize the benefits:

- Additive manufacturing and 3D printing technologies enable on-demand, just-in-time production of construction components with minimal material waste.

- Digital twins and simulations based on digital models allow for virtual testing and optimization of building designs, construction plans, and operational performance before physical implementation. This reduces real-world prototyping and experimentation costs.
- Integrated sensor technologies are being deployed to provide real-time monitoring and data-driven insights during construction, resulting in actionable analytics for optimization.
- Widespread use of drones and robotic automation is expected to substantially reduce manual labor requirements and injuries on construction sites in the future.
- AI and machine learning are finding a growing range of applications in construction from design optimization to progress tracking to quality control.
- Advanced materials like mass timber, cross-laminated timber, prefabricated components, and 3D-printed elements are enabling new efficiencies by streamlining assembly, expediting construction schedules, and reducing rework.
- Regulatory and policy support will be needed for companies to fully adopt emerging technologies at scale. Governments can play a key role in accelerating industry transformations through incentives, standards, and procurement policies that prioritize efficiency and sustainability.

Reduction in Waste and Resource Usage Through Prefabrication and Modular Construction

To translate all these efficiencies into meaningful potential cost savings for construction projects, the *World Green Building Council (WGBC)* recommends offsite, prefabricated, and modular approaches to construction, which significantly reduces waste and optimizes material resource usage compared to traditional ways.

Prefabrication involves manufacturing major building components or assemblies in controlled factory environments and then transporting them for assembly onsite. Modular construction takes a similar approach but with entire sections or modules constructed offsite and joined together onsite. The WGBC names several factors that lower the resource usage:

- **Less Material Spoilage Due to Exposure**: Prefabricated components are protected from weather damage and accidents on-site that result in trimming, rework, and landfilling of spoiled materials.
- **Tighter Tolerances Enable Less Material Overage**: The precision of robotic manufacturing and computer numerically controlled cutting minimizes the need for generous material overage to account for errors in cutting and fitting onsite.
- **Digital Modeling and Planning Optimize Material Quantities**: Computer-aided design and building information modeling software allow for precise material specifications and ordering that minimize leftovers.
- **Less Rework and Adjustments Required**: Strict quality control during factory production reduces the need for rework, adjustments, and repairs that generate waste.
- **Less Onsite Storage Needed**: Materials are delivered closer to installation time since production and transport schedules can be coordinated efficiently.

This translates directly to cost savings through lower material costs, easier material recycling, reduced logistics and storage costs, and less rework costs. The new methods shorten construction timelines, allowing projects to start generating value sooner and reducing the overall duration of resource consumption. This compressed schedule represents an additional potential source of cost savings through efficiency gains.

4.6.3 The Positive Impact on Asset Values and Risk Management

Enhanced Resilience and Lower Vulnerability to Climate Risks

Making urban developments and assets more resilient to climate risks can have a positive impact on asset values and risk management. When infrastructure, buildings, and communities are designed to withstand extreme weather events and long-term climate shifts, they become less vulnerable and can better maintain functionality. This is, according to the Urban Land Institute, what *enhances resilience*. For asset owners and investors, enhanced resilience translates to lower exposure to financial losses from climate-related disruptions. Assets that are resilient to floods, storms, heatwaves, wildfires, and other climate hazards experience fewer disruptions, less property damage, and fewer costs from business interruption. This stability helps maintain cash flow and revenue streams over time.

Lower vulnerability also reduces uncertainty about future conditions and performance. Resilient assets offer more predictable outcomes and require less contingency planning. Investors gain greater confidence in projected returns and asset values. Insurance costs may also decline for properties with high resilience ratings. That means that *proactively managing climate risks* through resilient design and adaptation strategies brings far more control over outcomes. Asset owners can mitigate risks they understand rather than reacting to unforeseen impacts. This shift to proactive risk reduction aligns with standard enterprise risk management practices that is directly correlated to the market value.

The Potential for Higher Property Values and Market Demand

CBRE in it's U.S real estate 2023 outlook gathered *substantial evidence* that sustainability and green building certifications yield financial benefits. For asset owners and occupiers, it happens through multiple pathways, which translates to higher property values, stronger rents, and stronger market demand overall. The size of this green value premium rises proportionally with the level of performance and arises from attributes fundamental to long-term investor preferences, tenant needs, and urban resource management. Several factors underpin the "green value premium":

- **Lower Operating Costs**: Advanced energy efficiency and waste reduction strategies in green buildings result in significantly lower utility and maintenance expenditures over time, with savings averaging 20–30% compared to conventional properties. These operational cost savings flow directly to the bottom line, enhancing cash flows and net operating income that drive property values. For tenants, lower rents or operating expenses effectively increase the value of occupied space.
- **Higher Rental Rates**: CBRE research finds that LEED-certified and ENERGY STAR-rated buildings command 5–15% higher effective rents on average. This premium stems directly from strong tenant demand for sustainable workplaces, while rent premiums rise with higher certification levels that signify greater performance. Leases that share financial benefits between owners and tenants further strengthen profitability.
- **Lower Capitalization Rates and Higher Cap Rates**: Numerous studies indicate that green-certified buildings attract lower capitalization rates from investors seeking stable cash flows and competitive returns within the risk profile. This translates to up to 16% higher property values according to CBRE. More stringent certifications that maximize efficiency and savings potential yield large cap rate differentials.
- **Reduced Obsolescence Risks**: Strategic sustainability investments allow properties to undergo "value regeneration" while lowering the risks of functional or economic obsolescence over time.

Efficiency gains improve competitiveness and resilience against utility price volatility, climate change impacts, and future environmental regulations that threaten asset performance. This risk mitigation premium grows in relevance with increasing physical and regulatory pressures.

- **Stronger Rental Growth**: CBRE evidence also shows that green-certified buildings experience lower vacancy rates and higher rental growth due to stronger demand from tenants seeking to reduce costs, attract workers, and improve their brand.
- **Enhanced Mortgage Terms**: Lenders are increasingly factoring in sustainability performance when underwriting loans, providing lower interest rates, longer terms, and higher loan-to-value ratios for green properties.
- **Wider Appeal to Investors**: Green buildings are able to access rapidly expanding sources of "sustainable capital" seeking investments that deliver social and environmental returns in addition to financial returns. This competitive differentiation increases valuations.
- **Greater Resiliency**: High-performance, low-carbon buildings prove more resilient against physical climate change risks and shifts in market preferences. This longevity supports stable cash flows, risk-adjusted returns, and asset values over the long run.

The Reduction of Regulatory and Reputational Risks

Proactive sustainability strategies can meaningfully reduce regulatory and reputational risks for real estate owners and investors, leading to stronger cash flows, stable returns, and higher asset values over time. Being able to anticipate and adapt to emerging environmental regulations helps minimize compliance risks, penalties, and disruption costs that otherwise erode financial performance. Transitioning properties to low-carbon resilience now reduces exposure to future regulatory costs.

The *Carbon Disclosure Project* (CDP), which collects environmental impact data from thousands of companies, finds these strategies do work in terms of risk reduction. Sustainability measures and transparency around impacts also build credibility and trust with stakeholders, minimizing reputational risks that threaten income growth and valuations. Being proactive in terms of managing risks through efficiency upgrades and impact disclosure represents a powerful risk mitigation premium for all involved. Safeguarding assets against looming pressures are rewarded by:

- **Credibility and Legitimacy**: Rigorous performance on sustainability issues provides a level of credibility that anticipates and responds to evolving regulatory and stakeholder expectations.
- **Lower Compliance Risks**: Proactivity that aligns with emerging public policies and tightening environmental regulations minimize noncompliance risks, penalties, and disruption costs that weigh on financial performance.
- **Reputational Benefits**: Transparency around impacts, certifications, and aggressive emission reduction targets build trust with stakeholders, minimizing risks to a property's reputation from inaction or greenwashing.
- **Stronger Demand**: Positive reputation strengthens tenant demand, business partnerships, government support, and investor interest.
- **Brand Differentiation**: Disclosure of initiatives and impacts differentiates a property's brand, enhancing its curb appeal, market positioning, and competitiveness in a way that commands a premium.

- **Access to Capital**: A proactive approach helps gain wider access to specialized green financing instruments, sustainable investors and ESG funds, lowering the overall cost of capital for properties, and increasing valuations.

4.7 Opportunities for Energy Savings and Revenue Generation

4.7.1 Energy Performance Contracting and Third-party Financing for Energy Efficiency Projects

Energy performance contracting and third-party financing models present opportunities to switch to energy efficiency with little to no upfront capital costs. They transfer the financial responsibility and risk of the project to an *energy service company* (ESCO) or other third-party, while allowing the client to benefit from energy saving and generated revenue.

In an energy performance contract, the ESCO conducts an audit to identify where exactly the opportunities lie. The team then designs efficiency upgrades like lighting, HVAC, controls, and building management systems, arranges project financing and only gets paid after the upgrades are installed, and delivers verified energy savings. The guaranteed energy savings cover the cost of the project and provide a revenue stream for paying the ESCO and financing costs. If the savings fall short, the ESCO is responsible for making up the difference. This transfers performance risk to the ESCO, removing it from the client. For the client, there are no upfront capital costs, no requirements, and no impact on their credit. They simply pay for the project over time from the verified energy savings. Any savings above what is needed to repay the ESCO provide an immediate return on investment for the client.

Third-party financing models operate similarly, except that they have a separate financing company which provides capital for the project rather than the ESCO. The ESCO still conducts the audit, implements upgrades, and guarantees savings. The financing company is repaid from the energy cost savings over an agreed term.

For clients, these financing approaches enable efficiency projects that would otherwise be unable to secure capital from budgets or traditional loans. They monetize underutilized efficiency opportunities that generate revenue through energy savings. And they do so with little to no operational disruption since implementation is managed by the ESCO.

4.7.2 On-site Renewable Energy Generation and the Potential for Selling Excess Energy Back to the Grid

The National Renewable Energy Laboratory, a US Department of Energy lab, focused on renewable energy research, argues that on-site renewable energy systems offer facilities opportunities for both energy cost savings and revenue generation.

By installing renewable energy systems, facilities can offset portions of their own energy consumption with lower-cost power generated on-site, reducing demand for higher-cost electricity from the grid, yielding significant energy cost savings over time. The National Renewable Energy Laboratory's research shows on-site renewables often produce electricity at a fraction of retail electricity rates.

When renewable energy generation exceeds a facility's instantaneous needs, most utility networks allow excess power to flow back to the grid through *policies like net metering or net billing*. The National Renewable Energy Laboratory finds this creates an opportunity for facilities to earn revenue by monetizing the excess renewable energy they produce but do not immediately consume. Under net metering, utilities typically credit facilities for exported power at a rate similar to the utilities' own wholesale power costs. The cash flows from selling this excess renewable energy can substantially offset the upfront costs of on-site systems.

In essence, it allows facilities to both reduce energy costs through self-generation of cheaper power and also earn supplemental revenue by selling surplus clean energy into the electric grid. Combined, these benefits enhance the value proposition and accelerate the payback periods for on-site renewable projects, in the lab's analysis.

4.7.3 Energy Efficiency Incentives, Grants, and Tax Credits Available for Building Owners

Various incentives, grants, and tax credits exist at the federal and state levels to encourage building owners to implement energy efficiency upgrades, according to the *Database of State Incentives for Renewables & Efficiency (DSIRE)*. These financial incentives aim to facilitate and accelerate efficiency improvements that save energy and costs while also reducing environmental impacts.

At the federal level, key programs include tax credits and deductions for commercial buildings that install energy efficient and renewable energy systems. For example, federal tax incentives exist for energy efficient commercial buildings, energy efficient new homes, geothermal heat pumps, and small wind turbines. However, much of the funding and strongest incentives for efficiency come from states utilities and local governments. DSIRE tracks over 900 state-level incentives that support energy efficiency in buildings, such as:

- **Rebates for High-Efficiency Equipment**: HVAC systems, lighting, controls, windows, insulation, and so on.
- **Low-Interest Loans for Efficiency Upgrades**: available through state programs and utilities.
- **Performance-Based Incentives**: paid per unit of energy or demand saved through approved measures.
- **Grants for Retro-Commissioning**: identifying and fixing inefficient operations and maintenance issues.
- **Property Assessed Clean Energy (PACE) Financing**: voluntary special tax assessments for efficiency projects.
- **Tax Credits or Deductions**: for efficiency improvements or certifications.
- **Expedited or Waived Permit Fees**: for projects achieving certain efficiency levels.

Many of these incentives aim to cover a portion of the upfront costs, overcome initial hurdles, and demonstrate the financial viability of efficiency, according to DSIRE. The ultimate goal is unlocking the longer-term savings and non-energy benefits that make a compelling business case for building owners.

A diverse network of financial incentives exists at multiple levels of government to encourage and support energy efficiency improvements in commercial buildings. These incentives offer opportunities for building owners to access funding, overcome financial barriers, and help manage initial costs—unlocking the broader benefits of efficiency that enhance long-term value and sustainability.

Imagine the transition to low-carbon, high-efficiency buildings as akin to discovering a hidden treasure for the modern age. The "X marks the spot" is not just environmental stewardship, but also a wellspring of financial benefits. Performance contracts act like a magic key, unlocking the treasure chest without requiring you to pay an upfront fee; they guarantee that the chest will refill itself over time, effectively covering its own opening costs. On-site renewable systems are like an unending supply of gold, not only reducing your energy bills but also turning excess power into sellable assets.

At its core, what we are talking about is a multi-layered strategy that turns buildings into veritable gold mines of energy productivity. The savings gleaned from efficiency measures are the nuggets you pick up along the way; they pay for your expedition. Revenue generated from on-site power and various incentives are the precious gems you did not even know were in the chest. And when traditional capital feels like a distant island, innovative financing tools build you a bridge. These are not just isolated opportunities; they are complementary gears in a machine engineered to make ambitious low-carbon projects not just viable, but self-financing. So, if you are wondering whether the juice is worth the squeeze, remember: the business case is not just about reducing your carbon footprint; it is about unlocking new realms of energy value that pay dividends.

4.8 Case Study – Emory University Campus Life Center

4.8.1 Project Overview

The Campus Life Center at Emory University, spanning 117,000 square feet, is a response to the increasing space demands on campus. It incorporates flexible and efficient dining services, technology and infrastructure upgrades, and expanded areas for student organizations and gatherings. Aligning with the University's commitment to sustainability, the project set ambitious performance targets.

4.8.2 Climate Analysis

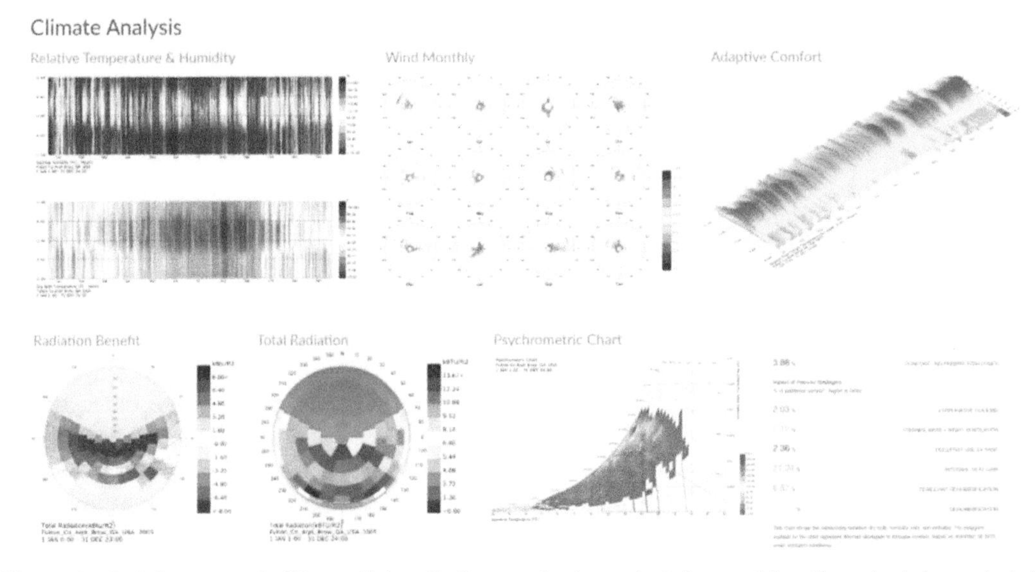

Climate Analysis in cove.tool of Emory University Campus Center project. Source: https://cove.tools/case-study/emory-university-campus-life-center/with permission of cove.tool.

A fundamental aspect of performance design is understanding the existing climate conditions. This involves conducting climate studies to inform decision-making.

4.8.3 Energy Analysis

Energy Analysis in cove.tool of Emory University Campus Center project.

Establishing target values is crucial for evaluating performance design options. Energy Use Intensity (EUI), measured as energy per square foot per year, helps compare different design iterations. Understanding these values aids in setting and achieving energy targets.

4.8.4 Daylight and Glare Analysis

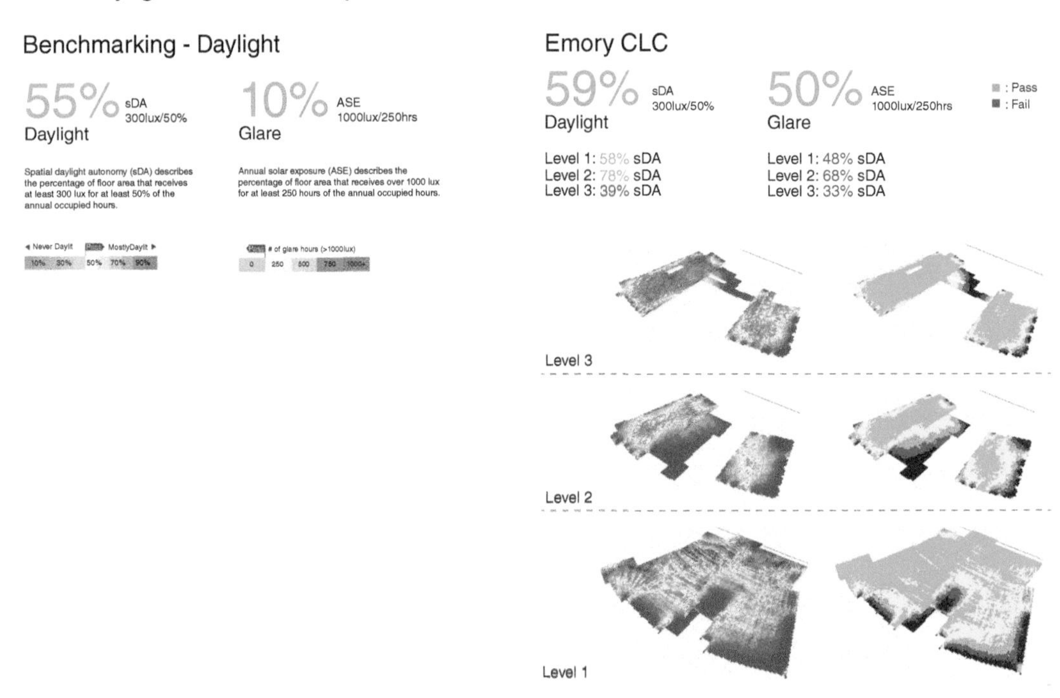

cove.tool daylight and glare analysis study of Emory University Campus Center project. Source: https://cove.tools/case-study/emory-university-campus-life-center/with permission of cove.tool.

Daylight quality is a key consideration in any building, especially in a campus center frequented by students. Studies link daylight to increased productivity and wellness. The team conducted comprehensive daylight and glare studies to optimize these aspects.

4.8.5 Facade Studies

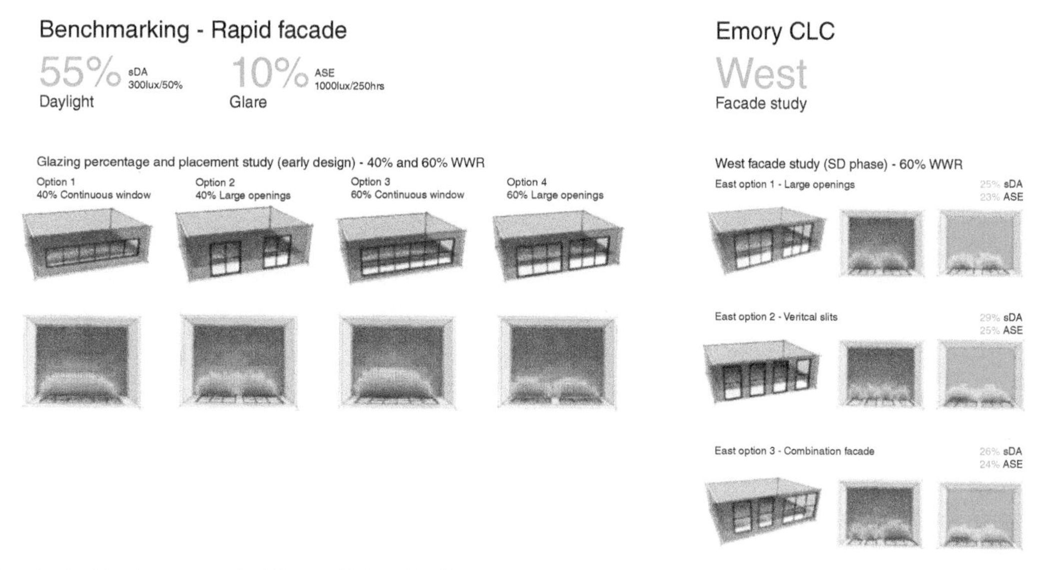

Facade Studies in cove.tool of Emory University Campus Center project. Source: https://cove.tools/case-study/emory-university-campus-life-center/with permission of cove.tool.

Rapid testing and comparison of various facade designs are essential. The team examined different glazing percentages, overhangs, fins, window sizes, and locations to determine the most efficient facade design.

4.8.6 Water Use Analysis

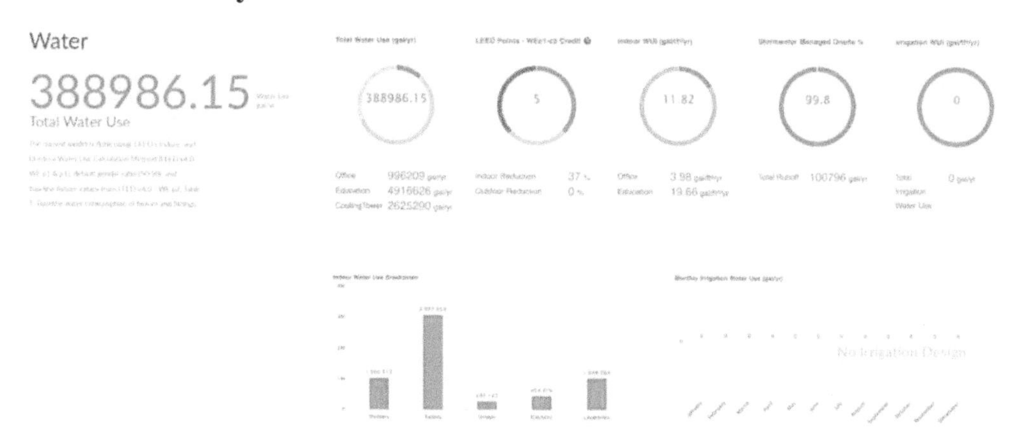

Studying the building water use (Internal and External). Source: https://cove.tools/case-study/emory-university-campus-life-center/with permission of cove.tool.

The project involved a thorough analysis of both indoor and outdoor water use. This included strategies for grey water reuse, stormwater capture, rainwater harvesting, and irrigation.

4.8.7 Cost Versus Energy Optimization

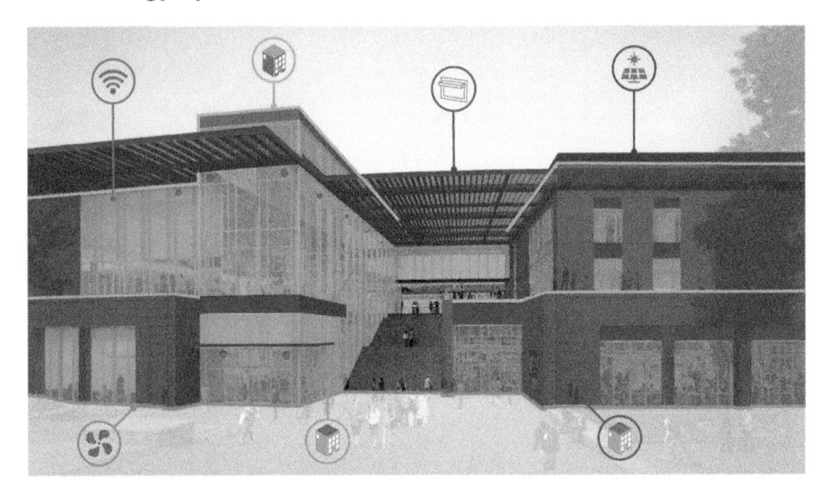

Cost versus energy optimization in cove.tool of Emory University Campus Center project.

With contractors involved early in the design process, cost was a key factor in decision-making. The project team focused on optimizing costs while maintaining energy efficiency. They examined multiple options for windows, insulation, HVAC systems, photovoltaic panels, shading strategies, and sensors, evaluating over 20,000 combinations to find the most cost-effective solutions. The team had a set budget that it exceeded in early design cost estimates. Through the cost optimization, the team cut approximately $1 Million off the 'tacked on' sustainability measures and was able to meet the goals.

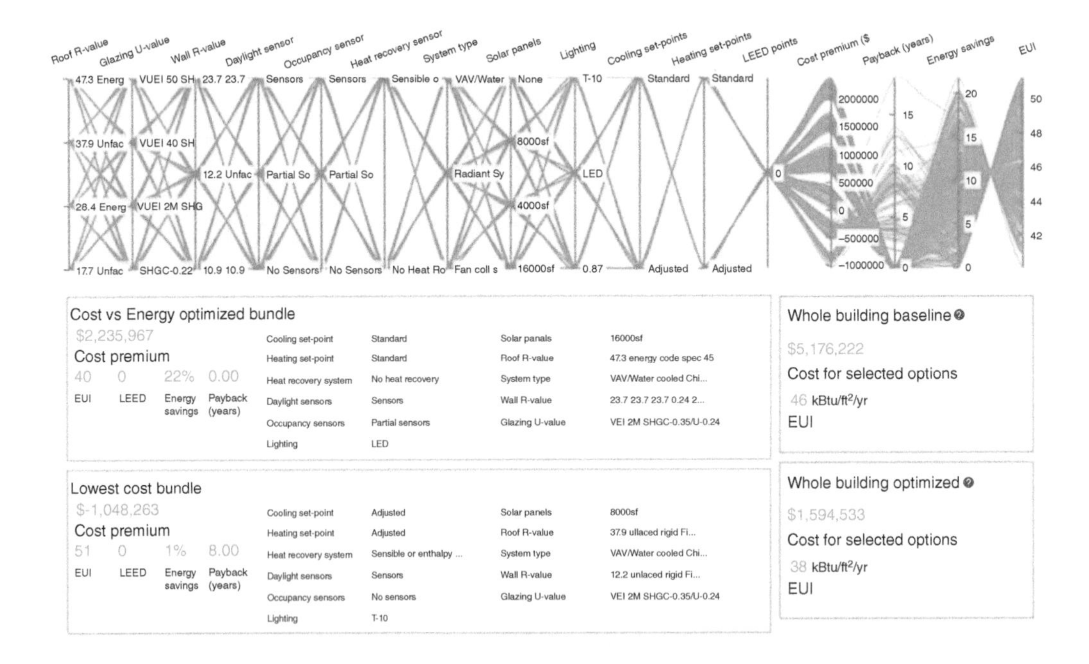

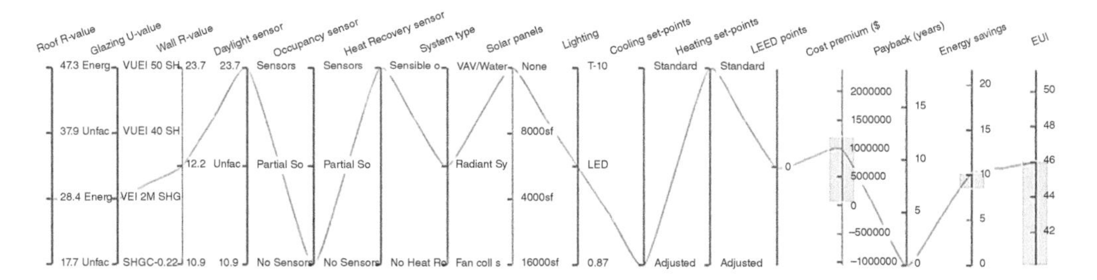

Source: https://cove.tools/case-study/emory-university-campus-life-center/with permission of cove.tool.

In conclusion, the design team employed a comprehensive approach to meet the project's sustainability and performance goals, balancing cost and energy considerations to achieve a highly efficient and cost-effective design.

4.9 Insights from Michael Beckerman, CEO at CREtech

As the founder of CREtech, the largest global conference and media platform focused on innovation and technology within the CRE sector, Beckerman is at the forefront of promoting climate tech as a crucial element of decarbonization. While "proptech" remains a dominant category in this sector, Beckerman

Source: CRETECH.

predicts that climate tech, which addresses decarbonization in the built environment, will attract significant venture funding and technological innovation in the coming years. He anticipates that within five years, climate tech investments in real estate will rival those in proptech (approximately US$20 billion annually), and within a decade, climate tech may even surpass proptech investments several times over.

Beckerman's personal and professional mission is to lead the decarbonization of the entire built environment. With CREtech's extensive global reach, trusted brand, and ability to convene stakeholders within the real estate ecosystem, it is uniquely positioned to drive concrete actions toward guiding the industry toward Net Zero. Through various platforms, including events, content, research, and digital media, CREtech is actively integrating sustainability into its messaging to the industry.

Source: CRETECH.

4.9.1 Trends and Opportunities for Building Owners

Within the commercial real estate (CRE) sector, there exists a prevailing sense of unpreparedness for the impending challenges posed by the climate crisis, according to Michael Beckerman. The real estate industry stands as the largest contributor to GHG emissions on a global scale, and projections indicate that decarbonizing it could require an investment of around $20 trillion. However, it is disconcerting to note that only approximately $100 million has been dedicated to decarbonization efforts within the industry to date. With mounting pressure from governments, tenants, residents, and financial stakeholders, the urgency for change is unmistakable.

Adding to this urgency is the realization that the physical risk to real estate assets, caused by climate-related factors, surpasses that of any other industry. Despite this alarming backdrop, there are encouraging signs in Europe, where a few early adopters within the CRE sector are recognizing the value not only in mitigating sustainability risks but also in reaping financial rewards by investing in climate technology solutions and sustainable building practices.

Beckerman believes that the tipping point for sustainability investments and practices in CRE will not necessarily be a moral shift but a financial one. He posits that when technology and venture sectors jointly support climate tech solutions that offer clear paths to achieving net operating income, the industry will witness a profound transformation. Research has demonstrated that green and clean buildings command higher rents and sales prices, setting the stage for a substantial divide between green and brown buildings in the marketplace. Brown, environmentally unfriendly buildings may soon become financially untenable.

4.9.2 Building Owners' Approach to Decarbonization

While building owners increasingly consider decarbonization as a means to reduce their properties' environmental impact, Beckerman notes that at this stage it is often viewed primarily as a regulatory compliance exercise and a public relations opportunity. For many, the economic viability of decarbonization initiatives must align with return on investment (ROI) and other financial objectives.

Source: yu/Adobe Stock Photos.

Beckerman delineates building decarbonization as a bifurcated market, addressing carbon emissions from building operations (comprising around 28% of all GHGs) on one hand and embodied carbon (accounting for roughly 11% of global GHGs) on the other. The former involves energy efficiency retrofits, including the installation of renewable energy systems like solar panels, wind turbines, and heat pumps, as well as measures such as improved insulation, energy-efficient lighting, and smart building controls. The latter focuses on sustainable construction methods, eco-conscious design principles, and environmentally friendly materials such as carbon-negative concrete and mass timber. Strategies may also encompass the offsetting of carbon emissions through investments in reforestation or renewable energy projects.

Beckerman acknowledges that while building decarbonization is in its early stages, it is making slow but steady progress, with decarbonization strategies increasingly factored into CRE strategies. He predicts that as the industry recognizes the financial opportunities inherent in green solutions, the pace of investment and adoption will accelerate dramatically.

In Beckerman's view, what motivates the CRE industry, venture capital, and startups to focus on decarbonization is less important than the collective effort to combat the climate crisis. The potential for climate tech to generate unprecedented wealth underscores the urgency of addressing climate change, with benefits accruing to all inhabitants of the planet.

4.10 Insights from VC Shaun Abrahamson, Managing Partner at Third Sphere

In the realm of decarbonization and its impact on building owners, Shaun Abrahamson, the driving force behind Third Sphere, has been diligently examining the trends and opportunities that lie ahead. He employs four distinctive frameworks to comprehend the intricacies of climate change opportunities and risks.

4.10.1 The Domino Effect

Shaun delves into the second and third-order consequences of climate risks. He illustrates how, for instance, a warehouse's value could plummet due to local flood management system failures, rendering roadways impassable. These cascading effects are increasingly significant as re-insurance firms grapple with more frequent once-in-a-century events like floods, hurricanes, and fires.

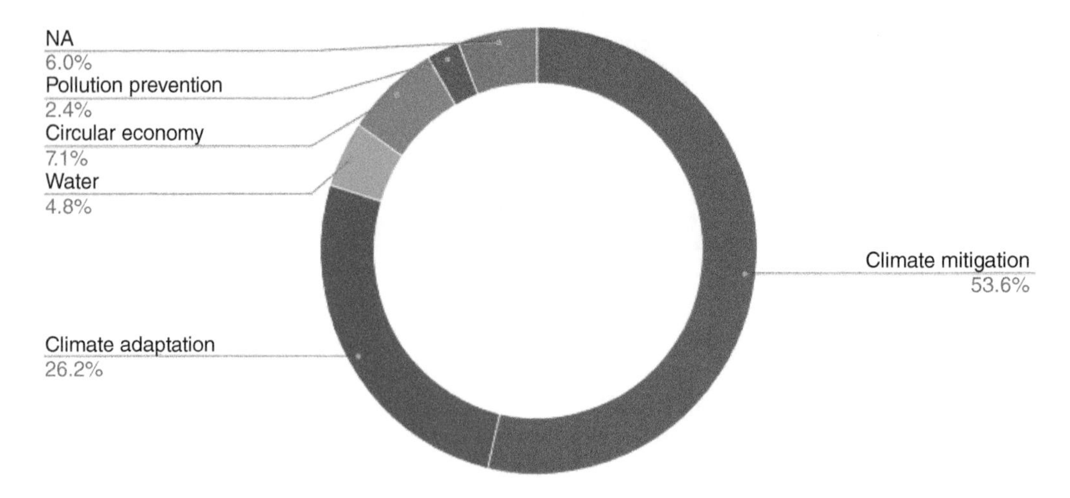

4.10.2 The Tech Revolution

Beyond the digital realm, Shaun highlights the transformative potential of physical technologies in the real estate landscape. While software has revolutionized aspects of building management and sales, the impending wave revolves around tangible technologies. Buildings may have a lifespan of 50 years, but key technologies like heating, cooling, and telecommunications typically span only a decade. Simultaneously, the cost of groundbreaking technologies such as solar modules and batteries has plummeted by 70% in a decade, ushering in new possibilities, from electric vehicles to cost-effective on-site power solutions.

Source: Shaun Abrahamson.

4.10.3 Beyond Environmental Goodness

Shaun underscores that sustainability initiatives are not solely for the betterment of the planet; they can also be financially prudent. Building owners and stakeholders stand to gain not only eco-friendly credentials but also economic advantages, such as cost savings and increased property values.

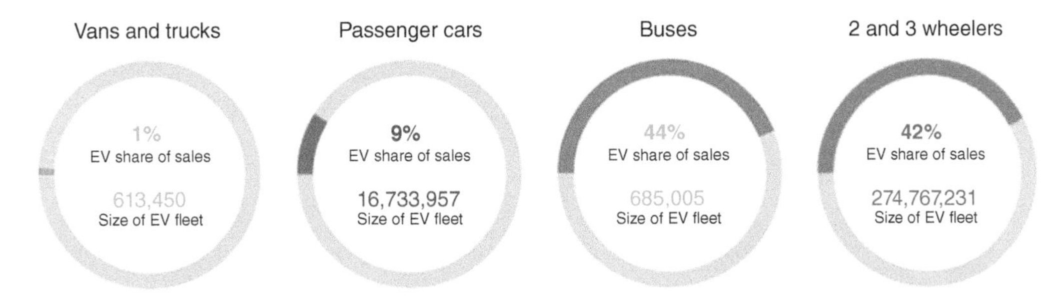

4.10.4 The Power of Collaboration

Recognizing the complexity of modern risks and opportunities, Shaun advocates for collaborative efforts. Coalitions and partnerships can unlock novel solutions and expertise, particularly in niche areas like backup infrastructure for essential utilities, facilitating the development of differentiated products and services.

Third Sphere, founded in 2013, emerged from a profound concern about climate risks and a recognition of untapped technological opportunities in the physical environment. Shaun's team has been dedicated to understanding the needs and aspirations of diverse stakeholders, from building owners to architects, engineers, utilities, insurers, investors, cities, and tenants. Facilitating introductions between entrepreneurs and potential customers and investors has become a hallmark of their approach, fostering connections that fuel innovation.

Their most successful endeavors have transcended the traditional "save the planet" narrative, resonating with customers seeking superior, more affordable, and efficient solutions. Subsidies offered by multiple countries for various sustainability initiatives have further propelled the adoption of these innovations.

Understanding that startups often require customers to embrace technological risks and uncertain investment returns, Third Sphere has spent the past five years developing a private credit platform. This platform transforms emerging technology hardware into leasable assets, complete with performance guarantees. This innovative approach has already catalyzed the adoption of technologies like electric bikes, EVs, residential batteries, and hybrid electric heating.

In the eyes of investors, the focus has shifted from whether decarbonization technologies exist to what impediments might hinder customers from choosing these eco-friendly alternatives over fossil fuels. Sales and marketing automation have emerged as invaluable tools, simplifying the process of evaluating and selecting low-carbon options for building owners and tenants. The elimination of "truck rolls"—on-site visits for assessments and permitting—presents numerous opportunities for time and cost savings.

Moreover, co-benefits play a pivotal role. On-site solar and storage, for example, transcend mere sustainability; they have become essential amenities for tenants, ensuring uninterrupted work and minimizing business disruptions.

In conclusion, Shaun's outlook is resolutely optimistic. The question is no longer whether the built environment will decarbonize, but rather how swiftly this transformation can be achieved.

5

The Role of Data-Driven Design in the Implementation Process

In the world of building and construction, data-driven design is emerging as the unsung hero in the epic quest to decarbonize our built environment. It is as if we have discovered a new secret weapon—one that has the potential to rewrite the rules of sustainable design. Traditional methods, reliant on intuition and experience, have had their day. Now, data-driven design is stepping onto the stage, wielding quantitative data, computational tools, and artificial intelligence (AI) as its instruments of change. The stakes? Meeting the stringent decarbonization targets that loom over buildings and infrastructure like a sword of Damocles.

Think of the data-driven approach as the Swiss Army knife of sustainability. This multi-tool leverages a goldmine of data—from historical projects to materials databases, simulations, and performance metrics. The outcome is a seismic shift from iterative guesswork to optimized low-carbon designs right off the bat. With Internet of Things (IoT) sensors and real-time monitoring, these structures do not just stand there; they teach us how to improve future buildings.

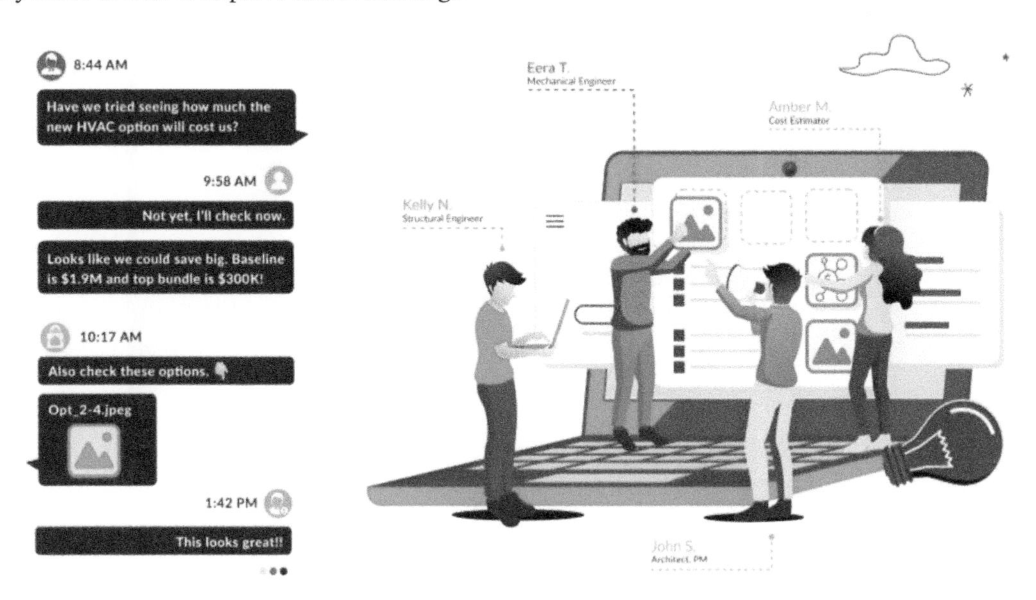

Real-time cross team collaboration with cove.tool.

Build Like It's the End of the World: A Practical Guide to Decarbonize Architecture, Engineering, and Construction, First Edition. Sandeep Ahuja and Patrick Chopson.

But this is not just about number-crunching; it is about a synergy of technologies. Building information modeling (BIM), AI-assisted design, extended reality (virtual reality—VR/augmented reality—AR/mixed reality—MR), and robotics are coming together, each method contributing to a harmonious productivity and quality boost. Pair these with cloud platforms and digital tools, and what you get is a collaborative innovation landscape that is still in its infancy but promises exponential growth. The caveat? We need new business models, workflows, and skillsets to catch up to fully tap into this potential. Imagine smart buildings that adapt and evolve, like living organisms responding to environmental stimuli. The longevity, resilience, and value of these structures could increase manifold, provided we learn to design this adaptability into their DNA from the get-go. And let 'us not forget, the benefits of these digital advancements must be inclusive, serving the aging and underserved communities as well.

So, what 'is holding us back from this utopian vision? Fragmentation—of data, standards, and tools. It is as if we have a full football team but everyone is playing from different game plans. The building information networking (BIN) model offers a potential solution, serving as a coach that can synchronize these disparate elements into a unified strategy for holistic decarbonization.

The revolution is within reach, but it is not just about the tools; it is about the vision. Achieving this transformation will require a cohesive effort across academia, corporate objectives, tools, and standards. It needs thought leadership, the kind that can champion this data-driven, human-centered approach and turn it from emerging trend to industry standard. Because in the end, progress is not just about technological leaps; it is about creating a future where the rewards of this revolution are spread equitably across society.

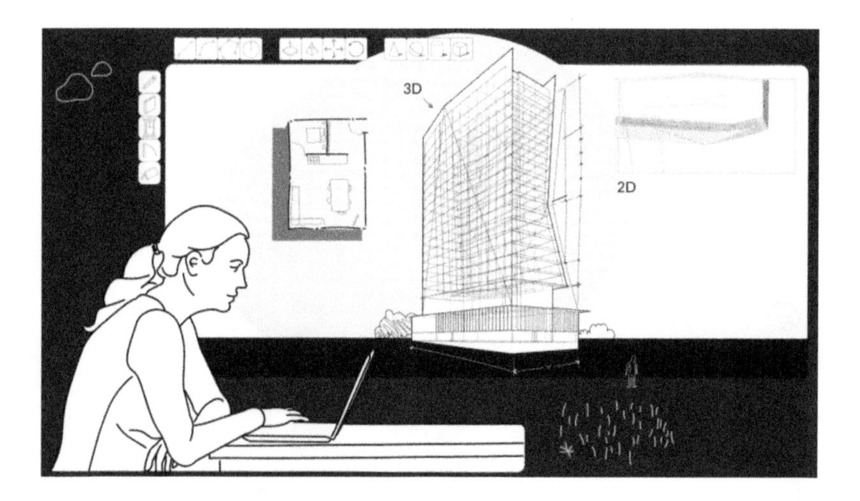

Integrating technology in the design workflow at the early stages.

5.1 Current Challenges in the Building Design Process

5.1.1 The Limitations of PDF and Email-based Workflows

As we stand on the brink of a technological revolution in building design and construction, the industry's traditional workflows are increasingly resembling relics of a bygone era. The reliance on PDFs and emails for disseminating design information is akin to using a horse-drawn carriage in the age of electric vehicles—charming, perhaps, but *grossly inefficient* and fraught with challenges. This section dives deep into the multiple layers of complexity and inefficiency that characterize these outmoded methods, from lack of interoperability to the enormous financial costs of fragmentation.

The building and construction industry's affinity for PDFs and emails as tools of trade might seem convenient, but it is effectively stifling progress. This mode of operation is akin to using an abacus in a world of supercomputers. It is not merely archaic; it is a formidable barrier to embracing the technological advancements that are just around the corner.

PDFs and emails serve as static murals in an art gallery where the demand is for dynamic, interactive installations. These 2D, "read-only" portrayals lock design data into an uneditable state, forcing various stakeholders to manually re-enter information into their systems. This is not just laborious—it is a recipe for human error, leading to inconsistencies and discrepancies that could have ripple effects on the entire project.

Think of updating PDFs and emails as akin to a never-ending game of whack-a-mole. Every time a change is made, a new set of PDFs and emails must make the rounds, requiring each stakeholder to manually incorporate these changes into their isolated design files. It's an exhausting loop of repetitive tasks that not only increases the margin for error but also inflates project costs by racking up unnecessary man-hours.

Added to that, utilizing PDFs and emails for design information is like looking at a complex jigsaw puzzle one piece at a time. You might understand each individual element, but you miss the bigger picture. Consequently, there is a gaping hole in the comprehension of how a change in one area impacts the overall design and coordination. This lack of a holistic view can lead to costly misunderstandings, resulting in construction delays and expensive rework.

In a scenario where multiple stakeholders have their hands on individual design elements, maintaining an authoritative version becomes an exercise in futility. It's akin to a group of people each editing a separate draft of a screenplay simultaneously—with no one knowing which version will make it to production. The result is confusion, wasted time, and mistakes rooted in outdated or conflicting information.

Relying on PDFs for design is like trying to understand the universe through a child's telescope; it severely limits your scope. These 2D drawings are fundamentally incapable of supporting advanced simulations that could provide insights into critical performance factors such as energy efficiency, structural integrity, and construction sequencing.

Most traditional design workflows focus on capturing what is planned rather than what actually gets built. This is like a film director only reading the script and never watching the final cut of the movie.

These methods fall woefully short in recording real-time field decisions, construction changes, and variations, ultimately failing to capture the building as it stands upon completion.

The fragmentation issue is not just an inconvenience; it is a financial sinkhole. A 2021 McKinsey report alarmingly highlighted that data and practice fragmentation in the building industry leads to an eye-watering US$284 billion in missed profit opportunities each year. To put it bluntly, this is not just an operational inefficiency; it is a financial catastrophe that is begging for immediate redressal.

Simply put, the era of PDFs and emails as the linchpin of design workflows is over. The future calls for a migration to digitally connected workflows anchored on interactive, 3D models. Not just a technological shift but an imperative transition. One that promises to enable real-time collaboration, automated updates, comprehensive performance analyses, and accurate "as-built" state capturing. It is a holistic overhaul that could very well redefine how buildings are conceived, engineered, and constructed in the modern age.

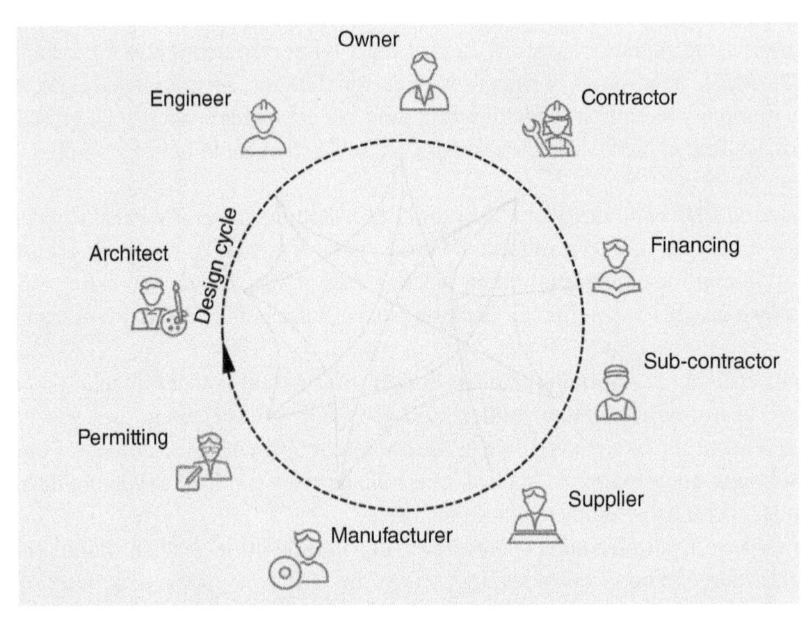

Diagram showcasing the inter-connected workflows of various professionals in a building design and construction process.

5.2 Digital Technologies in Decarbonization Strategies

As we traverse deeper into the 21st century, the construction industry is undergoing a radical transformation, shaking off its historical inertia to embrace a suite of digital technologies. What we are witnessing is not just an incremental shift; it is a full-scale digital metamorphosis driven by remarkable gains in productivity, efficiency, sustainability, and coordination across the entire lifecycle of construction projects. The marquee players in this transformation range from AI and digital twins to robotics and blockchain, each serving as a catalyst for reengineering core construction processes.

5.2.1 Digital Workflows and Design Optimization

The advent of generative technologies such as AI, digital twins, and BIM has brought about a seismic shift in design workflows. Imagine an architect's drafting board upgraded into a dynamic virtual sandbox. Here, project designs can undergo a series of tests and optimizations even before a single brick is laid. The generative algorithms within AI can sift through myriads of design permutations, making it possible to identify low-carbon, high-efficiency solutions that would be inconceivable through manual calculations. Gone are the days of wasteful prototypes and costly rework; the entire design process has evolved from a linear, high-carbon pathway into a circular, virtual, low-carbon marvel.

By rapidly analyzing thousands of data points from historical records, current conditions, and future projections, these algorithms can fine-tune schedules, optimize material orders, and create realistic cost forecasts. Technologies like radio frequency identification (RFID) tags offer real-time monitoring of material and equipment logistics, facilitating lean workflows, and just-in-time deliveries. The end result? A radical improvement in resource allocation, drastic reductions in waste, and a more predictable project timeline.

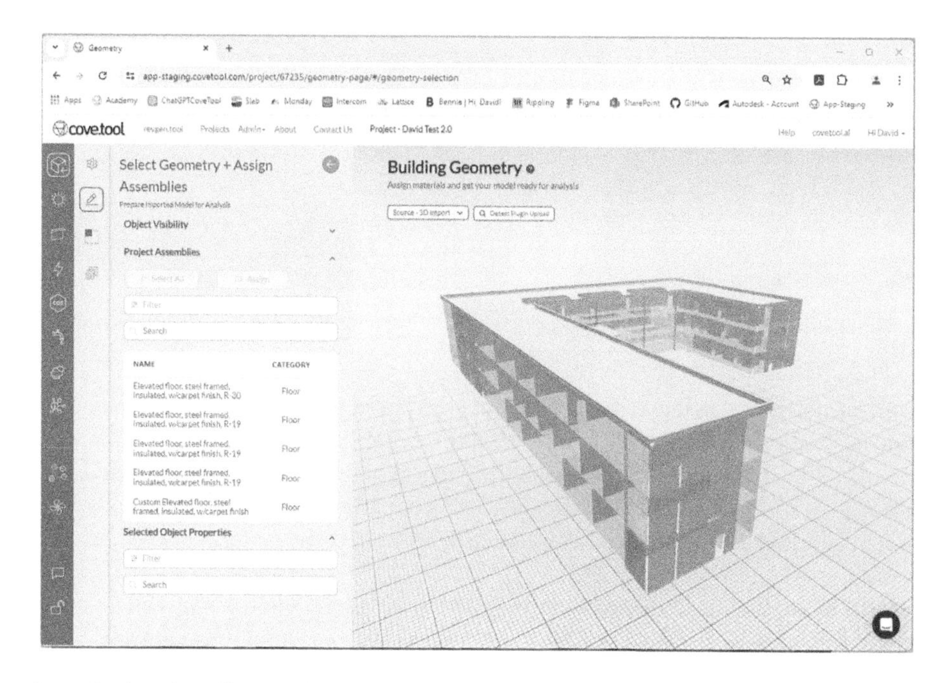

Having a dynamic virtual sandbox can be effective is rapid prototyping and decision making.

The on-site construction landscape is also getting a technological facelift. Robotics, drones, 3D printing, and augmented/virtual reality are taking over tasks that were traditionally manual and tedious. Picture a bricklaying robot that not only works faster than a human mason but also with surgical precision. Drones equipped with advanced sensors can perform site surveys remotely, offering real-time progress tracking without the need for manual inspection. These technologies are not merely productivity boosters; they are also safety enhancers, eliminating the need for human intervention in potentially hazardous situations.

Waste minimization is a perennial challenge in construction, and here too, digital technologies are making their mark. Wearable devices, sensors, and digital twins serve as the industry's new nervous system, providing real-time data that can be used to optimize material orders and minimize rework. Imagine a construction site where every piece of material, every man-hour, is utilized with maximum efficiency. Digital tools can also analyze waste streams, opening up new avenues for material recycling and reuse, thereby putting a significant dent in the industry's traditionally high waste levels.

Leveraging technology is a critical part of moving the dialogue forward. Source: Angelo/peopleimages.com/Adobe Stock Photos.

The era of working in isolated email or document silos is fast becoming obsolete. Today, cloud-based platforms are facilitating a paradigm shift toward real-time, collaborative work environments. Technologies like collaborative BIM and blockchain-enabled ledgers allow instantaneous access and sharing of up-to-date design and construction data among all project participants. This real-time exchange is not just a boon for coordination; it is a potent antidote to delays and costly rework induced by poor communication.

Emerging technologies are also breathing new life into construction supply chain management. RFID tags and AI-powered platforms offer an unprecedented level of real-time visibility and integration between planning, logistics, and suppliers. Meanwhile, blockchain technology ensures a secure and transparent information exchange. Collectively, these tools bring about just-in-time deliveries, customization options, and resilience within the labyrinthine supply networks of the construction industry.

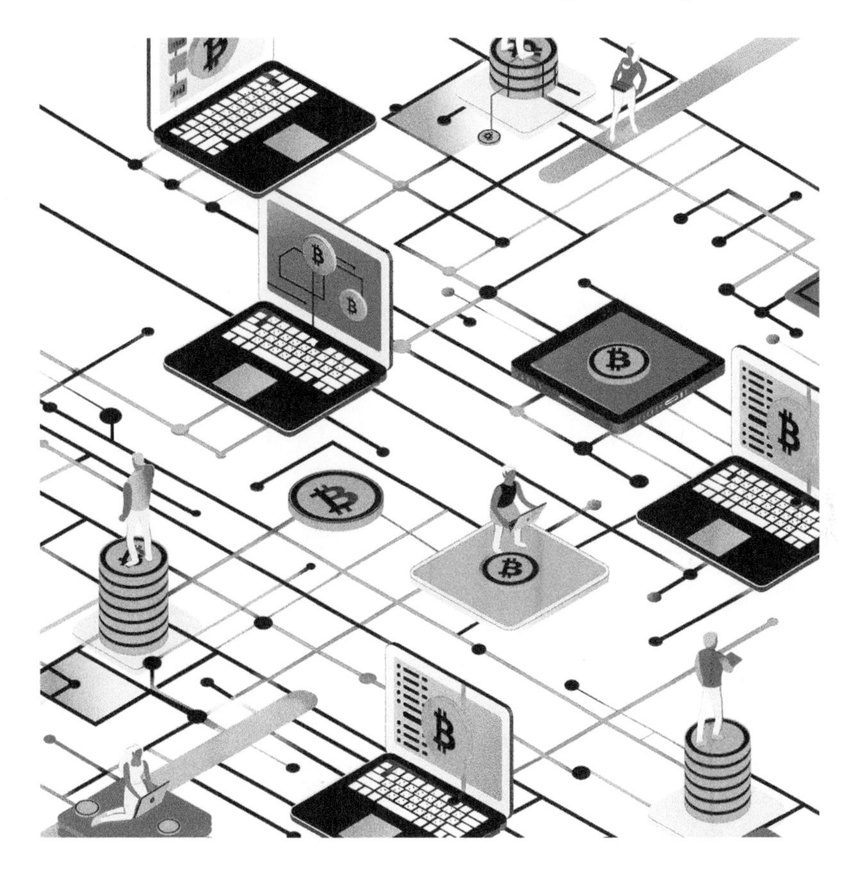

Creating workflows for secure and transparent information exchange.

While the digital transformation is in full swing, it is still a work in progress. Early adopters are set to gain a significant competitive edge as digital fluency becomes an industry standard. However, technology alone is not a silver bullet. The human element—skills development, ethical considerations, and organizational culture—needs to evolve in tandem with technological advancements. A siloed focus on new tools will yield limited benefits.

The construction industry's digital transformation will reach its full potential only through an ecosystem-wide approach. This entails collaborative efforts across companies, government agencies, educational institutions, and communities to spearhead a transition toward digitally-enabled, sustainable, and human-centric built environments. The promise of technology is immense, but its realization is contingent upon people-centric governance and a clear sense of purpose.

Leveraging software to arrive at project outcomes and decision faster.

Source: bernardbodo/Adobe Stock Photos.

5.3 The Impact of Rapid Digitalization in Construction

5.3.1 The Increasing Use of AI, Drones, Robotics, Mobile Apps, and Cloud Storage in Construction

Researchers have noted the increasing use of emerging technologies like AI, drones, robotics, mobile apps, and cloud platforms *within the construction industry*. While they have been around for a while, we are now further down the road in technological development, having figured out many issues with interfaces and user experience (UX), which allows for a wider adoption. This was especially important for industries such as construction that require detail-oriented focus, precision, and a low risk of technical failure.

Design team professionals using emerging technology to improve their workflows. Source: Halfpoint/Adobe Stock.

The benefits are now in most cases guaranteed, and if strategies are correctly executed, these tools will help solve many challenges:

AI and machine learning are enabling more automated solutions for scheduling, design optimization, procurement, and risk analysis. However, fully leveraging these technologies will necessitate augmenting rather than replacing human expertise. Researchers have also warned of the potential negative socioeconomic impacts that must be carefully managed.

Drones equipped with sensors and AI are automating tasks like inspections, monitoring, and 3D modeling for improved safety, speed, and accuracy. There are some technical challenges, such as limited battery which restricts scalability of drone-based solutions, but there are already innovative on-going research projects to tackle this.

Industrial robotics deployed for material handling, 3D printing, and assembly show potential for boosting productivity through less labor and higher precision. The implications for jobs and livelihoods are an active area of research, with many studies highlighting the need for retraining programs and policy changes.

Mobile apps for construction workflows offer benefits through real-time information access in the field, but yet again, less than optimal UX currently limits the impact. This is a subject to UX/UI development, while the main subject is investigation of intuitive navigation.

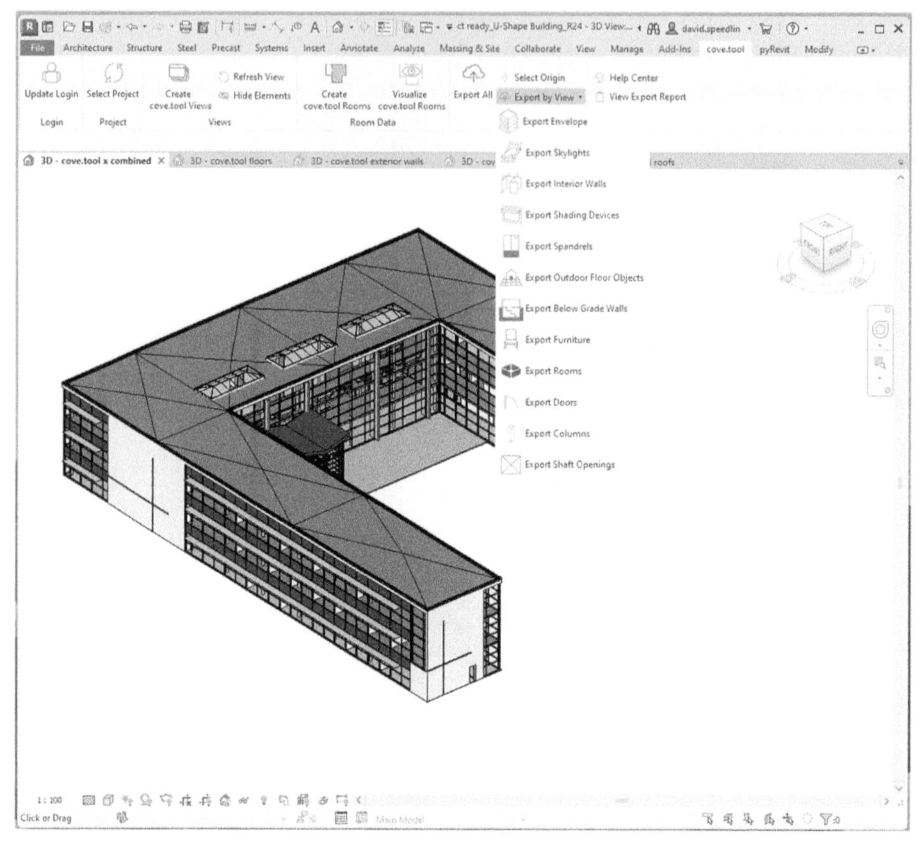

Web based and ai powered platforms can be interconnected through plug-ins and open APIs to create a seamless workflow.

Cloud platforms enable interconnected models and databases, facilitating better coordination across distributed teams. By providing on-demand access to computing resources like data storage, applications, and services via the internet, they offer a way to integrate disparate software, data formats, and tools used by the various stakeholders involved in projects. When combined with cloud-based BIM platforms and document management systems, they allow real-time visibility into project status.

Digital technologies sign considerable gains for construction, yet research stresses that holistic, inclusive approaches considering the entire ecosystem will be critical for realizing transformation at scale. Embedding issues of human-centered design, skills development, integration challenges, and uneven access across the industry appear as common themes regarding the path forward. Further longitudinal studies tracking real-world adoption and outcomes over time currently help identify optimal strategies for leveraging technology in equitable ways.

5.3.2 The Need for Professionals to Adapt and Acquire the Necessary Technical Skills

As digital transformation progresses, the skills and competencies required of professionals within the industry are also undergoing significant change. Acquiring technical skills in areas like data analytics, coding, and digital design will be considered a basic requirement and absolutely critical for those in the construction sector to remain relevant, productive, and employable in the future.

Focusing on education to adapt and acquire the key technical skills is a big part of our growth. Source: EwaStudio/ Adobe Stock Photos.

Traditionally, construction workers and professionals acquired skills through on-the-job training and experience rather than formal instruction in technical subjects. But this model is becoming insufficient as digital tools and processes become core competencies. The ability to use and integrate emerging technologies effectively will differentiate high-performing workers and firms in the coming years. Traditional "hands-on" skills will, of course, still be needed to be complemented with higher-level technical fluencies.

The skills most in demand are expected to center around digital design and modeling:

- BIM
- Data analytics applied to real-time project and asset data; programming, automation and robotics
- AI and machine learning
- Additive manufacturing
- Extended reality—XR (VR/AR/MR)
- Sensors, drones, and IoT devices.

For architects and engineers, skills in software like CAD, Revit, AutoCAD, and SketchUp will have to be complemented by literacy in programming languages like Python, Javascript, C++, and others. Beyond technical skills, higher-order cognitive abilities around digital literacy, adapting to change and collaborative problem-solving across distributed teams will also be essential to thrive within a digital construction ecosystem.

Skills gaps currently exist within the industry due to a reluctance by some firms and workers to invest in training, a lack of accessible options for reskilling and retraining, and an aging demographic among the current workforce. Younger entrants to the sector tend to have a stronger foundation in digital skills.

Source: Lazy_Bear/Adobe Stock Photos.

To address this, companies will need to provide more opportunities for ongoing technical and digital upskilling, but governments and educational institutions have to play a role by developing targeted digital skills training programs, certifications, and apprenticeships. Integrating digital curricula into undergraduate construction programs will also better prepare future professionals.

5.4 The Influence of Modular Construction Techniques

Modular formwork panel system. Source: Tomasz Zajda/Adobe Stock Photos.

The construction industry is undergoing a transformative shift, thanks to the rising prominence of modular construction methods. Long beset by challenges like escalating costs, labor shortages, and inefficiencies, the sector is finding a lifeline in modular construction—a methodology that takes significant portions of building projects off-site for fabrication, assembly, and meticulous quality control before modules are shipped for swift on-site installation. This is not just a tweak to existing processes; it is a radical rethinking that has captured the industry's imagination.

Modular construction is to traditional building what an assembly line was to handcrafted manufacturing—a game-changer in efficiency and speed. By shifting the chaos of on-site fabrication to the controlled environment of factories, modular construction leverages streamlined production lines, standardization, automation, and lean manufacturing principles to fast-track project timelines. How fast? Well, some modular projects have been clocked at a staggering 50% quicker completion rate compared to their conventional counterparts. Moreover, the factory setting eliminates the uncertainties of weather-related delays, a notorious productivity-killer on traditional construction sites.

One of the most compelling arguments for modular construction is its ability to deliver superior quality. In a factory, every cut is precise, every angle is exact, and quality control is not just an afterthought but an integral part of the manufacturing process. The result is a product with fewer defects and higher reliability—a level of quality control that is almost impossible to achieve on a bustling, weather-beaten construction site.

Waste reduction in modular construction is not incremental but revolutionary. By employing digital design, precision fabrication, and just-in-time logistics, modular construction can slash waste by up to 80% when compared to traditional methods. It is a difference that impacts not just the bottom line but also the environmental footprint of the entire project.

The modular methodology does not just cut corners; it cuts carbon emissions too. With a combination of efficiency, waste reduction, and digital optimization, modular buildings can boast up to 40% lower embodied carbon emissions than their traditionally constructed counterparts. Plus, the modular framework is highly conducive to the integration of sustainable materials and renewable energy systems, making it a win-win for both builders and the planet.

While modular construction offers tantalizing advantages, it also poses a challenge to the industry's conservative norms. To fully harness its benefits, firms will need to overhaul their technologies, business models, and even their entire value chains—a transformational leap that not all established players may be prepared to take.

The march toward automation and off-site fabrication does raise valid concerns about the potential displacement of skilled trades. Industry leaders must stay ahead of the curve by initiating training programs and updating educational pathways to align with the changing landscape.

Realizing the full spectrum of benefits that modular construction offers necessitates a tectonic shift in the industry's operational ethos. This means cultural transformations within companies, tighter integration among partners, and an overhaul of archaic regulations that hinder modular methodologies. For those willing to lead this transition, the rewards are lucrative—cost savings, quality improvements, and the repositioning of construction as an avant-garde, tech-savvy sector.

The modular revolution also calls for a new cadre of project leaders who are not just familiar with construction but are also tech-savvy and systems-oriented. These are individuals who must wear multiple hats:

- *Champion of Change*: They must vociferously advocate for the modular shift, guiding both firms and partners through the transformation maze.
- *Strategic Visionary*: A piecemeal approach will not cut it. These leaders need a comprehensive strategy that integrates every phase from design to operation.
- *Cross-Disciplinary Collaborator*: The modular process is a web of interconnected workflows, requiring a mastery of diverse disciplines and expertise.
- *Systems Thinker*: Rather than thinking linearly, these leaders focus on interdependence, standardization, and continuous improvement.
- *Technology Vanguard*: Fluency in enabling technologies like BIM, automation, and data analytics is non-negotiable for optimized modular construction.
- *Calculated Risk Taker*: Pioneering change involves taking measured risks and learning from both successes and setbacks.
- *Process Innovator*: These leaders must be adept at continually refining digitally-enabled processes for even greater efficiency and effectiveness.

The modular construction revolution is not just a fleeting trend but a seismic shift that has the potential to redefine the construction industry's operational DNA. This metamorphosis promises a spectrum of benefits—cost-efficiency, speed, quality, and sustainability—that traditional methods could only aspire to achieve. However, this transformation does not come with an autopilot mode. It necessitates committed leadership, a willingness to innovate, and an openness to disruptive technologies and methodologies.

The leaders who rise to this challenge will not just be conventional project managers. They will be visionaries who understand the symbiotic relationship between technology, sustainability, and human capital. These are the individuals who will not merely manage change but spearhead it, armed with a holistic, lifecycle perspective that integrates every element from design and manufacturing to on-site assembly and operation.

The ripple effects of this transformation will extend beyond individual projects or companies with the potential to reconfigure the entire construction ecosystem. This will involve a synergetic effort that brings together stakeholders across the board—from companies and contractors to policymakers and educational institutions. For those firms and leaders willing to embrace this change, the reward is not just short-term gains but long-term industry leadership. They will be the standard-bearers for an advanced, tech-savvy, and sustainable construction industry that is not just a part of the future but is actively shaping it.

5.5 Insights from Dennis Shelden, Associate Professor at Rensselaer Polytechnic Institute

Dennis Shelden's journey through the realm of decarbonization and building technology is a fascinating narrative filled with innovation and disruption. As an architect and technologist, Shelden has been at the forefront of reshaping the way we think about buildings, sensors, data, and sustainability. In this case study, we delve into his insights and experiences, exploring the evolution of building technology and its pivotal role in the decarbonization revolution.

5.5.1 A Transformative Journey: From Georgia Tech to RPI

Dennis Shelden's journey into the world of building technology began at Georgia Tech, where he spent years honing his skills and expertise. However, in 2020, he embarked on a new chapter, moving to Vermont for 18 months before joining Rensselaer Polytechnic Institute (RPI). At RPI, he took charge of the Center for Architecture, Science, and Ecology, a program with a significant reputation in the field.

Shelden's role at RPI allowed him to co-direct an initiative focused on the intersection of energy, the built environment, and smart systems. He explained, "It's like one of the IRIs at Georgia Tech. Energy, built environment, and smart systems, but it's specifically about energy and the built environment system." This shift marked the beginning of his deep dive into sensor technology.

5.5.2 Sensors: The Game Changers

Shelden's fascination with sensors became a driving force in his work. Traditional building retrofitting occurred once every few decades, but lightweight sensors were changing the game. He highlighted the emergence of "disruptive technologies," such as Chiclet-sized sensors from Norway that could send signals through concrete and had a 15-year battery life. These sensors could be strategically placed throughout buildings, forming ad hoc sensor networks.

Shelden explained, "There's this whole kind of things about building ad hoc sensor networks that then tie into the smart home thing." These sensor networks would not only revolutionize building management but also play a crucial role in decarbonization efforts.

5.5.3 A Water-Sensing Startup: Addressing ESG and Carbon Concerns

Shelden's passion for sustainability led him to venture into a startup focused on water sensing. While not a high-end technology endeavor, it held immense potential for addressing environmental, social, and governance (ESG) concerns. Shelden emphasized, "There is a carbon component to it, although it's not as big as you would think. But there's definitely like an existential water problem now."

This venture highlighted the broader shift toward addressing environmental challenges in the built environment, where even seemingly modest technological advancements could contribute to decarbonization efforts.

5.5.4 Occupant Behavior Modeling: Bridging the Gap

A significant challenge in building technology and decarbonization efforts has been the disconnect between simulations and real-world building performance. Shelden described the emerging field of occupant behavior modeling as a potential solution. Traditional simulations had fixed schedules and lacked the ability to adapt to human behavior within buildings.

Shelden explained, "So how do you overlay that? And then how do you start incorporating kind of an understanding of, like, okay, let's look at what people are actually doing and see if we can incorporate that into the simulation." This approach aimed to bridge the gap between theoretical building performance and real-world occupant behavior.

5.5.5 Data Standards and Semantic Web: Enabling Collaboration

In his continued work with BuildingSMART International, Shelden was passionate about developing open data standards for the building industry. He envisioned a future where data communication in the industry would transition from outdated methods to modern web services.

Shelden highlighted the challenge of dealing with multiple data schemas and emphasized the importance of creating data frameworks that allowed interoperability despite differences.

5.5.6 Decentralized Building Data: A Paradigm Shift

Shelden introduced the concept of decentralized building data, drawing inspiration from the gaming and computer graphics world. Instead of relying on monolithic files like IFC, he envisioned an ecosystem where entities were identified with universal identifiers (URIs). These entities could then be augmented with data packages, fostering agility, and flexibility.

He emphasized, "Don't tell me what you think a universal view of a wall is. I don't care... everything else has to be kind of agile, sort of, collections of stuff that people are, you know, it ought to be completely open to people to define."

5.5.7 The Path Forward: Collaboration and Automation

Shelden concluded the conversation by touching on initiatives related to recycling building components, automated tagging, and reverse engineering building compositions. He also highlighted the need for collaboration and orchestration in distributed systems to validate and query building data effectively.

As our conversation came to an end, Shelden's passion for transforming the building industry and advancing decarbonization efforts was evident. His journey from Georgia Tech to RPI had been a transformative one, and his vision for the future of building technology was nothing short of revolutionary.

5.6 The Potential of Emerging Technologies

5.6.1 The Use of Augmented Reality Capabilities and Building Information Modeling (BIM)

AR is one of the emergent technologies which, when combined with BIM, *shows much promise* for revolutionizing how construction projects are executed. But first, let's lay down some definitions.

Source: Iokanan Pro / Adobe Systems Incorporated.

AR refers to technology that layers computer-generated enhancements atop views of the real-world environment in real time. Using devices like smartphones, tablets, or AR glasses, AR applications overlay digital assets like images, videos, sounds, and data onto a user's view of the physical world. AR uses digital overlays—combining virtual objects with the real world—to enhance workflows across the construction lifecycle. AR applications for construction include:

- Remote assistance using AR glasses that overlay digital data onto live video feeds, allowing experts to visually guide and assist workers in the field.

- Layout and inspection with AR markers that visually indicate the precise location of architectural or structural elements, improving accuracy, and reducing errors.
- Visualization of 3D BIM models overlaid onto the physical worksite to aid coordination, previsualizing construction sequences, and identifying potential issues or clashes before they arise.

In construction, AR applications are used to enhance productivity, quality and safety through improved visualization, planning and communication capabilities. But wide-scale adoption faces hurdles, mostly related to knowledge gaps among traditionally trained workers, usability challenges, and high upfront costs related to both hardware and software. But AR is a great promise, especially when combined with BIM, which digitally captures key attributes of physical components throughout a building's life cycle in an interactive 3D model.

AR and BIM technologies show much potential for improving construction processes when combined effectively. AR can visualize 3D BIM models directly on construction sites as workers compare digital overlays of designs with the real environment. AR tools are able to access relevant data and metadata stored within BIM models to provide additional context-specific information, instructions, and guidance to these workers, while BIM models can be updated live based on information collected through AR applications used for as-built documentation or progress monitoring—feeding data back into the model to keep it accurate. AR also facilitates of construction sequences planned and scheduled in 4D BIM simulations, enabling workers to identify potential issues in AR overlays before beginning physical work. Automation of repetitive tasks through digital workflows that guide automated processes planned in BIM models are also enabled. During operations, AR interfaces provide building operators real-time access to relevant BIM model data to support maintenance. Combined AR/BIM solutions promise benefits like improved coordination, clash detection, scheduling, tracking, as-built BIM, automation, and facilities management, though a lack of interoperability currently limits their synergies.

In ideal conditions, these two technologies would blend seamlessly. However, there is still a lack of interoperability between AR applications and BIM platforms that currently limits the potential synergies between them. Overcoming technical barriers that limit the interoperability between these systems will be critical to fully realizing this combined transformative potential.

5.6.2 The Increasing Investment in Technologies like 3D Printing, Robotics, and Modularization

The 3D printing technology enables printing of construction materials layer-by-layer to create standardized or customized components. It shows much promise for improving how construction elements are fabricated through benefits like optimizing material usage, simplifying workflows, and enabling mass customization. The technique can print building components like walls, floors, and structures directly on-site in a single step, potentially reducing logistics. Computer-assisted design of 3D printed parts aims to yield lower material costs and waste through optimizing material usage while facilitating architectural customization possibilities that are difficult with traditional methods.

Utilizing 3D printing ain building design and construction process. Source: stokkete/Adobe Stock Photos.

Issues currently limiting the impact of those technologies include high costs of durable 3D printing materials for large-scale structures, lack of scalability as most applications remain prototypes rather than commercially viable, and absence of regulations and standards hindering adoption. Fully realizing 3D printing's potential will require investments in R&D for more affordable, printable concrete materials suited for large structures, engineering 3D printing platforms tailored for construction applications, pilot projects using 3D printed parts in commercial buildings to validate impacts, and training programs for workers to gain proficiency in design and production. Balanced approach considering implications for existing labor and smaller suppliers would maximize industry-wide benefits from increased 3D printing adoption for construction. But all three—modular construction techniques, 3D printing, and robotics—show promise for transforming how buildings are delivered. Scale-up will require greater support to speed up the transition and start reaping the maximum benefits.

5.7 Building Information Network (BIN): A New Paradigm

5.7.1 Introduction to BIN and Its Advantages over BIM

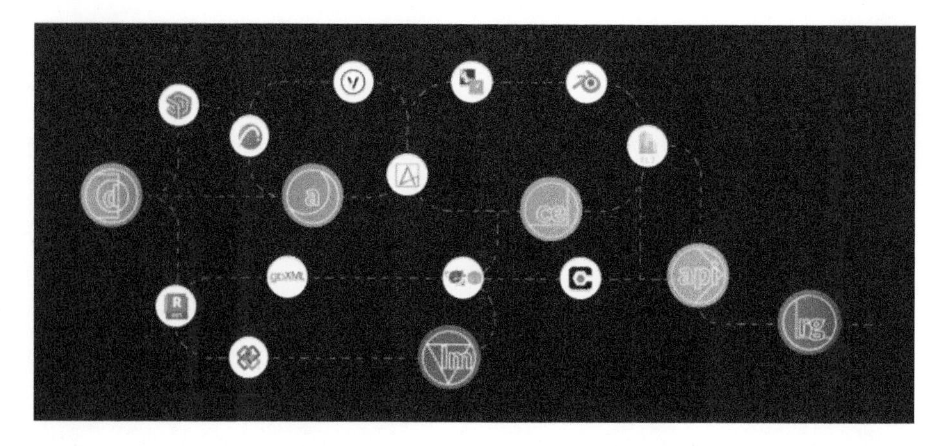

Thinking through an interconnected web of ai, autmation, and web-based technology vs point based solutions.

While BIM has been instrumental in advancing digital collaboration during the design and construction phases, it has been largely confined to these stages, with limited capacity for real-time data integration. This is where BIN makes its groundbreaking entrance, aiming to extend the digital capabilities of BIM into a real-time continuum during the design and construction process.

Leveraging technology for finding optimal building design and construction outcomes. Source: Peter/Adobe Stock Photos.

BIN is not merely an evolution of existing technologies; it is a transformative approach that interlinks a multitude of real-time data sources to create a unified and dynamic digital ecosystem. This ecosystem is active and relevant throughout the design and construction phases of a project. By harmonizing data streams from IoT sensors, architectural software, materials databases, and even worker workflows, BIN creates a living, breathing digital twin of the project. This affords stakeholders an unparalleled real-time view of the project's status, challenges, and performance indicators.

Finding the right approach, generative design ai vs rule based automation. Source: Wagner/Adobe Stock Photos.

Traditionally, keeping tabs on a project's progress involved manual inspections, static reports, and a plethora of meetings. BIN upends this model by enabling real-time remote monitoring via IoT technologies like RFID tags, indoor positioning systems, and drones. This gives design teams and contractors the ability to visually track, in real-time, key project metrics like schedule adherence, material usage, and even quality control, all within a live 3D digital environment.

Design flaws and construction errors are not just costly; they can derail entire projects. BIN is envisioned as an AI with real-time data monitoring that can track various metrics such as material health, carbon, energy use, daylight, glare, and other constraints that affect the design. This allows for early detection of potential issues like material or design inefficiencies, providing the opportunity to rectify problems before they escalate into costly delays or redesigns.

Even during the construction phase, BIN's data-rich environment can offer invaluable insights. By continuously comparing sensor data with design benchmarks, BIN can highlight discrepancies almost instantaneously. This enables design teams to make immediate adjustments, ensuring that the construction aligns closely with the original design parameters and performance expectations, thereby streamlining the entire commissioning process.

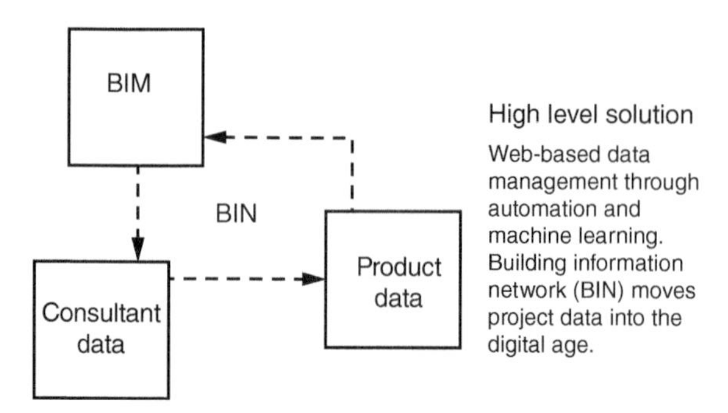

BIN fosters a collaborative environment where real-time data is readily available to all stakeholders. This is especially valuable during the handover from design to construction teams, as it ensures that all parties have access to the same up-to-date data, minimizing information lapses or misunderstandings. This data continuity not only streamlines the handover but also lays a strong foundation for agile and informed construction processes.

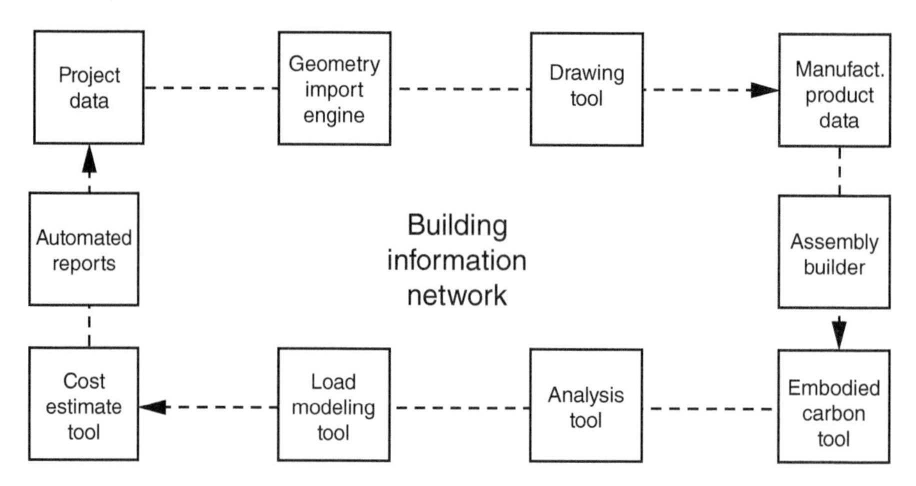

Building information network visual.

One of the most innovative aspects of BIN is its ability to feed real-time data back into the design process. Whether it is performance data from recently implemented design elements or workflow efficiencies from the construction phase, this feedback loop allows for dynamic adjustments to the design itself. This makes the design process less static and more of an evolving entity, responsive to real-time inputs.

BIN represents more than just an incremental improvement in building design and construction but a way to integrate AI into our workflows. By seamlessly integrating real-time data throughout the design and construction phases, BIN eliminates the data silos and blind spots that have traditionally limited project understanding and efficiency. While the concept is still in the adoption phase, its potential to revolutionize the industry is clear. Through BIN, stakeholders gain access to a continuously updated, holistic view of a project, allowing for unprecedented levels of collaboration, foresight, and dynamic problem-solving. BIN will not just change how we manage building projects but fundamentally alter how we conceive, design, and execute them.

5.8 Applying the GitHub Model to Building Design

5.8.1 The Concept of a "Main Branch" in Building Design

In the rapidly evolving landscape of building design and construction, it is important to explore innovative approaches that can enhance efficiency, reduce risks, and improve outcomes. Using lessons from the software industry, we can bridge the gap between design intent and how the project is realized by offering insights that architects, engineers, contractors, and building owners can leverage. One such insight is the concept of a "main branch," a cornerstone of software development that holds immense potential for transforming how building projects are conceived, designed, and executed.

For those unfamiliar with the world of software development, GitHub is a platform used for version control, allowing multiple people to work on projects simultaneously. Think of it like a highly sophisticated set of building plans that can be edited by various specialists at the same time. At the heart of this system is the "main branch," akin to the original, stamped set of architectural drawings. This main branch represents the most stable, dependable version of the project, serving as the foundation from which all changes or "branches" emerge.

Just as an architect would not make changes directly to a stamped set of drawings without due process, in software development, alterations are not made directly to the main branch without rigorous testing and review. This principle can be directly applied to the field of building design, where the process is inherently iterative, involving multiple drafts, adjustments, and stakeholder inputs. Here, the "main branch" would represent the approved, baseline design. Any modifications or experiments would be branches off this main design, to be merged back only after thorough evaluation.

So, what would a "main branch" in building design entail? At its core, it would be a fully vetted, coordinated design that has received all necessary approvals from owners, engineers, and other stakeholders. It would be a "release-ready" version of the design, meaning it could actually be built as-is with no further changes. Just like in software development, any alterations would be made in separate "branches," ensuring that the main design remains stable and reliable at all times.

The advantages of maintaining a stable main branch in building design are many. Contractors bidding on the project have a clear, unambiguous baseline to work from, enhancing the accuracy of cost estimations. Project schedules can be developed with greater certainty, as they are based on a stable design. It also simplifies the work for engineers, who can finalize their contributions confident that the foundational design is unlikely to change arbitrarily.

A stable "main branch" in building design is not just about efficiency but also a potent tool for risk mitigation. When contract administrators have a defined project scope based on a stable main branch, it eliminates much of the ambiguity that can lead to contractual disputes. Changes and alterations are rigorously tested and evaluated in separate "branches" before they are merged into the main design. This creates a documented, formal process for changes, adding a layer of legal protection for both owners and contractors.

In the construction world, accountability can sometimes be as complex as the structures being built. The "main branch" concept offers a way to streamline this. Each change made to the main design is meticulously documented and can be reviewed by all stakeholders. This ensures that if an issue does arise, it is straightforward to determine who approved what change and when. Such transparency can significantly reduce disputes and finger-pointing during the project or even after its completion.

There are two big decarbonization benefits of a "main branch" approach. By providing a stable baseline design, architects and engineers can more confidently integrate sustainable features without worrying about these elements being altered or removed in subsequent design changes. This facilitates the implementation of energy-efficient systems, green materials, and other low carbon features from the get-go. Second, the rigorous testing and vetting process for new "branches" ensures that any proposed changes do not compromise the sustainability features already in place. Thus, a "main branch" approach not only simplifies the inclusion of sustainable elements but also helps in maintaining their integrity throughout the project's life cycle.

As technology continues to evolve, the integration of software development methodologies into building design is likely to deepen. One intriguing possibility is the application of AI in automating the review and vetting process for changes to the "main branch," thereby speeding up project timelines. Machine learning algorithms could also predict the impact of design changes on various parameters like cost, timeline, and sustainability, offering data-driven insights before changes are finalized. We could see a future where every change to the "main branch" is immutably recorded, providing an unprecedented level of transparency and accountability.

Compliance with building codes and engineering requirements is a non-negotiable aspect of construction. Here, too, a stable "main branch" proves invaluable. Because only thoroughly vetted and coordinated changes are merged into the main design, stakeholders can have high confidence that the project meets all compliance criteria. This reduces the risk of costly last-minute changes or code compliance issues, ensuring that the project stays on schedule and within budget.

In a world where the demands of building design and construction are increasingly complex and fraught with risk, the concept of a "main branch" offers a revolutionary approach to managing these challenges. By adopting this disciplined strategy, which has already proven its worth in the software development realm, architects, engineers, contractors, and building owners can achieve higher levels of coordination, accountability, and quality in their projects. The time has come for the building industry to take a page out of the software development playbook and consider the manifold benefits of implementing a "main branch" approach in building design.

5.8.2 Revolutionizing the Architectural, Engineering, and Construction (AEC) Industry with the GitHub Approach: A Paradigm Shift in Design Change Management

The architectural, engineering, and construction (AEC) industry, historically marked by its methodical pace and risk-averse nature, stands on the precipice of a transformative shift. The catalyst? The GitHub model, a beacon of collaboration, transparency, and agility, which has the potential to revolutionize the way design changes are managed in the building industry.

Source: ponsulak/Adobe Stock Photos.

Consider the apparently straightforward task of modifying a door in a building's design. While seemingly innocuous, such a change can trigger a domino effect, necessitating adjustments in spatial layout, structural considerations, and aesthetic harmony. Enter the GitHub-inspired methodology. By conceptualizing the building's blueprint as a "main branch", each proposed alteration—whether it pertains to the door's dimensions, material, or design—can be tracked as a distinct "branch." This approach allows stakeholders to examine the repercussions of each modification in isolation, ensuring the primary design remains unaffected. Once an alteration is evaluated and approved, it seamlessly integrates into the main plan, providing real-time updates to all involved parties.

Now, let us extrapolate this model to a more intricate scenario: altering a building's wall structure. Such a modification carries extensive consequences, impacting load-bearing aspects, thermal properties, and more. In a GitHub-inspired system, this change would emerge as an individual branch, permitting a collaborative examination by engineers, architects, and other stakeholders. This inclusive process does not just uncover potential conflicts or compatibilities with existing structures but also spurs innovation, encouraging the adoption of sustainable materials and energy-efficient designs.

A pivotal benefit of this model is its intrinsic accountability. In an industry where minor modifications can bear significant legal and safety ramifications, tracking who proposes which change and when is critical. The GitHub approach offers a transparent audit trail, alleviating liability concerns and fostering a milieu where cross-disciplinary collaboration thrives, unencumbered by traditional bureaucratic constraints.

Embracing the GitHub model marks a radical departure from the AEC industry's conventional methodologies. This adoption heralds an era of enhanced agility, efficiency, and cooperative spirit, mirroring the evolution observed in software development. By adopting these principles, the building industry can not only refine the management of specific design changes but also redefine the paradigm for executing complex, multidisciplinary projects. The potential for sustainable design, innovative architectural solutions, and AI integration in this new framework is boundless, inviting professionals to not only adapt to but also shape the future of the AEC industry.

5.9 Decarbonization Strategies in the BIN Framework

In the rapidly evolving landscape of building design and construction, the BIN has emerged as a transformative force. It brings real-time data analytics to the core of the industry, enabling architects, engineers, contractors, and owners to make immediate, informed decisions that enhance sustainability and achieve decarbonization goals. This shift from traditional, slower workflows to a more agile, data-driven approach is revolutionizing how projects are executed.

Low carbon timber structure. Source: ungvar/Adobe Stock Photos.

Real-time data has become a linchpin for effective decarbonization. It provides a continuous feedback loop that allows for quick adjustments in design and construction. Unlike traditional methods that rely on periodic assessments, this real-time capability offers an agile and dynamic approach to reducing carbon footprints. This is particularly vital for managing built systems, which often involve complex and energy-intensive processes. Take an HVAC system that suddenly shows an increase in energy consumption, for example. With BIN's real-time analytics, project teams can quickly diagnose and rectify the issue, saving money and reducing carbon emissions.

But BIN's capabilities are not limited to managing existing systems; they also make it easier and less risky to implement new sustainable technologies. If a newly installed solar panel array falls short of expectations, real-time analytics can identify the issue, enabling immediate corrective action. This feature has the added benefit of reducing the financial risks associated with adopting new technologies, providing a more conducive environment for innovation.

Understanding the complex trade-off within building design and construction is critical to a low carbon and low cost output. Source: chokniti/Adobe Stock Photos.

Perhaps one of the most challenging aspects of sustainable building design is managing the complex trade-offs that come with it, such as selecting eco-friendly building materials without compromising structural integrity. BIN's real-time data analytics offer a solution by allowing for a dynamic comparison between different building materials in terms of their performance, structural implications, and carbon footprint. This is invaluable for balancing building performance with carbon reduction goals. For instance, the energy efficiency of various HVAC systems can be weighed against their upfront carbon costs, providing a nuanced approach to meeting sustainability objectives.

One of BIN's most compelling features is its ability to integrate multiple dimensions of a construction project into a unified, real-time framework. This enables a holistic approach to decarbonization, where each aspect of the project—from material selection to energy use—is continuously optimized to meet sustainability goals. Such real-time integration also allows for unprecedented flexibility and adaptability in project management. Teams can respond to challenges and opportunities as they arise, rather than being constrained by a static plan. This agility can result in more effective decarbonization efforts and significant cost savings over the life of a project.

But BIN's impact extends beyond mere data collection by fostering a culture of proactive sustainability. By providing a platform for real-time, informed decision-making, it encourages all stakeholders to consider sustainability as an integral part of their daily operations. The long-term implications of this shift could be revolutionary for the construction industry. By embedding real-time analytics and data-driven decision-making into its DNA, BIN has the potential to make sustainable construction the norm rather than the exception.

The integration of BIN and real-time analytics into the building industry shifts our understanding of design and construction. It combines the best of data science and sustainable design to create a more agile, responsive, and sustainable approach to building design and construction. As the industry continues to evolve, the adoption of BIN could well be a tipping point, paving the way for a future where sustainability is not just an aspiration but a reality.

Incorporating renewable energy can help with operational carbon use but needs to be integrated in a cross functional way. Source: ktsdesign/Adobe Stock Photos.

5.10 Case Study: How a Change in Wall Structure can Impact the Entire Building Design

In a BIN framework, the roles of all participants—architects, engineers, contractors, and building owners—become even more integrated and collaborative compared to a BIM framework. While BIM facilitates some early coordination through 3D modeling, BIN takes this a step further by establishing an actual information network connecting all parties.

Within the BIN, all project stakeholders can access and contribute information in real time. Plans, models, specifications, schedules, issues, and resolutions are shared openly on the network. This essentially breaks down barriers between traditional project silos by providing a single source of truth visible to everyone.

This is a level of transparency and connectivity that allows the roles of each party to become highly interdependent. Architects and engineers must design with constant input from contractors on constructability and from owners on functional needs. Contractors identify issues as they arise and work with design teams to find solutions within the BIN. Owners become actively involved in day-to-day decisions by tapping into up-to-date information on the network.

Everyone essentially takes on a "network role" where they contribute information relevant to their expertise, while also accessing and incorporating information from others on the BIN. Project responsibilities become a shared effort across parties, rather than distinct responsibilities in silos. The ultimate goal of a BIN framework is to leverage real-time connectivity and transparency to optimize coordination, reduce waste, and manage changes dynamically. Projects are essentially "designed while being built" through constant and immediate feedback circulating within the network. The BIN framework fundamentally shifts roles by establishing an open information ecosystem where architects, engineers, contractors, and owners work together interdependently as a networked team. Distinctions between traditional responsibilities become blurred as everyone takes on the shared goal of optimizing the project through the BIN.

5.11 Case Study: How a Simple Change (Like Changing a Door) can be Efficiently Managed in a BIN Framework

Manage Operational Complexities: The increasingly intricate web of decisions in building design and decarbonization calls for real-time, data-driven tools. This is where AI, particularly within a BIN, can be a game-changer. However, its application is akin to wielding a double-edged sword. A study by Boston Consulting Group found that when applied for creative tasks, generative AI like GPT-4 can boost performance by an astonishing 40%. On the flip side, its application in business problem-solving led to a 23% decrease in performance. It is important to understand the boundaries between human capabilities and the machine just like any other tool. The stage is set for AI in BIN, where its capabilities could either unlock new dimensions of sustainability or introduce a new set of problems.

One of the most promising avenues for AI within BIN is in the realm of managing complexity. The balancing act between building product selection, building performance, structural calculations, and carbon emissions is a daunting task. AI algorithms can sift through the labyrinth of trade-offs, offering data-backed suggestions that can drastically reduce the carbon footprint while optimizing performance metrics. These algorithms are not just about "doing more with less"; they are about "doing better with less," helping architects and planners make decisions that are aligned with decarbonization goals.

AI is not a silver bullet; it serves as an "invisible assistant." This dynamical learning and adjustment capability is invaluable but requires prudent human oversight. AI can navigate through heaps of data and offer solutions, yet it lacks the nuance and critical thinking that human experts bring to the table. The warning from the BCG study about the risks of over-relying on AI should serve as a cautionary tale. While AI can be a powerful ally, it cannot and should not replace human judgment, especially in matters as complex and multidimensional as building design and sustainable urban planning.

Another less discussed but equally critical facet is the impact of AI on organizational creativity and the diversity of thought. The BCG study highlighted a 41% reduction in the diversity of creative ideas when generative AI was used. This poses a significant risk in the context of BIN, where innovation and diverse perspectives are key to solving complex sustainability challenges. A homogenized thought process might offer incremental improvements but could potentially stifle groundbreaking innovations.

The frontier of AI competence is not static. It is continually evolving, requiring firm leaders to stay informed of new capabilities on a monthly basis. What seems beyond reach today could well be within grasp next month. As the capabilities of AI expand, so too will its role within BIN frameworks, making the relationship between AI and human decision-making an ever-evolving dynamic. Do not just focus on what AI can do now but keep an eye out for what it might be able to do next. This evolving landscape necessitates that we not only adapt but also anticipate, continually revisiting our strategies and assumptions.

The integration of AI within BIN is an exciting but cautious step forward. It offers the promise of greater efficiency and more effective decarbonization strategies, but it also comes with its own set of challenges and limitations. As we stride into this data-driven future, the key will be to wield this double-edged sword with the right balance of enthusiasm and caution, ensuring that we harness its full potential without losing sight of the human elements that make our buildings not just sustainable, but also livable.

Computationally driven design and construction solutions have been moving the discussion forward in the last decade. Source: Denys Nevozhai / Unsplash.com / Public domain / https://unsplash.com/photos/low-angle-photography-of-architecture-building-izC9Yob6DGM.

Data driven decision making is key for complex buildings. Source: Claudio Colombo/Adobe Stock.

5.12 The Challenges Amid Rapid Industry Change and Industrialization

Various Factors: Project delays, arguably the most visible of the challenges, are often the tip of the iceberg. When innovative, low-carbon technologies make their debut, they are usually shrouded in uncertainty due to limited real-world performance data. While these technologies promise a greener future, they are often nascent and untested, complicating project timelines. Firms find themselves in a catch-22; they want to adopt new technologies to meet decarbonization goals but are wary of the risks involved.

Moreover, the bureaucratic labyrinth of approval processes adds another layer of complexity. Government agencies, unaccustomed to new technologies, may take longer to evaluate and approve them. This is not just a matter of ticking boxes; it is an issue that can potentially stymie innovation and discourage firms from adopting greener technologies.

The workforce transition, often overlooked, is a silent time-consumer. As the industry evolves, so too do the skillsets required. Construction firms are discovering gaps between the skills they need and those their current workers possess. Retraining a workforce is not an overnight task; it is a long-term investment that can extend project timelines significantly.

Supply chain disruptions can throw a wrench into the most well-laid plans. As the industry moves toward low-carbon materials, supply chains are often unprepared to meet the new demand, resulting in limited availability and subsequent delays.

5.12.1 The Slippery Slope of Scope Creep

Scope creep is an insidious challenge that often begins innocuously. As a project progresses, our understanding of its requirements may evolve. For instance, what starts as a straightforward energy retrofit might gradually expand to include water-saving features, sustainable landscaping, and more. While these additions are commendable, they extend the project's scope and can lead to a slew of issues, from delays to budget overruns.

Another contributing factor is the iterative nature of design in the digital age. With technologies like BIM allowing for rapid prototyping, firms may be tempted to tweak designs continually, leading to a bloated scope that spills into the construction phase.

Underestimating complexity is another pitfall. Projects often involve intricate systems integration, merging traditional construction with innovative technologies and natural strategies. Inexperience in handling this level of complexity can result in an expanded scope as firms scramble to meet their goals.

Budget overruns often follow in the wake of project delays and scope creep, but they have their own unique triggers. The financial aspects of adopting new, low-carbon solutions are still a gray area. Limited commercialization means higher costs, which are often volatile. Coupled with uncertain life-cycle expenses, budgeting becomes an exercise in educated guessing.

Addressing workforce challenges further strains budgets. Upskilling an existing workforce or hiring new talent with the necessary expertise is an investment that can exceed initial budget estimates. And let us not forget that scope creep, in expanding the project's range, naturally inflates its cost, often beyond what was originally budgeted.

5.12.2 Strategies for Navigating the Complexity

Confronted with these challenges, the industry cannot afford to be passive. Firms need proactive strategies to navigate the complexities of this transitional period. For instance, building strong relationships with approval agencies can accelerate the permitting process. Workforce development is another area requiring attention; targeted training programs and apprenticeships can fill the skills gap more efficiently. Supply chain resilience, often an afterthought, should be a focal point, achieved through diversification and local sourcing where possible.

Improving project management practices can preempt many issues, catching them before they snowball into larger problems. Leveraging digital tools like virtual design and construction can enhance predictability and reduce uncertainties. Innovative contractual models that share risks and rewards among partners can incentivize flexibility and adaptability, key attributes in an ever-changing landscape.

In terms of digital transformation, continuous learning is the way forward. The rapid pace of change requires a shift from traditional training models to just-in-time learning methods, such as online courses and simulations. Micro-credentialing can validate evolving skill sets, and a focus on cultivating attributes like curiosity and adaptability will be just as important as technical expertise.

The construction and building industry is at a critical juncture, shaped by the twin forces of decarbonization and digital transformation. While the road ahead is fraught with challenges like project delays, scope creep, and budget overruns, these are not insurmountable. Through strategic planning, continuous learning, and a willingness to adapt, firms can navigate these complexities. The end goal—a more sustainable, efficient, and innovative industry—is well worth the effort.

Up skilling in the digital era is necessary for the construction industry. Source: jjfarq/Adobe Stock Photos.

5.13 Change Management in the Digital Era

5.13.1 The Importance of Change Management in the Implementation of New Technologies and Practices

Change management is critical in the successful implementation of new technologies and practices entering the construction industry as it transitions toward decarbonization. The effective aspects of change management in this context are communicating the need for change, developing transition strategies, but also addressing resistance likely to be encountered on the way.

Change management is extremely important for successfully implementing new technologies and work practices. Effectual change management necessitates transparently communicating the need for change by articulating a compelling vision for how novel technological and decarbonization strategies will augment project outcomes, mitigate risks, and unlock novel opportunities. It entails explaining how jobs, skill requirements, and working styles will evolve, emphasizing gains in interest, autonomy, and flexibility rather than solely losses. Showing concrete benefits to clients through boosted quality, lowered costs, and sustainability will be of value in the future, just like engaging stakeholders through multiple channels to collect feedback, address concerns, and build buy-in.

Developing transition strategies and pilot-testing the changes on a small scale is a good choice before organization-wide rollouts. This way, change management tools can address both the technical difficulties, but also the components beyond the technical realm that would necessarily have a big impact on a large scale, namely emotional, behavioral, and cultural components of transition. It goes hand in hand with learning how to address opposition constructively by acknowledging valid concerns and eliminating obstacles to gain cooperation and participation. Giving opportunities for input, feedback, and two-way dialogue will ameliorate skeptical views, build trust, encourage focus on pain points, and catalyze momentum. By strategically communicating the need for change, everyone can navigate the structural, cultural and behavioral dimensions of transformation.

5.13.2 How Effective Change Management can Facilitate the Transition to a BIN Framework and the Adoption of Decarbonization Strategies

BIN represents a significant departure from traditional design and project management processes, requiring many changes. Similarly, decarbonization demands a tremendous effort in adaptation. Both transformations involve shifting incentives, processes, capabilities, and ways of thinking entrenched in existing organizational culture.

Source: Claudio Colombo / Adobe Systems Incorporated.

Communicating a compelling vision for an integrated BIN approach will be crucial to gaining support for transitioning from siloed strategies to a holistic framework that maximizes outcomes. Leaders must clearly explain to stakeholders how built system upgrades, technological solutions, and natural climate solutions working together synergistically through an integrated BIN framework will enable deeper reductions in embodied and operational carbon. The benefits of a coordinated, systems-based approach must be articulated in a vision that inspires buy-in for the new, holistic way of thinking required from all stakeholders across sectors. Effectively communicating the transformative value proposition of integrated network frameworks will therefore be essential to facilitate their adoption over traditional disjointed strategies.

Developing transition strategies will require piloting integrated BIN solutions on a smaller scale. Pilot projects can demonstrate benefits, provide lessons, and build support for wider adoption. New cross-functional teams trained in a BIN mindset and metrics assessing outcomes across BIN dimensions will be needed, along with partnerships between complementary organizations that can enable holistic BIN solutions. Institutional structures may require realignment to incentivize systems-level thinking. Pilots can inform strategies that embed an integrated BIN approach through reshaped teams, partnerships, metrics, and incentives.

Addressing this issue involves reframing common concerns. Fears about job losses can be countered by spotlighting the new roles emerging at the intersections of systems that combine and integrate diverse expertise. These are the opportunities the BIN framework creates. Tradeoffs and risks within single strategies can be recast as surmountable when multiple complementary levers are coordinated through a holistic approach. Frustration with another "initiative" can be quelled by demonstrating how the BIN framework provides a unifying structure that maximizes and subsumes existing decarbonization efforts, positioning them for synergistic impact. An integrated system-level view reframes common concerns and exposes opportunities resistance as obscured. The BIN lens ultimately reveals a pathway forward by shifting fears into concrete actions and synergies.

The first cornerstone in this transformative journey is human capital development. Training programs that instill a BIN mindset are crucial; they must go beyond mere technical skills to foster cross-disciplinary thinking that breaks down silos. Alongside this, communities of practice serve as fertile grounds for the cross-pollination of ideas, bridging professionals from diverse BIN domains. These communities not only disseminate best practices but also act as collaborative think tanks, solving challenges through collective wisdom. The key to this is knowledge transfer, a process that allows for the rapid dissemination of innovations and solutions across different sectors within the BIN framework.

A shift in organizational structures and metrics is imperative. Job roles must evolve to encompass a multidisciplinary approach, valuing skills like collaboration, adaptability, and systems thinking. Traditional metrics, too, need an overhaul. Instead of evaluating outcomes within isolated dimensions, a more holistic yardstick involving life cycle assessments, triple bottom line accounting, and systems analyses is necessary. This extends to the organizational leadership, who must be adept at communicating a holistic vision and championing cross-functional initiatives. Furthermore, institutions must re-evaluate their incentive structures to reward this kind of holistic, multidisciplinary thinking, acknowledging that cultural shifts often begin at the top echelons of an organization.

The broader ecosystem cannot be ignored. Incremental, iterative approaches like pilot programs offer a pragmatic path to test assumptions and build experience. Yet, external forces—policymakers, investors, and clients—hold significant sway in accelerating this transition. Through mandates, financial incentives, and procurement practices that value coordination across BIN dimensions, these stakeholders can exert a transformative influence. Amid all this, technology serves as an enabler but not a panacea. While AI, automation, and digital twins offer robust tools for systems-level analyses and coordination, they are complements to, not substitutes for, the human capacity for holistic, conceptual thought crucial for true systems-level transition toward decarbonization.

These three pillars—human capital, organizational change, and external influences—constitute a comprehensive strategy for transitioning to an integrated BIN framework, setting the stage for a future where sustainability and innovation are intrinsically linked.

5.14 Conclusion

Source: MaciejBledowski/Adobe Stock.

This chapter delves into data-driven design, a complex but transformative discipline with the power to set us on the path to decarbonization. Data is more than just numbers on a screen; it is a new way of seeing the world. But this perspective is incomplete without human insight to contextualize it within a broader societal and moral framework.

Data-driven design is not an island but part of a larger archipelago that includes technological solutions and natural resource strategies. Fixing the system—not just its individual components—is the task at hand. While data can act as our navigational tool, the role of the navigator falls to us. We need to use this tool not just to make informed decisions but to make decisions that are in alignment with long-term sustainability and ethical considerations.

As we chart the course for the future, our decisions need to extend beyond the immediate horizon. We are not just building for today; we are influencing the architecture of tomorrow. AI and BIN frameworks are powerful tools that offer us the scaffolding for this future. But they are tools, nothing more. The responsibility for using them wisely falls squarely on our shoulders.

To truly unlock the potential of these tools, we must cultivate conditions ripe for transformative change. We need to foster systems thinking and ethical governance. Collaboration needs to replace competition, and a shared commitment to planetary stewardship must become our guiding principle. Our approach must continually optimize outcomes as conditions change and our understanding improves. We also need to scale and commercialize innovative technologies, but do so within a system context to reveal interactions, dependencies, and opportunities. As we navigate this complex landscape, we must never lose sight of social and equity dimensions. The data we gather and the insights we glean must be used to benefit all, not just a select few.

The challenges ahead are not just technical but adaptive, requiring a reorientation of our values and institutions. The most potent force for change is not any particular technology; it is the collective will to use that technology in a manner that benefits not just the present but the future as well.

5.15 Insights from Tristram Carfrae RDI, Deputy Chair at Arup

Tristram Carfrae holds significant roles at Arup, where he serves as one of the Deputy Chairs and holds the esteemed title of Arup Fellow. He presides over the Digital Executive and champions the pursuit of excellence across all aspects of the organization. With a background in structural engineering, Tristram's expertise has played a pivotal role in the creation of numerous award-winning structures, including the iconic Water Cube designed for the Beijing 2008 Olympic Games.

"Little Island" in Manhattan, New York City, New York. Source: creativefamily/Adobe Stock.

Carfrae's distinguished career is underscored by his recognition as a Fellow of the Royal Academy of Engineering. He was honored with the title of Royal Designer for Industry (RDI) in 2006, received the prestigious Institution of Structural Engineers' Gold Medal in 2014, and was granted IABSE's International Award of Merit in Structural Engineering in 2018.

5.15.1 Decarbonization: A Huge Topic

In the world of decarbonization, Carfrae is a visionary whose journey through the evolving landscape of technology and data trends is nothing short of remarkable. When asked about what he saw on the horizon, Tristram admitted, "Huge, how about huge?"

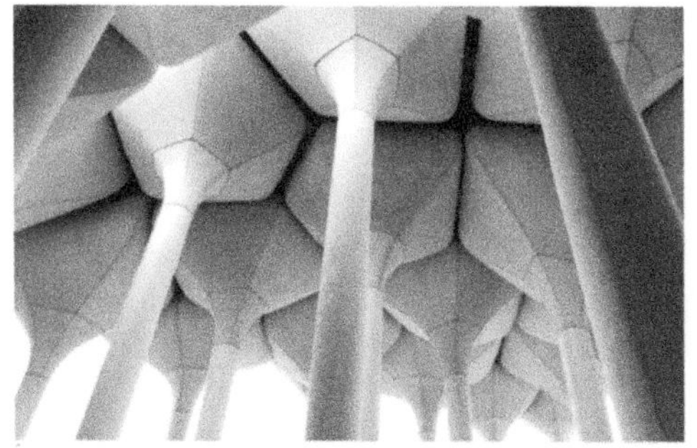

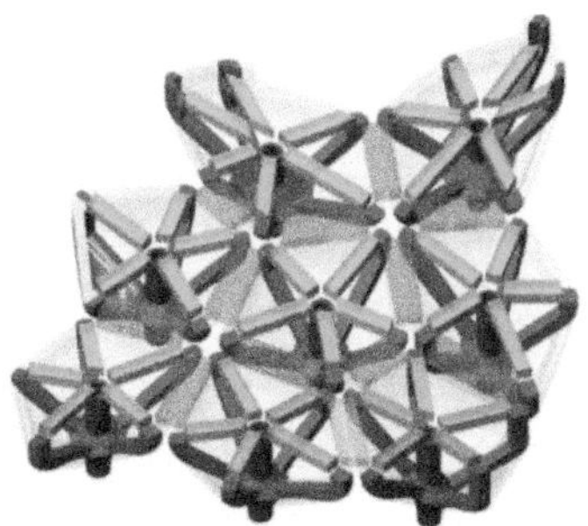

The typical pot is 4.6 m tall and composed of four to six "petals." It sits atop a central column head which transitions to a cylindrical pile.

To create the doubly curved shape of the pots, rigid foam blocks were shaped using automated machinery.

The enormity of the topic was undeniable, especially when reflecting on the past seven years of what he affectionately referred to as "digital transformation." It all began with a vision to harness the power of emerging technologies—social media, mobile, big data, AI, cloud computing, VR, AR, and the concept of user-centered design. It was a time when doors to uncharted territory swung open, making professionals across industries take a step back to recalibrate.

5.15.2 Data in Design, Construction, and Delivery

One element that had caught many off guard was data. In a world where design, construction, and delivery often existed in silos, the concept of gaining insights directly from data felt foreign. Architects and engineers had grown accustomed to crafting meticulous design instructions and construction documents, but once the project was handed off, they seemed to distance themselves from the outcome, leaving it to fate. Carfrae noted, "We never went back and measured; we didn't obtain feedback. It was as if designers preferred to remain cocooned in their creative vacuum."

Data, however, had other plans. It promised a future where the built environment could be closely monitored, where the performance of buildings and the patterns of people's lives could be quantified and analyzed. It offered a new perspective on construction, renovation, and the impact of environmental changes. For Tristram, the possibilities were tantalizing.

He delved into a story of a confidential project involving 5000 buildings facing unexpected seismic activity. Armed with photogrammetry and satellite imagery, they made educated guesses about the buildings' possible materials and structures. They then subjected these virtual structures to a barrage of simulated earthquakes, producing a probabilistic assessment of their likely performance. It was a unique application of data-driven insights, one that could potentially save lives and resources.

But data's influence extended beyond this project. Carfrae shared a tale of Shanghai, a city grappling with stormwater drainage issues. Using satellite imagery and Earth observation data, they allocated land use patterns pixel by pixel, crafting a stormwater model that could inform decisions about green spaces and waterways. The result? A more absorbent city, cost savings, and improved quality of life.

Shanghai urban drainage.

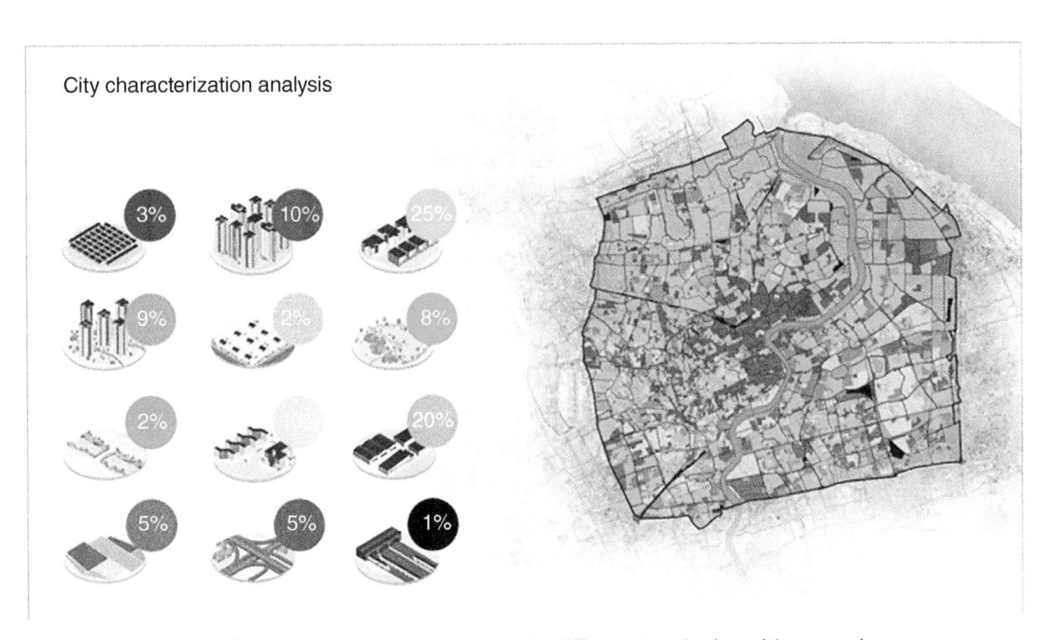

Machine learning categorized the masterplanning area into different typologies with respective green infrastructure usage known as the Shanghai masterplan.

He marveled at the ability to extract actionable information from data, not just in hindsight but in real time. A UK railway project in the works, traversing rural natural heritage areas, had benefited from satellite imagery. Identifying biodiversity hotspots and areas with less diversity helped determine the optimal route.

5.15.3 From Man to Machine

Yet, as technology accelerated, so did Carfrae's ponderings. He spoke of trust and the challenge of relinquishing control to machines. Drawing inspiration from an astronaut's wisdom, he underscored the need for humans to validate data-driven decisions. However, he also emphasized the importance of diverse verification methods, never placing absolute trust in one approach.

Carfrae then shifted the conversation to decarbonization, a topic that has soared in relevance. He discussed the evolving landscape, highlighting the shift from operational to embodied carbon as a critical concern. Measuring and understanding carbon emissions across the life cycle of buildings became a key goal. Material passports and the circular economy emerged as potential solutions, heralding an era of sustainability.

5.15.4 Engineers, Designs, and Change

The conversation circled back to engineers and their pivotal role. Carfrae painted a compelling picture: engineers and designers standing at the precipice of change, holding the power to shape the future of carbon emissions in construction and design. Buildings account for a significant portion of global emissions, and engineers were uniquely positioned to tackle this challenge.

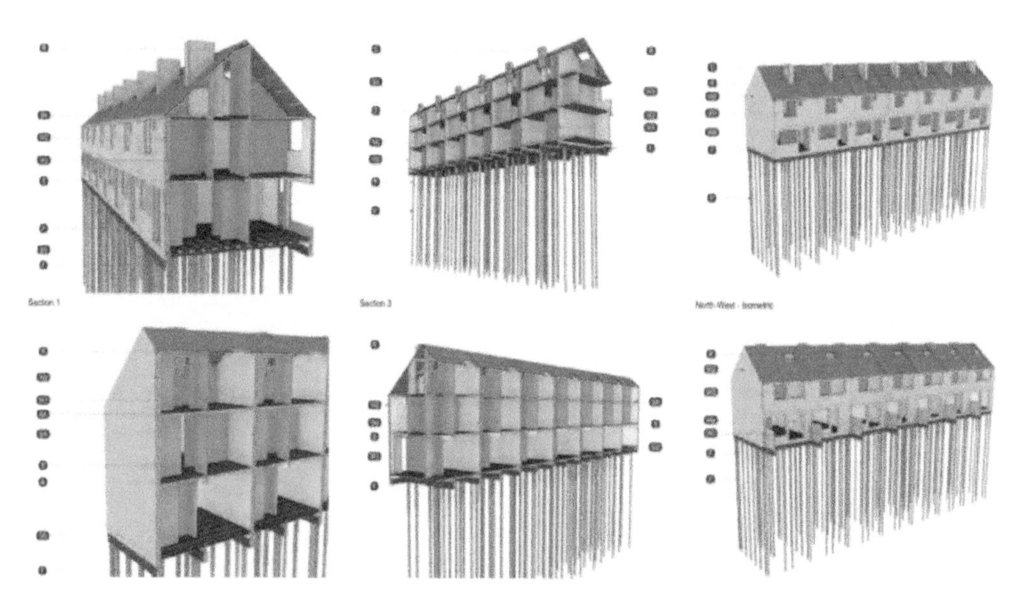

All building elements are tagged in the BIM program using a consistent naming convention.

Carfrae's message was clear. The journey ahead was vast and uncertain, but the fusion of technology, data, and engineering had the potential to steer humanity toward a sustainable future. With each story shared, a vision of a world where data-driven insights and sustainable practices could reshape our cities and our lives came into sharper focus. It was a journey worth embarking upon, for the sake of the planet and all its inhabitants.

6

Decarbonizing Buildings: "The Low Hanging Fruit"

In the grand endeavor of curbing our carbon emissions, the spaces we inhabit—our homes, workplaces, and community buildings—stand out as a domain ripe for transformation. This chapter delves into the array of straightforward and economical design choices that can profoundly decrease the energy demand and carbon footprint of our built environments, thereby bolstering the well-being, comfort, and productivity of occupants.

We will traverse the spectrum of actionable interventions applicable throughout a building's lifecycle, from nascent design sketches to deep retrofitting initiatives. The strategies discussed hinge on the principle of decarbonizing energy consumption at its origin. This involves leveraging passive design, material efficiency, behavioral adjustments, and systematic efficiency improvements, with particular attention to the manipulation of microclimates, thoughtful orientation, and the optimization of roofing and facade systems.

The journey to substantial carbon abatement does not necessarily hinge on avant-garde innovation or exorbitant financial outlays. It is anchored in established, yet frequently overlooked strategies that are ripe for broad application. With the right incentives, policy frameworks, and the engagement of building users, these approaches can scale up to become a formidable force in the fight against climate change.

This chapter aims to illuminate these accessible solutions and demonstrate their considerable, yet often unexploited, potential to enact change. Through illustrative case studies, actionable advice, and broad impact assessments, we will reveal how these "low hanging fruits" hold the promise of leading us toward a more sustainable and carbon-conscious future in building design and operation.

6.1 Understanding the Basics of Building Science

6.1.1 Defining Key Terms and Concepts

The building science turns on an understanding how energy and matter interact within and around our structures. The discipline is a confluence of the physical sciences—physics, chemistry, and biology—and their principles govern the performance of a building and its attendant systems. For architects and engineers committed to the decarbonization of the built environment, a robust understanding of building science is indispensable to allow the design team to achieve the project goals using data rather than guesswork.

Build Like It's the End of the World: A Practical Guide to Decarbonize Architecture, Engineering, and Construction, First Edition. Sandeep Ahuja and Patrick Chopson.

Consider the building envelope—walls, roof, and foundation—which delineates the boundary between the controlled interior environment and the natural elements outside. As a professor of mine, Rich Cole, AIA, liked to say, "The process of building is putting sticks in the mud and putting sticks across those sticks until you cannot put your hand through it." It is here that the exchange of air, water, and energy is managed. The deftness with which a building's envelope conducts this exchange is a key measure of its performance. The envelope must be meticulously integrated and sealed to maintain indoor comfort and minimize energy losses, pivotal in the journey toward sustainability. To know how to design and construct low carbon envelopes, let us dive into the principles underpinning the science of buildings.

Thermodynamics is the cornerstone of building physics, providing a framework to understand energy dynamics within buildings. It says that energy, in its myriad forms, is conserved and merely transitions from one state to another. The implications for buildings are profound; they are designed to find a thermal balance with their surroundings, striving for equilibrium but also conserving energy wherever possible.

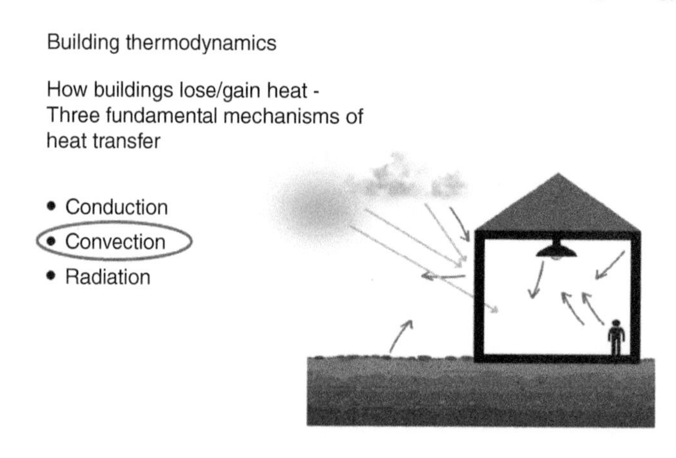

Conduction involves the direct transfer of heat between objects in physical contact. The rate of conductive heat transfer depends on the temperature difference, contact area, and conductivity of the materials. Insulating materials like foams, fibers, and composites have low conductivity, impeding conductive heat transfer. Common building insulation includes spray foam, batts, rubber membranes, and aerogels. Higher *R*-values indicate more resistance to conductive heat flow, with a higher *R*-value corresponding to better insulation performance. Placing insulation strategically in building assemblies helps reduce energy demand for heating and cooling by slowing interior heat loss in winter and heat gain in summer, lowering operational carbon emissions. Common areas of insulation include exterior and interior walls, floors, attics, crawl space, and foundations.

Convection involves heat transfer between a fluid and a surface via the movement of that fluid. Convection currents within building cavities carry heat from one part of a structure to another, impacting thermal performance and comfort. We can use strategies like air sealing and foam insulation in cavities to help disrupt convection currents and slow heat transfer while reducing air leakage. This lessens conductive and convective heat loss, lowering energy demand.

Radiation accounts for a significant portion of total heat transfer within buildings. All surfaces above absolute zero emit infrared radiation, with hotter surfaces emitting more radiation that is absorbed by colder objects. Radiant barriers and air gaps impede radiant heat transfer, while highly emissive surfaces also help dissipate heat via radiation. Utilizing daylighting, shading devices, and high-reflectance surfaces are great ways to help moderate radiant heat gains and losses.

Moisture management is a critical concern, as water in its vapor and liquid forms can traverse building materials, potentially causing damage and promoting mold growth. Water is the enemy of buildings and why studying strategies to combat this are essential skills for the entire design and construction team. Techniques to master include flashing, coping, vapor barriers that restrict vapor diffusion, materials that allow buildings to "breathe," and proper active ventilation to control indoor humidity. Air barrier systems are essential in maintaining a distinct boundary between the exterior and interior, ensuring that air movement is controlled and conditioned space is preserved. These barriers, when applied correctly, contribute significantly to a building's thermal efficiency and comfort levels.

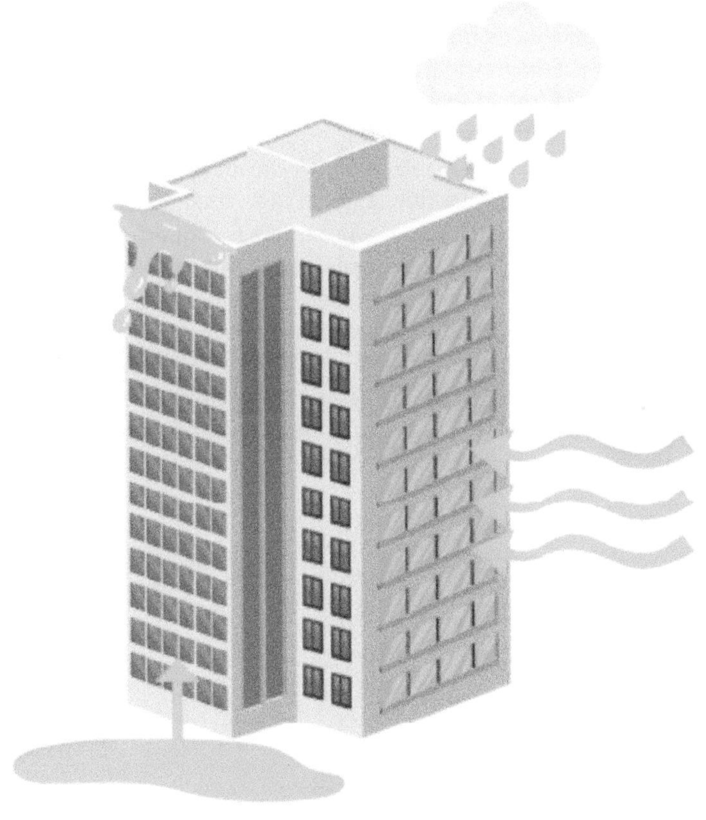

Water gets into buildings through (1) bulk pathways, (2) capillarity, (3) being carried as vapor through air, and (4) vapor diffusion. Bulk water has by far more volume than the other three, and capillarity is second—and so on down the line. Source: Adapted from https://www.greenbuildingadvisor.com/article/four-ways-that-water-gets-into-buildings.

Thermal bridging occurs when materials with different thermal conductivities come into direct contact, creating a pathway for heat to flow across the building's envelope, circumventing the insulation. This not only compromises the insulation's effectiveness but also leads to increased energy costs and reduced comfort for occupants. Think of thermal bridges as leaks in a ship's hull; just as water will find its way through any breach, so too will heat escape through these thermal shortcuts. To combat this, architects and builders must employ meticulous design and construction practices. This may include the use of thermal break materials that interrupt the path of heat flow or the strategic placement of insulation to cover potential bridging spots, ensuring the continuity of the thermal barrier.

Ventilation is another critical aspect of building science, serving the dual purpose of maintaining indoor air quality and regulating internal temperatures. However, the methods employed can swing the pendulum of energy efficiency widely. Natural ventilation harnesses environmental conditions to circulate air within the building, but its effectiveness is subject to the whims of the weather and may not always be reliable. Mechanical ventilation systems offer more control and consistency, yet they consume energy and can be costly to operate. The balance lies in choosing or designing a ventilation strategy that provides the best of both worlds: adequate fresh air intake with minimal energy expenditure. The implementation of energy recovery ventilation systems can recover the thermal energy from exhaust air, reducing the energy burden of heating or cooling the incoming fresh air.

In the broader context of decarbonization, the strategies become even more multifaceted. Operational carbon, the emissions produced from the energy used to heat, cool, and power buildings, can be significantly reduced by optimizing the efficiency of heating, ventilation, and air conditioning (HVAC) systems. The adoption of electric heat pumps is a prime example, capable of providing heating and cooling in a more sustainable manner compared to traditional fossil fuel-based systems. On-site renewable energy generation, such as solar photovoltaic panels or wind turbines, can offset a building's energy demand from the grid, further cutting operational carbon emissions.

Embodied carbon, on the other hand, encompasses the emissions from the entire lifecycle of building materials—from production to transportation to installation and disposal. Selecting materials with lower embodied carbon, such as sustainably sourced wood, low carbon concrete, or recycled content products, can make a substantial difference. Moreover, the design itself plays a pivotal role: optimizing the building's form, orientation, and envelope for passive solar heating, natural ventilation, and daylighting can substantially reduce the need for active systems and the carbon footprint associated with them.

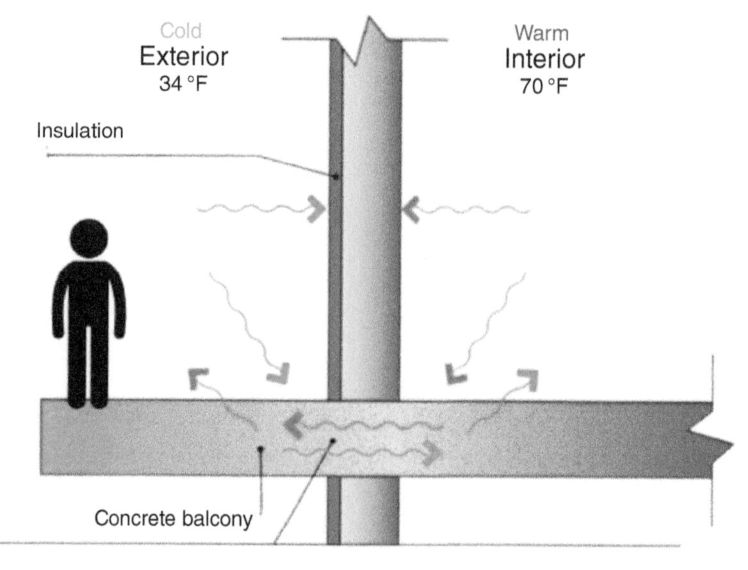

Importance of thermal bridging. Source: Adapted from Nogap / https://www.nogapinsulation.com.au/what-is-thermal-bridging-and-why-is-it-important/.

When these building science principles are synergized with decarbonization strategies, a transformative shift in the built environment is within reach. Architects, engineers, and builders, armed with a scientific and systematic approach, are well-positioned to drive this change, crafting buildings that not only minimize carbon emissions but also enhance performance and occupant well-being throughout the building's life.

As we forge ahead, it is essential to delve deeper into the specifics of insulation, thermal mass, and daylighting. These elements serve as critical instruments in our decarbonization orchestra, each playing a distinct role in the harmonious performance of sustainable building design. Let us turn our attention now to these components, understanding their individual and collective contributions to energy-efficient and low-carbon building strategies.

Studying the impact of building design during the early design phases.

6.1.2 The Principles of Heat Transfer, Insulation, and Thermal Mass

Building on our understanding of thermal bridging and ventilation, we shift focus to the role of insulation, thermal mass, and daylighting in building science and their contribution to decarbonizing buildings.

Insulation is a critical component of building design, pivotal in minimizing energy loss and, by extension, the carbon footprint of a building. It serves as a protective layer that slows down the unwanted transfer of heat through the building envelope—be it resisting the cold during the winter or the heat during the summer. The effectiveness of insulation is measured in *R*-values in the United States and in *U*-values in the rest of the world, with higher *R*-values and lower *U*-values indicating greater resistance to heat flow. Proper insulation leads to reduced energy demands for heating and cooling systems, which translates to lower operational carbon emissions. This conservation of energy is not only environmentally responsible but also economically beneficial, as it lowers utility costs over the building's lifespan.

Insulation must be seen not just as a material that fills gaps and wraps buildings, but as a strategic investment in energy conservation and long-term sustainability. By carefully selecting insulation materials and considering their placement, designers can ensure that every square inch contributes to a building's thermal efficiency. The choice of insulation—be it mineral wool, batt, or panel—must be informed by an understanding of the local climate, the building's specific thermal needs, lifecycle, human health, embodied carbon, and the environmental impact of the insulation material itself. This practice goes beyond merely meeting code requirements. It is also about anticipating future energy scenarios and ensuring that the building can stand resilient in the face of rising energy costs and stricter environmental regulations.

Thermal mass, typically embodied in materials like concrete or brick, acts as a thermal battery within the building structure, absorbing and storing heat energy and releasing it over time. It helps in stabilizing indoor temperatures by reducing the peaks and troughs associated with external temperature fluctuations, leading to a more comfortable indoor environment. When used thoughtfully, thermal mass can reduce the reliance on mechanical heating and cooling systems, contributing to a building's energy efficiency and carbon reduction. However, it is a double-edged sword; without proper insulation, thermal mass can become a conduit for energy loss and become very uncomfortable. Hence, it is crucial to strike the right balance between insulation and thermal mass to harness its full potential for passive temperature regulation.

Daylighting leverages natural light to illuminate building interiors, reducing the need for artificial lighting and the energy it consumes. The art and science of daylighting involve the strategic placement of windows and reflective surfaces to maximize the penetration of sunlight while avoiding glare and excessive heat gain. It is a delicate balance that requires a deep understanding of the building's site, climate, and function. Effective daylighting can significantly reduce a building's operational carbon footprint by cutting down on electricity use during peak daylight hours. Moreover, access to natural light has been shown to improve occupant well-being and productivity, showcasing the multifaceted benefits of this strategy.

Each of these components—insulation, thermal mass, and daylighting—plays a unique role in the decarbonization narrative of buildings. They are not standalone solutions but interdependent elements that need to be carefully coordinated in the building design process. By integrating these strategies with a holistic understanding of building science, professionals in the field can create structures that are not only more energy-efficient and sustainable but also places where people thrive. In the following sections, we will delve deeper into these strategies, examining their practical application and the tangible benefits they offer in the quest for a decarbonized built environment.

6.1.3 Explaining the Science of Daylighting and its Energy Benefits

Daylighting stands as a vital part of sustainable building design, offering a harmonious blend of aesthetic enhancement and energy efficiency. This delicate balance is achieved through a science that understands and applies the principles of views, photometry, and thermal physics, all aimed at harnessing natural sunlight to illuminate indoor spaces. By reducing the dependence on artificial lighting, buildings can see a significant decrease in electricity usage and, subsequently, a reduction in operational carbon emissions.

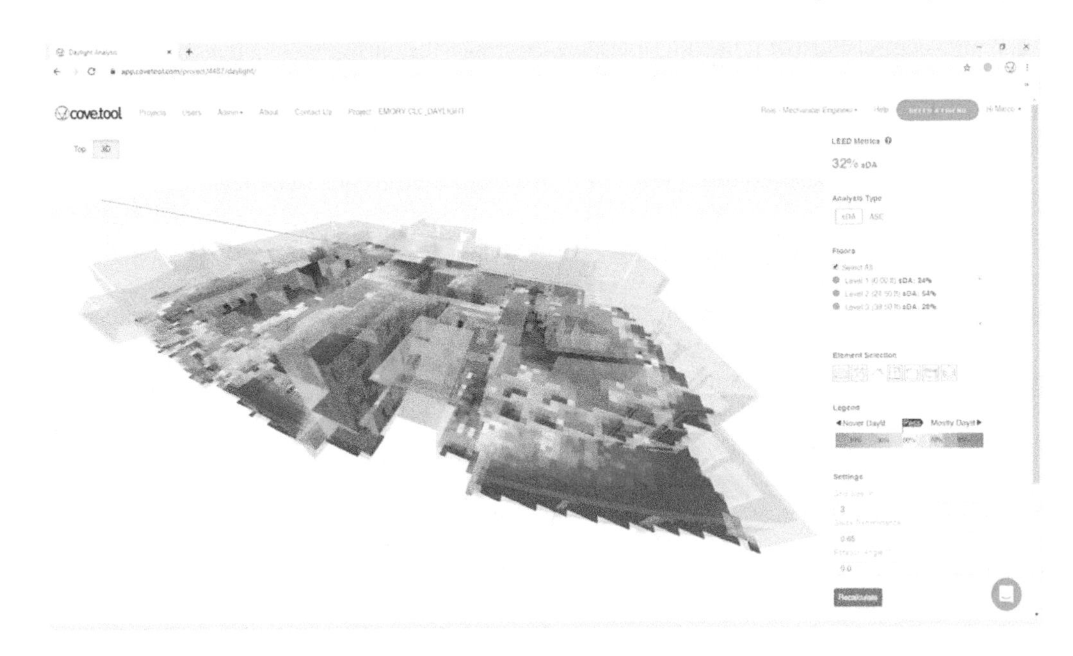

cove.tool's daylight simulation method follows the guidelines required of the BD+C LEED v4.0 IEQ.

The journey of daylight into a building is organized through carefully placed apertures—windows, skylights, and other openings that invite light in. The quality and quantity of this natural illumination hinge on a multitude of factors: the size and orientation of these apertures, external shading devices, the building's geographical positioning, and its internal configuration. To optimize daylight while mitigating unwanted heat gain, design strategies employ larger windows facing the equator or southern directions, complemented by shading solutions like overhangs and louvers that temper direct sunlight and reduce glare, enhancing the visual comfort of occupants.

Interior finishes also play a pivotal role in daylighting. Light-colored ceilings and surfaces act as canvases, reflecting and diffusing light deeper into the building. The strategic use of reflective materials and the placement of windows relative to the interior work planes are essential, ensuring that illuminance levels are tailored to both the task at hand and the well-being of occupants. Target illuminance levels, typically ranging from 250 to 750 lux for general office tasks, allow for an environment that is conducive to both productivity and energy conservation.

Yet, the bounty of daylight comes with a caveat—excessive exposure can lead to overheating. The science of daylighting, therefore, involves a careful calibration of the glazing area relative to the floor space, with glazing technologies such as low-emissivity coatings stepping in to filter out infrared radiation while welcoming visible light. Overhangs and light shelves are not mere architectural features but critical elements that modulate thermal gains and enhance comfort.

Automation brings daylighting into the realm of smart design. Photosensors linked to electric lighting systems and motorized shades provide dynamic adjustments to the indoor lighting environment, responding in real time to the sun's intensity and the building's illumination needs. Occupancy sensors and dimming controls further refine this interplay, creating a responsive system that maximizes natural light while minimizing energy use.

To encapsulate the essence of effective daylighting, one must consider the synergy of three essential elements: (i) the geometrical precision that maximizes natural light penetration, (ii) the interior layout that fuses light with thermal comfort, and (iii) the automated controls that fine-tune the balance between natural and artificial light. When these components are integrated thoughtfully, the building not only reaps energy benefits but also becomes a space where access to natural light supports the circadian rhythms and well-being of its occupants.

Daylighting extends beyond the technicalities of window placement and light reflectance. It embodies a design philosophy that views light as a vital, shaping force within a building. The interplay of light and shadow, the modulation of brightness and contrast, and the color of light at different times of day—all contribute to the experience of the space. Daylighting design must balance these aesthetic considerations with energy savings, ensuring that spaces are not only well-lit but also energy-efficient. Advanced glazing technologies, automated shading systems, and integrated lighting controls can all enhance the effectiveness of daylighting strategies, leading to buildings that adapt to their occupants' needs and the rhythms of the natural world.

By viewing insulation, thermal mass, and daylighting not as isolated elements but as part of a comprehensive approach to building design, we lay the foundation for truly sustainable architecture. This approach respects the complex interdependencies of the built environment and harnesses the full spectrum of building science to create spaces that are both environmentally responsible and deeply human-centric. As we continue to confront the challenges of climate change, the principles of building science provide a roadmap for creating buildings that not only minimize carbon emissions but also enhance the quality of life for their occupants.

6.2 Site Selection and Building Orientation

6.2.1 How Site Selection and Building Orientation can Significantly Affect a Building's Energy Consumption

The art of placing a building on a site and orienting it precisely is akin to setting a sail to the wind; it requires an intimate understanding of the local environment to harness its natural energies effectively. The decision of where to plant the foundations of a building can profoundly influence its energy consumption and efficiency. Such site selection must be undertaken with a discerning eye toward the area's microclimate conditions, which include the lay of the land, native vegetation, proximity to water, and potential obstructions—all factors that shape the ambient temperature, airflow, and light penetration.

Microclimate conditions offer a palette of natural potential. A site's topography can create pockets of shade or channels for breezes, significantly influencing the thermal comfort within a building. Likewise, the presence of water bodies can modulate the local climate, offering cooling effects. Prior to breaking ground, these characteristics can help forecast the energy demands of the future structure, signaling whether the site is predisposed to passive solar heating or natural cooling.

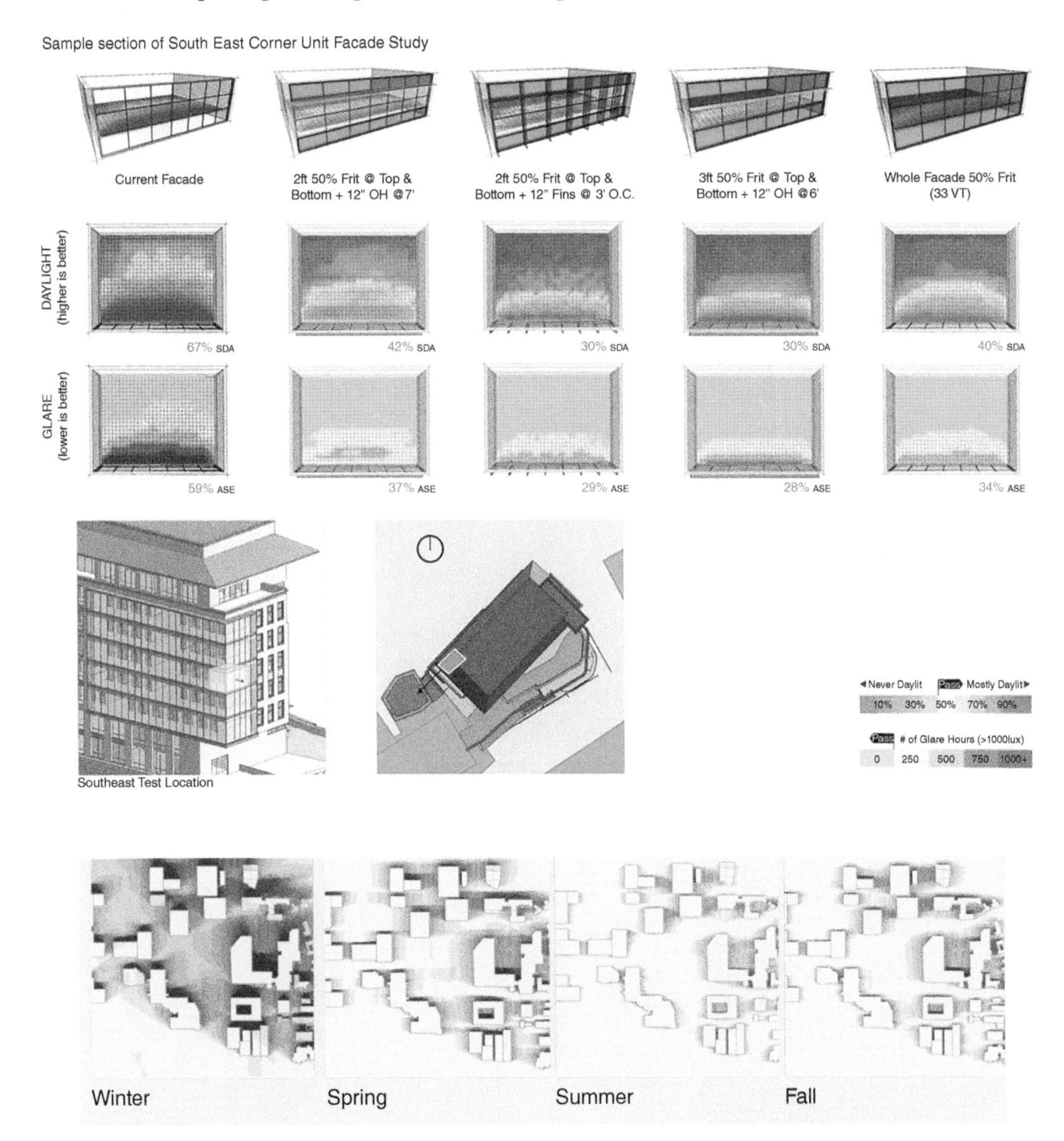

The orientation of a building relative to the sun and wind dictates its interaction with the elements. In temperate zones, positioning a building's longer axis along an east-west line, with a broad side facing south, allows it to soak up the winter sun, reducing the need for artificial heating. Conversely, minimizing exposure to the intense summer sun from the south can mitigate the risk of overheating. This strategic orientation takes advantage of the sun's arc, using it as a natural thermostat to warm spaces in the cooler months and keep them temperate when it is warmer.

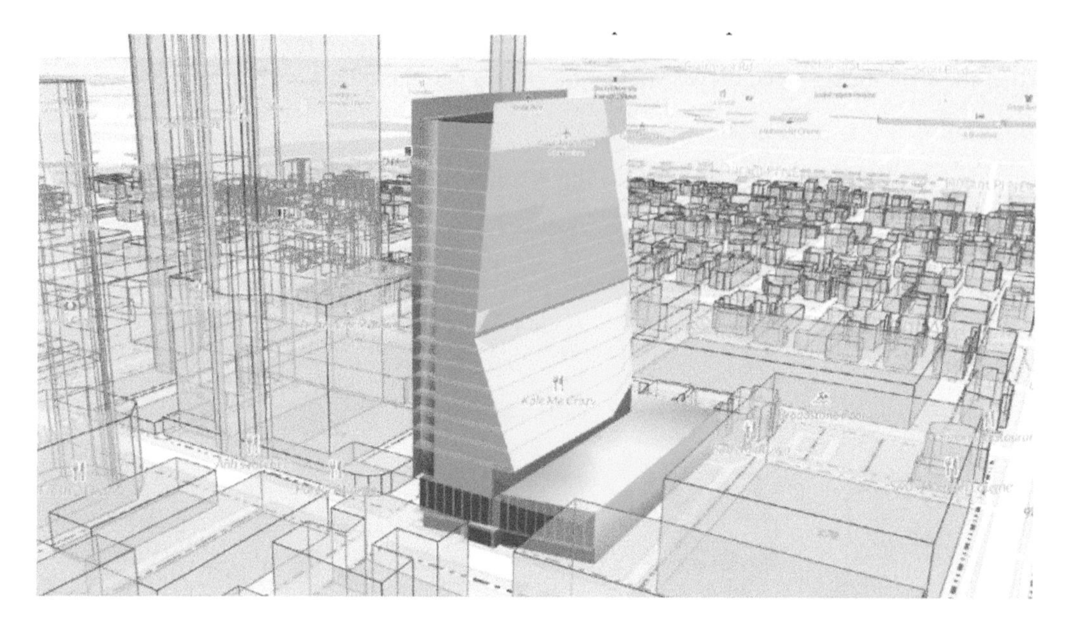

Hours of Sun study of a highrise office building in cove.tool.

The building envelope—comprising its facades, roof, and foundation—is the mediator between the indoor and outdoor environments. By using high-performance materials that insulate and breathe, the envelope can become a sophisticated boundary that minimizes unwanted heat exchange. High-tech glazing, robust insulation, and thoughtfully designed shading are just a few tools that can be employed to bolster the envelope's defensive capabilities against energy loss.

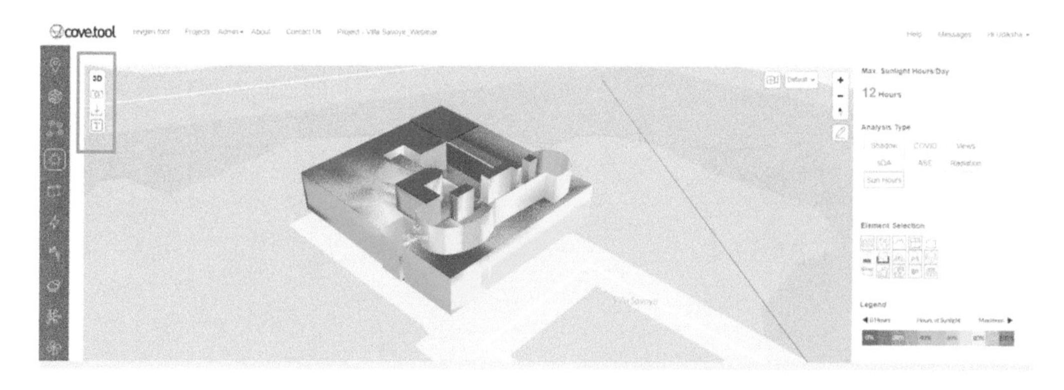

Radiation study of a low rise K-12 project in cove.tool. Source: Cove Tool, Inc.

Designing with a sensitivity to the site's intrinsic factors can have the most profound impact during the early stages. Analyzing the microclimate, prevailing winds, and solar access, and then marrying these findings with an intentional orientation and envelope design, can yield a building that leans into the natural energy flows of its environment. By doing so, we invite the sun, wind, and light to serve not just as passive contributors to the building's performance but as active players in reducing energy consumption.

6.2.2 How to Optimize Orientation for Sun Exposure and Wind Direction to Facilitate Natural Ventilation and Daylighting

Optimizing for sun exposure and wind direction is not a new concept in architecture, but its applications in modern design are ever evolving. To make the most of these natural resources, we must take a comprehensive approach:

For sun exposure, we start with an analysis of the site's microclimate, including its topography and surrounding features. The goal is to maximize solar heat gains in winter, warming the building naturally, while avoiding excess heat in the summer. In the northern hemisphere, southern exposures should be wide to welcome diffused light and warmth, while northern exposures should be minimized to reduce heat loss. The building's shape also matters; compact forms reduce surface area and heat transfer, while taller, narrower shapes can enhance natural ventilation through the stack effect.

Harnessing the power of the sun begins at the foundational level—selecting the right site. This initial step is a thorough examination of the microclimate, including the local topography, vegetation, and nearby water bodies, all of which play a part in the building's future relationship with the sun. These elements shape the site's temperature profile, influence wind patterns, and determine the availability of daylight, affecting the potential for natural ventilation and lighting. Careful site analysis can uncover opportunities to leverage these natural resources for energy savings and comfort.

Orienting a building is a deliberate act, aligning it with the celestial dance of the sun across the sky. By positioning the building's long axis on an east-west trajectory and directing the majority of the window glazing and roof surface toward the south, architects capture the low winter sun to warm the building, while shielding it from the higher, more intense summer sun from the north. This intentional alignment optimizes the building's thermal performance, harnessing natural warmth when needed and providing shade when it is not.

Southern exposures are a canvas upon which the sun paints with a light that is both bright and diffuse, bathing spaces in even illumination exposures not only light up interiors but also serve a dual function in the colder months, preheating ventilation air before it circulates within, contributing to a comfortable indoor climate while reducing the demand on heating systems.

The eastern and western facades of a building, embraced by the sun at dawn and dusk, are prime real estate for windows that invite daylight to stretch deep into the building for most of the day. These same openings are strategic points for capturing cross breezes that enhance natural ventilation, reducing the reliance on mechanical systems for fresh air.

Northern exposures in the northern hemisphere are more reserved in their interaction with the sun. They receive a less direct kiss from the sun's rays, making them ideal for minimizing glazing to reduce heat loss. These facades can also play a role in natural ventilation strategies, where exhaust openings can be placed to facilitate the escape of warm air, contributing to efficient stack ventilation.

The shape of a building is more than an aesthetic statement—it is a thermodynamic decision. Compact forms minimize the building's surface area, thereby reducing heat exchange with the outside environment. Tall, slender designs can enhance the stack effect, where warm air rises and draws in cooler air at the base, driving natural ventilation through the structure.

The building envelope is the barrier and the gateway between the indoors and the outdoors, requiring a thoughtful approach to its design. The size, type, and placement of fenestrations—windows, doors, and other openings—must be carefully calibrated to admit the right amount of daylight while managing solar heat gains and preventing the loss of conditioned air. High-quality insulation, responsive shading devices, and materials that can breathe naturally support this delicate balance.

Incorporating elements like sawtooth roofs, clerestories, and monitor roofs into the building's design can dramatically enhance the penetration and diffusion of daylight. These features, alongside green roofs and strategically placed solar panels, can provide additional shading, contributing to a building's passive climate control strategies.

Turning to wind exposure, understanding the prevailing winds at a site is like knowing the currents of a river before setting sail. Architects and planners must discern the dominant wind patterns and orient the building's openings and forms to catch these natural currents, providing fresh air and cooling breezes during occupied hours.

Breezeways are not accidental; they are created by positioning buildings to harness the wind, funneling it through the site to ventilate spaces naturally. These can be formed between existing structures or as part of the new design, creating corridors of air that flow through the building's openings.

Windbreaks, such as trees or constructed barriers, can temper the force of the winds, redirecting and modulating them to improve the airflow around the building. These features act as shields, but they can also be shaped to channel breezes into the building, serving both aesthetic and functional roles.

The building's form, when designed with the wind in mind, becomes aerodynamic. Streamlined, narrow shapes that align with the prevailing wind patterns create less resistance and turbulence, allowing the wind to glide past and through the building, facilitating airflow and enhancing natural ventilation.

Openings are critical for cross ventilation—they must be placed thoughtfully to harness the wind without creating discomfort. By positioning windows, doors, and louvers on the leeward side, opposite to where the wind is coming from, buildings can draw in breezes and allow air to move freely from one side to the other.

Features like wind towers, which can be as decorative as they are functional, capture and direct breezes into the building, providing cooling and ventilation. These traditional elements can be modernized and integrated into new designs, proving that old wisdom can complement new technologies.

The stack effect is an age-old principle that remains relevant in modern building science. By creating vertical spaces within the building, such as atria or lightwells, architects can set the stage for warm air to rise and draw in cooler air from below, fostering a natural ventilation cycle.

Facade permeability is a key factor in a building's ability to breathe. By using materials and construction methods that allow for the walls to have a degree of openness, especially on the windward side where the wind first makes contact, buildings can become more harmonious with their environment, allowing for natural temperature regulation.

Seasonal variations in wind patterns are a reality that designers must contend with. A building that is adaptable, with adjustable shades and ventilation options, can respond to these changes, maintaining comfort throughout the year.

Finally, the design process is iterative and should be informed by empirical evidence. Wind tunnel testing and computational fluid dynamics simulations offer valuable insights that can refine wind-assisted ventilation strategies, ensuring that the final design is not just theoretically sound, but practically effective.

Operable windows when designed well can be very effective for energy use reduction and indoor comfort. Source: as/Adobe Stock.

Each element—from microclimate considerations to the shape of the roof—interacts with others, creating a complex web of design decisions that collectively impact a building's energy performance. Documenting and tracking the performance metrics such as temperatures, airflow, and lighting levels can provide valuable feedback, enabling a continuous process of optimization and improvement in the pursuit of energy-efficient, harmoniously designed buildings.

6.3 Passive Design Strategies: Harnessing the Power of Nature

6.3.1 Introducing the Concept of Passive Design Strategies

Passive design strategies embody the essence of sustainable architecture, invoking the age-old wisdom of working with the natural environment to foster comfortable indoor conditions. The aim is to minimize reliance on mechanical systems such as air conditioning, artificial heating, and electric lighting, by capitalizing on the free and abundant resources of sunlight, wind, and thermal stability offered by the Earth itself. The first step in passive design is a conscientious site analysis—embracing the inherent characteristics of the environment to establish a symbiotic relationship with the natural forces at play. When harnessed efficiently through innovative yet often simple measures, these strategies can significantly reduce a building's energy consumption, lower operational costs, and enhance the well-being of its occupants.

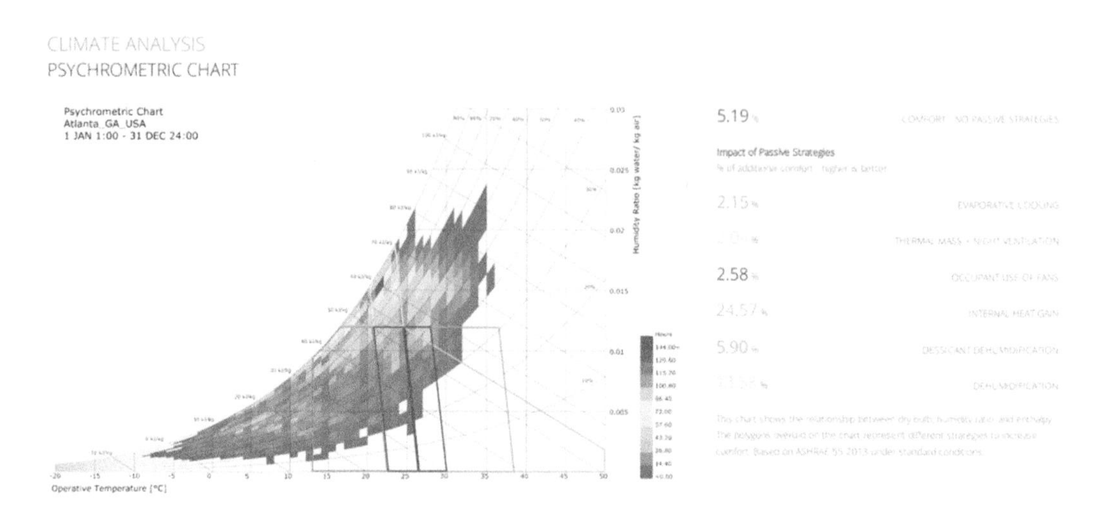

Passive strategy, or passive design strategies, are building design methods using ambient energy sources to cool, heat, shade, and or ventilate a space. In place of mechanical support systems, or purchased energy, passive strategies are used to keep buildings comfortable and thermally excel in climates where design can out-performs technology.

At its core, passive design sings a harmony with the environment. It is about dialing down the hum of machines and amplifying nature's whispers of wind, warmth, and light. It is crafting spaces that sip energy lightly, reduce emissions, and cradle health and productivity in their sunlit arms. Strategic site choice, savvy building orientation—these initial moves dictate a building's rapport with its surroundings. With a design that rises from the earth, aligned with the climate's cadence, the result is architectural poetry in motion—a building at one with the ecosystem.

As outlined in the previous chapter, site selection and orientation are critical first steps, establishing a building's responsiveness to environmental resources. Carefully choosing locations with optimal microclimates characterized by breeze channels, shade, and solar access lays the groundwork for passive strategies. By incorporating them at a building's outset, optimized from the ground up for natural climatic inputs rather than technological outputs, passive design offers architecturally elegant yet highly effective sustainable solutions centered on human well-being and ecological wisdom. When done right, the result is a building that appears to operate effortlessly, in perfect harmony with nature.

6.3.2 The Different Types of Passive Strategies and Their Benefits

With definitions being set, now we will look at the details of specific passive strategies and systems that minimize energy demands by working with natural forces. The key types of passive techniques will offer details on their technical procedures and benefits. After carefully selecting a site and setting up the orientation of the building, which belong to the factors outside of the building, we will now look at what can be optimized on and within the building in order to be passively "in tune" with the environmental setup.

6.3.3 Building Envelope

The building envelope represents the critical interface between the controlled indoor environment and the variable conditions outdoors. It is the building's first line of defense against climate fluctuations and a key player in its energy performance narrative. The design of this envelope encompasses the selection of materials, the assembly of structural components, and the integration of openings, all engineered to optimize thermal comfort while minimizing energy consumption.

In this section, we will examine the multifaceted role of the building envelope as it balances the act of shielding interiors from extreme weather while harnessing natural forces to maintain a comfortable indoor climate. As we do so, we will appreciate its complexity as more than a mere barrier but as a dynamic, responsive system that contributes to a building's efficiency, cost-effectiveness, and occupant well-being. Each aspect, from the walls and roofs to the fenestration and insulation, works in concert to achieve the dual objectives of sustainability and livability.

Glazing Elements: Windows are not merely transparent openings but complex systems engineered to minimize thermal transfer while maximizing daylight. The application of double or triple glazing, equipped with low-e coatings and filled with inert gases, is a technological innovation that curtails heat loss and gain. They admit light beneficially, without the accompanying unwanted heat flow. Complementing these are exterior shading devices—including overhangs and shutters—which tactically obstruct excessive solar heat in summer, curtailing the need for energy-intensive cooling, and admit low-angle winter sunlight to naturally warm interiors, reducing heating demand.

Fins and overhangs. Source: Duda Paine Architects.

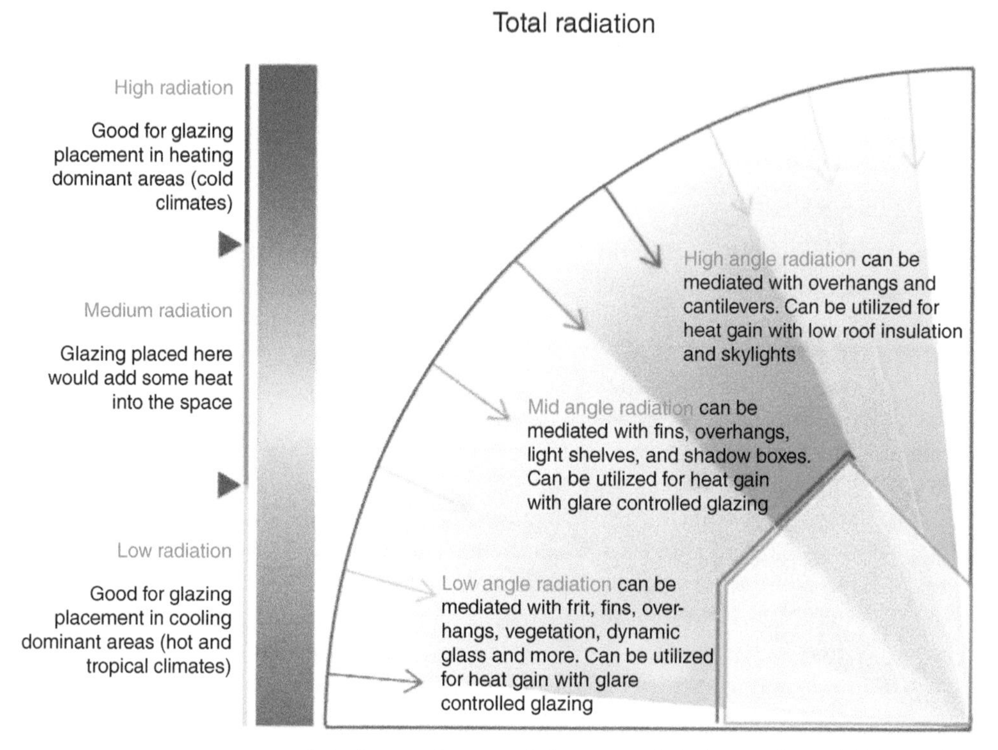

This diagram shows a radiation map projected onto a dome which is used to determine the best placement for glazing and possible radiation mediation strategies. Use the color, angle, and direction of radiation to identify what strategy you can use to optimize thermal comfort and ideal placement

Insulation and thermal mass materials: The role of insulation is to thwart heat transfer through the building's skin. Using materials with high R-values and low U-values provides substantial thermal resistance, diminishing heat loss and gain. Options such as spray foam or rigid panels offer superior insulation compared to traditional materials like fiberglass batts. In parallel, thermal mass materials, such as concrete or brick, absorb and store solar heat during the day, releasing it gradually, thereby stabilizing interior temperatures and lessening the need for mechanical temperature control.

Insulation examples.

Ventilated facades: They provide a highly effective passive strategy for convectively cooling the building's envelope and reducing heat gains. Double-skin facade systems with an air gap between the outer cladding and the main wall structure allow natural airflow through this channel. Operable vents located at the top and bottom admit cross breezes that flow through the air gap, cooling the facade by convection. The airflow is driven by positive pressures from phenomena like the stack effect arising from temperature differences and wind pressure from prevailing breezes. As warmer air in the gap rises and exits through higher vents, cooler outside air is drawn in through lower vents, inducing cross ventilation through the facade air gap. The natural ventilation of the facade minimizes conductive and convective heat gains, reducing overheating risks and cooling demands from the building interior. Ventilated facades thus harness environmental resources through natural ventilation and cross breezes rather than mechanical conditioning to keep the building envelope cool.

Permeable building fabrics: materials with small, interconnected pores and passages that allow air to permeate the building envelope provide an additional pathway for passive natural ventilation and cooling. Permeable building fabrics like rammed earth, adobe, straw bale, and other natural materials constructed with a porous cellular structure promote airflow infiltration that circulates within wall cavities and assemblies. Tiny gaps and holes permeating these permeable envelopes enable outside air to infiltrate indoors, replacing warm interior air that exits through higher openings. The internal convection circulation driven by temperature and pressure differences—known as infiltration ventilation—supports passive space cooling, offering an additional natural ventilation strategy that helps reduce mechanical cooling loads. Integrating permeable envelope materials as part of a holistic passive design can improve a building's overall "breathability" where natural airflow enters, circulates within and exits the structure to help regulate temperatures.

Various renewable energy technologies: They can be seamlessly integrated into the building envelope to help offset a portion of the structure's mechanical loads. Building-integrated photovoltaics (BIPVs) such as solar panels and thin-film solar cells embedded directly into cladding materials like roof tiles or facades generate clean electricity that displaces a percentage of the building's power consumption. Meanwhile, active solar thermal systems, including solar hot water collectors and solar air heaters, incorporated into the envelope capture heat for uses like domestic hot water or space heating, reducing mechanical demands for those functions. When thoughtfully designed as part of a holistic passive strategy, these integrated technologies work symbiotically with other envelope elements like glazing, shading, insulation, and natural ventilation. Together, they aim to minimize a structure's total mechanical energy demands through a mix of passive design responses and renewable energy generation at the building's own envelope scale.

Vertical shot of big panels of solar batteries along the high building's wall in the city. Source: EKH-Pictures/Adobe Stock.

Ventilation

Effective ventilation is not merely a component of design but the very breath of a building, ensuring comfort and enhancing air quality with minimal environmental impact. Natural ventilation is a key strategy that exploits environmental conditions to cool and circulate air within buildings. From the simplicity of cross ventilation to the calculated design of wind-induced systems, and the natural physics behind stack and buoyancy ventilation, we will dissect these passive cooling strategies that are essential in creating energy-efficient, high-performing buildings. These methodologies showcase how intelligent design can align with nature's forces, forging a sustainable path forward in modern construction.

Passive Cooling

- *Cross ventilation*—It is a highly effective passive cooling strategy for ventilating buildings naturally. It occurs when prevailing winds are captured through openings oriented to admit cross breezes that move through and across interior spaces. To harness cross winds, windows, vents and louvers are typically located on the leeward, or downwind, side of a building relative to the direction from which breezes are blowing. The building's overall form can also be optimized to minimize air turbulence and resistance to cross winds as they pass through, further facilitating natural ventilation. Cross ventilation demonstrates a simple yet powerful passive design technique for ventilating buildings and improving occupant comfort using only natural wind forces.

- *Stack ventilation*—It harnesses the natural occurrence of stack effect to drive convective air currents within and through buildings. Stack effect occurs as warm air, being less dense, rises and exits a building through higher openings while cooler, denser air enters at lower levels, creating vertical convection currents. To maximize this stack effect, vertical shafts, atria, and courtyards are often integrated into building designs. They utilize the vertical distance and differences in air density as warm air rises upward through tall interior volumes. Taller interior spaces and larger vertical openings within a building likewise raise the pressure differentials that draw in cross breezes and naturally ventilate interior environments.
- *Wind-induced ventilation systems*—They harness prevailing winds to naturally ventilate buildings through the use of features that capture and funnel air into structures. Wind towers, wind scoops, and wind catchers—vertical funnels or ducts oriented to face the direction of prevailing breezes—capture incoming winds and channel that air flow downwards. As wind-driven air moves through these funnels and ducts, it induces vertical and horizontal air movement within buildings. Automated vent systems that incorporate adjustable dampers can further optimize airflow regulation based on changing climate conditions, maximizing ventilation effectiveness.

Ceiling metal ceilings with a dome for ventilation and removal of ammonia on a modern cow farm. Source: 99paginas/Freepik.

- *Buoyancy ventilation*—It relies on differences in air density and temperature between interior and exterior environments to naturally ventilate buildings. The chimney effect leverages the tendency for hot air, being less dense, to rise and exit a structure through higher openings while denser, cooler air is drawn in at lower levels. Vents, chimneys, and flues that exhaust warm air from the top of interior spaces help create the negative pressure that draws in fresh, cooler air from outside. As heated air rises and exits through these vertical vents, it sets up convection currents that continuously cycle fresh air into buildings. The ability of high vents to exhaust warm air, aided by the upward movement of heated air due to its lower density, demonstrates the basic mechanics behind buoyancy ventilation. By simply incorporating higher vents or chimneys within building designs, structures can harness differences in air density driven by temperature variances to passively induce air movement and refresh interior environments without mechanical assistance.

Several additional considerations are important for maximizing the effectiveness of natural ventilation strategies. Seasonal variations in wind patterns and outdoor temperatures require design flexibility and adaptive strategies within natural ventilation systems. Integrating features like shading devices, evaporative cooling towers, and ground coupled heat exchangers can help maximize temperature differentials between interior and exterior air, further enhancing buoyancy-driven air currents. Computer simulations and physical scale models can evaluate and optimize natural ventilation proposals before finalizing building designs, identifying potential issues, and enhancing performance.

Daylighting

Daylighting transcends mere illumination; it represents a harmonious blend of functionality and sustainability within architectural design, aimed at optimizing natural light's penetration and its ensuing benefits. It is about finding multifaceted strategies that leverage the sun's free energy and light to enhance indoor environments while curbing reliance on artificial lighting and cooling. The interplay between building orientation, fenestration, shading devices, and spatial planning are integral to mastering daylighting's delicate balance of light and shadow. These design elements serve not only to illuminate but to breathe life into spaces, creating interiors that are not only visually comfortable but also energy efficient, encapsulating the very essence of sustainable design innovation.

- *A building's orientation and placement of fenestration*—They are critical design factors for maximizing natural daylight admission while minimizing overheating risks. Proper orientation of facades and windows involves positioning based on a location's solar path diagrams and seasonal variations in sun angles. In northern latitudes, orienting larger windows and openings toward the south typically admits the most winter daylight while avoiding higher summer sun angles that pose overheating risks. The lower sun angles from the south in winter months allow sunlight to penetrate deep into interior spaces, improving illumination levels without requiring much active shading. Meanwhile, the higher northern sun angles in summer tend to overheat interiors unless designed to exclude or mitigate solar gains.
- *The placement and sizing of windows and skylights*—Within a building layout, they are critical design factors for adequately illuminating interior spaces with natural daylight. Windows are strategically located based on an interior's daylight zones, with larger fenestrations provided in primary living and working areas to maximize illumination levels. Window dimensions likewise consider solar altitude angles at different times of day and year in order to optimize the penetration depth of daylight into

interior spaces. By accounting for changes in the sun's path and angle of incidence, window apertures can be sized to maximize daylight admission during core hours of occupation when illuminance levels are most needed. Horizontal band windows oriented toward the best exposures often provide better penetration of daylight deep into floorplans, while narrower and taller windows near building perimeters maximize privacy.

- *External shading devices*—They add to the benefits of natural daylight while minimizing unwanted solar gains and overheating. Fixed shading elements like overhangs, louvers, light shelves, and exterior fins are positioned to block direct sunlight during peak hours when it could cause excessive heat gain and glare, while still transmitting useful diffuse daylight into interior spaces. Operable and adjustable external shading features offer even greater flexibility, allowing control of solar penetration based on seasonal and diurnal variations in the sun's path and intensity. Thoughtfully designed exterior shades optimize solar access for daylighting while preventing glare and overheating that compromise occupant comfort and building performance. The key lies in differentiating between useful diffuse daylight and undesirable direct solar radiation, and shaping the building facade accordingly.
- *Interior layout and space planning*—A building's interior layout and space planning should strategically locate key functional areas to maximize access to natural daylight. Floor plans ideally center workspaces, congregate areas and primary activity zones under optimal daylight zones closest to exterior fenestration where the most illumination is available. Perimeter zones adjacent to windows and skylights typically receive the highest illumination levels from natural daylight, while deeper core zones located farther from the building skin require artificial lighting to meet target lux levels.
- *Light wells and clerestory windows*—They provide strategic means of introducing natural daylight into core zones and deep plan depths that typically receive little exterior illumination. Vertical light wells and light shafts funnel daylight down into interior spaces located further from perimeter zones, illuminating core areas that would otherwise require artificial lighting. Similarly, higher clerestory windows admit diffuse top lighting into deep plan spaces, providing more uniform illumination to interior zones distant from exterior walls. The vertical shafts and higher windows act as passive "light pipes" that deliver daylight deeper into floor plans, minimizing the footprint of artificially lit zones. Often these strategies can be a source of energy loss if the design team is not careful to pay attention to the U-value and SHGC of the glass in the skylight or monitor. There is often also a glare penalty, so proper shading and orientation is critical to the success of this strategy. Together, light wells and clerestory windows demonstrate design strategies for improving daylight penetration within compact or opaque floor plans that inhibit natural illumination of deep interior spaces through conventional horizontal fenestration alone.
- *Rooftop daylighting systems*—They provide an overhead approach for introducing natural light into interior spaces. Features like sawtooth rooflines, light scoops, and light tunnels channel exterior daylight down into floor plans by capturing light at high angles and redirecting it vertically. Flat or sloped skylights and roof monitors likewise maximize ceiling heights to diffuse downward lighting across wide areas. By collecting light where it is most available and redirecting it downward, rooftop daylighting geometries offer a useful complement to conventional horizontal fenestration strategies for maximizing natural illumination throughout interiors. Be careful to consider the constraints similar to skylights.

- *Integrated technologies*—They can help optimize the performance of all mentioned natural daylighting strategies within building designs. Daylight-responsive dimming systems offer an automated means of fine-tuning artificial lighting levels based on the amount of natural light available. Photo sensors mounted in key locations detect changes in illuminance levels from daylight admission and signal dimmable electric light fixtures to raise or lower output accordingly. Dimming systems are a complementary technology that supports passive daylighting strategies. As natural illuminance levels rise from clear skies or oriented fenestration, electric light output is reduced to minimize energy use while maintaining overall target lux levels. Conversely, as clouds block direct sunlight or daylight diminishes later in the day, artificial light levels are increased to compensate. Integrated together, natural and electric illumination sources are balanced in real time to optimize lighting quality while minimizing energy consumption for artificial lighting.

Overall lighting and facade designs must account for the dynamic range and variability inherent in natural light sources. Static building orientations and fenestration alone are often insufficient without complementary shading strategies, diffusion techniques, and lighting controls that adjust for continually changing daylight qualities. Therefore, the ideal approach is to combine passive daylighting approaches and active lighting technologies to provide a balanced lighting solution that harnesses the benefits of natural illumination while avoiding visual and thermal discomfort. A thorough consideration of natural light's variabilities and potential drawbacks, alongside strategic design of shading elements, lighting controls, and material finishes, allows buildings to maximize the productivity gains and energy savings offered by well-executed daylighting strategies.

Roofing

Integrating green roofs into architectural designs not only encapsulates a commitment to aesthetics but also embodies a functional approach to environmental stewardship and resource management. A striking feature of green roofs is their remarkable ability to manage stormwater. According to the Environmental Protection Agency (EPA), green roofs have been found to reduce rainstorm runoffs by an impressive 75%. This is accomplished through the natural process where the plants in our green roof systems absorb rainwater via transpiration and release it through evaporation. Meanwhile, the vegetation and substrates act as biofilters, retaining harmful pollutants that would typically be destined for local waterways or treatment facilities, thus offering a dual benefit of water retention and purification.

Roof of a modern building. Source: Tomas Bazant/Adobe Stock.

Even better, the economic incentives for adopting green roofs are becoming increasingly tangible. Many cities, recognizing the long-term benefits of sustainable design, are offering tax credits and grants. By participating in programs such as LEED and BREEAM, developers and property owners are not only able to capitalize on financial benefits but also contribute positively to social, economic, and environmental systems.

The acoustical benefits of green roofs add another layer of urban comfort. In the bustling heart of city life, where noise is a constant, green roofs can provide a much-needed buffer, reducing noise pollution by up to 15 decibels. This level of sound abatement can make a perceptible difference in the urban soundscape, akin to dampening the sound from a lawn mower to that of a vacuum cleaner.

While green roofs offer a myriad of benefits, their success and longevity hinge on effective maintenance. Especially amidst the evolving challenges of climate change, such as heatwaves and flooding, the care for plant life on green roofs becomes paramount. Maintenance practices must adapt to ensure that the vegetation remains healthy and resilient, capable of withstanding extreme weather events and continuing to perform their critical ecological functions. Regular care, including proper watering, weeding, and monitoring for pests and diseases, ensures that green roofs continue to deliver their full range of benefits, from stormwater management to temperature moderation and biodiversity enhancement.

The importance of maintenance underlines the green roofs' role as dynamic systems that interact continuously with their environment. When carefully designed, implemented, and maintained, green roofs are not just a feature of a building; they are active participants in the building's life and the broader ecosystem, adapting and responding to the changing climate and urban landscape.

6.3.4 Cool Roofs

Cool roofs are a great example of passive cooling technology, championing high solar reflectance and thermal emittance as their primary mechanisms for combating heat absorption. These roofing systems are strategically engineered to reflect a substantial portion of the sun's energy, diverting it away from the building envelope. As a result, surface temperatures can be significantly lower—by as much as 60°F—than those of conventional roofing materials.

This considerable temperature disparity is not merely a superficial benefit; it translates directly into reduced interior temperatures and, subsequently, a diminished reliance on air conditioning systems. By lessening the heat transfer into buildings, cool roofs play a pivotal role in curbing the energy consumption dedicated to cooling interiors, fostering both economic and environmental dividends.

On a macro scale, the collective implementation of cool roofs can act as a countermeasure to the pervasive urban heat island effect. This phenomenon, which leads to elevated urban temperatures, is mitigated as cool roofs reflect more sunlight and emit more heat than traditional roofing, thereby reducing the accumulation of heat in urban areas.

A multitude of materials are available to construct cool roofs, each with unique properties tailored to specific needs. White or light-colored membranes, typically fashioned from thermoplastic olefin (TPO), polyvinyl chloride (PVC), or modified bitumen, serve as the reflective surface that repels solar energy. Meanwhile, materials with high emissivity ensure that any absorbed heat is not retained but released, further amplifying the cooling effect. It is important to balance the embodied carbon and chemical properties of any material, especially roofs, since they typically are replaced every 20–30 years.

The integration of cool roofs with other environmentally considerate technologies, such as solar panels, enhances their efficiency. Solar panels can exploit the expanse of a cool roof to generate clean energy while simultaneously benefiting from the cool surface, which can improve their performance. Similarly, green roofs installed atop roofing membranes can extend their ecological and thermal regulation benefits to the entire roof assembly. A multidimensional strategy that combines shade, evaporation, embodied carbon, energy usage, and reflectivity is vital to maximize efficiency and comfort.

Tool for welding and application of PVC and TPO synthetic membrane. Source: Doralin/Adobe Stock Photos.

6.4 Case Study: High Performance Tower by TVS Architecture and Interior Design

6.4.1 Introduction

Looking beyond conventional wisdom requires using data to prove innovative ideas can work especially in the concept stage. With a building performance strategy led by *Paul McKeever, AIA*, the confidential multistory commercial tower sits in a dense urban neighborhood. Passively cooled with a unique central atrium, the common areas harness the climate avoiding energy use for ventilation, heating, and cooling. The design team used cove.tool to quantify the energy and cost savings. Armed with this data, the team successfully pitched the concept to the owner by justifying the return on investment in an unconventional design.

The first step to justifying the concept was identifying the baseline energy consumption for a comparable building and then setting goals for this project. cove.tool played a decisive role by making that analysis iterative and simple. This enabled the team to quickly test out ideas and fine-tune them, reaching a cumulative savings of 50.2%.

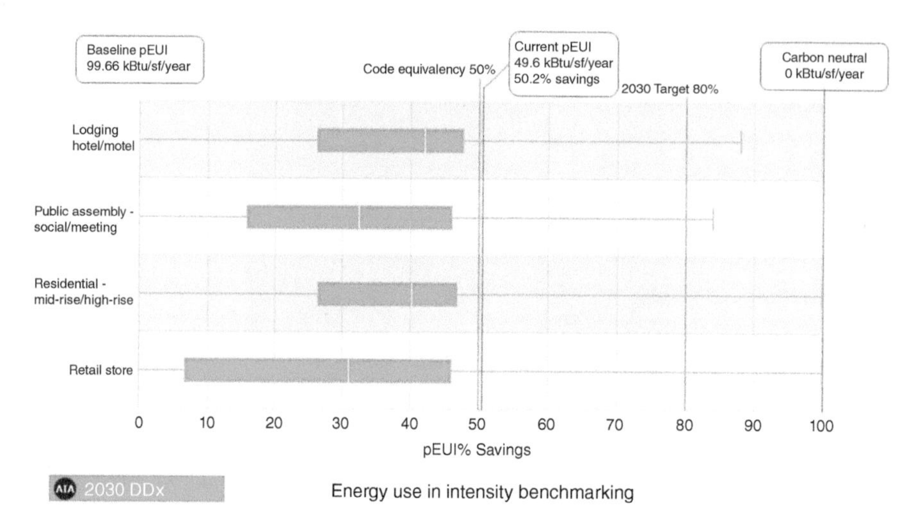

Energy use intensity benchmarking from AIA.

6.4.2 Process

tvsdesign approached the project from multiple angles to get holistic benefits from an optimized design solution. They analyzed the key aspects of space conditioning, daylight, window-to-wall ratio (WWR), and use of renewable energy all in-house giving the team a better understanding of the constraints and opportunities.

The focal point of the design was a central atrium with a seasonal strategy for heating/cooling and natural ventilation. Using cove.tool, the design team assessed how much energy could be saved by enhancing the stack effect and allowing for passive strategies like night cooling. Not only did the space serve well during the cooling season (warm weather), but also in the heating season (cool weather) when it can be used as a closed sunspace to trap the heat indoors. All these intuitive design strategies were quantified and assessed for cost versus energy consumption, making a good business case for the client.

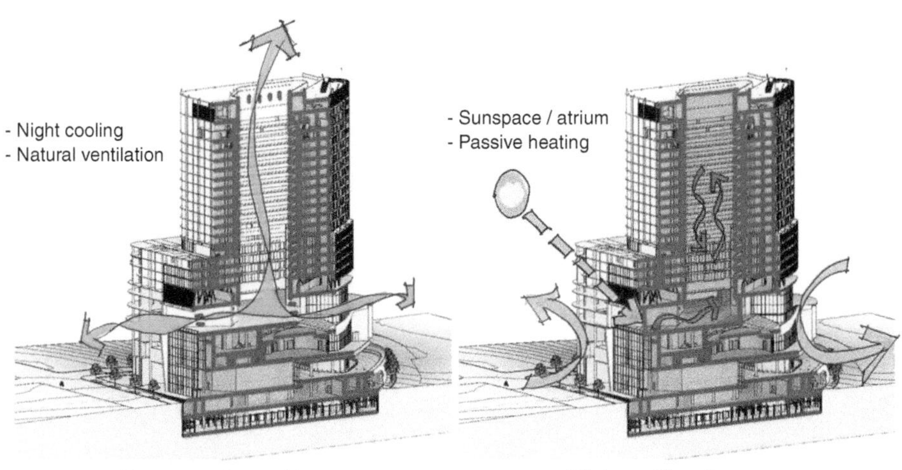

Passive cooling and heating.

The benefit of an integrative tool is that the team could do rapid prototyping, saving them many hours in modeling different iterations. As recommended best practice by the *AIA Architect's Guide to Building Performance*, a parametric study was done for different values of the WWR along with the use of photovoltaic panels (PV) to compare their combined impact on the project's energy use intensity (EUI). Using cove.tool's *facade guidance tool*, the team swiftly modeled variations of a typical bay with different glass types, shading strategies, and infiltration rates.

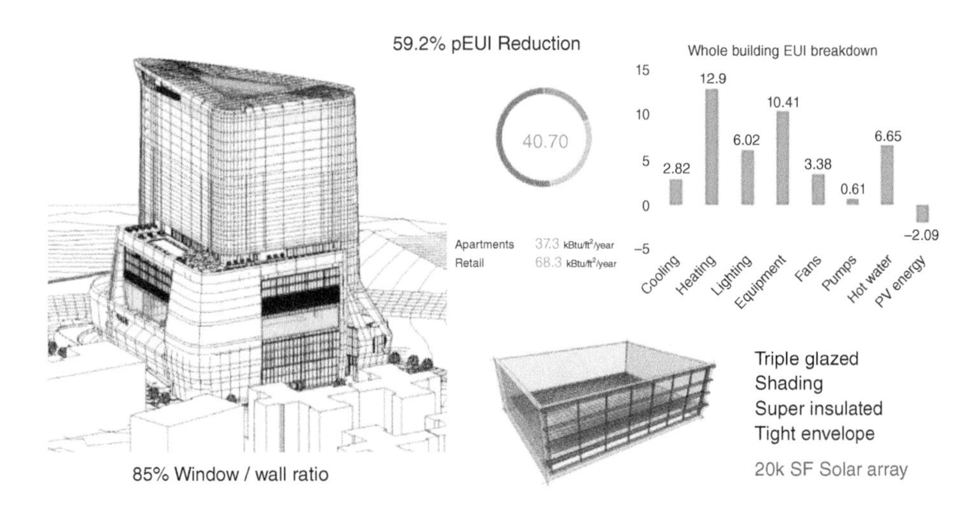

Window to wall ratio (WWR) and photovoltaic panels (PV).

The project benefited from the *Sun Hours* and *Solar Radiation* studies to optimize the facade details and location of photovoltaic panels. When combined with the iterative WWR studies, the design team was able to customize the fenestration and glass type.

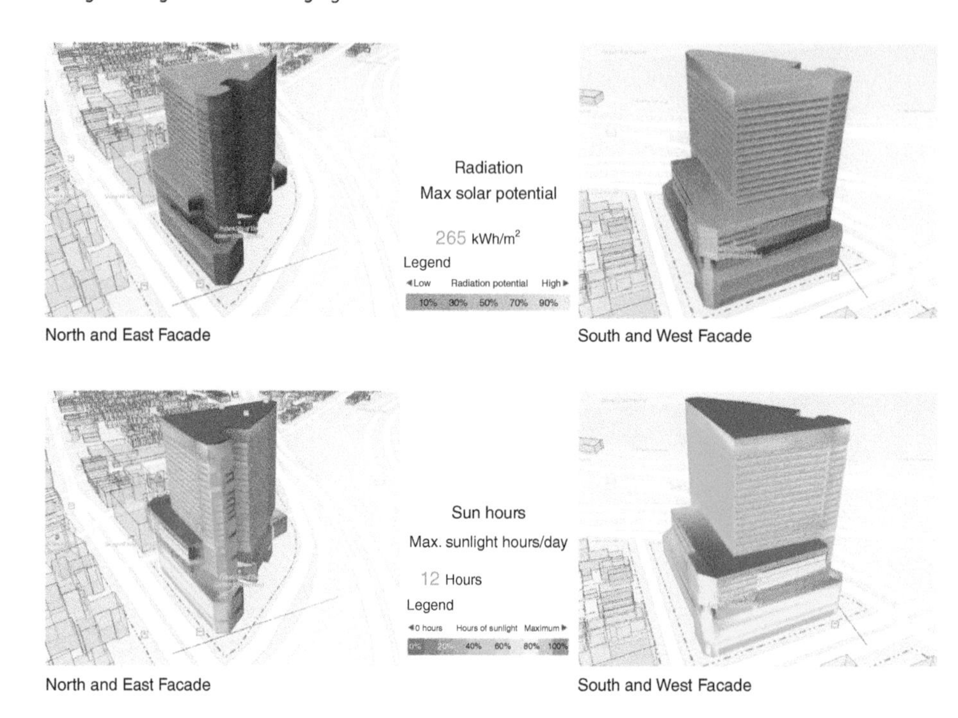

Sun hours and radiation studies.

Using these analyses, the design team enhanced facade details for each orientation, selectively adding details like fritted glass and recesses for openings. The *full floor plate daylight analysis* was very useful to identify the interior spaces that needed transparent glazing to allow more daylight penetration, while limiting the glare (*ASE analysis*).

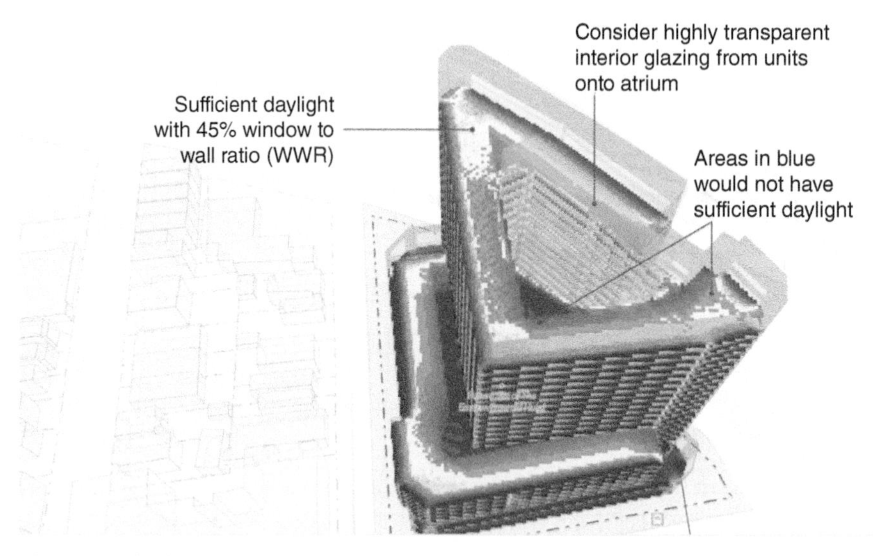

Enhanced façade details for each orientation - selectively adding details like fritted glass and recesses for openings. The full floor plate daylight analysis was very useful to identify the interior spaces that needed

The design team advanced from a baseline *EUI* of 74 kBTU/sf/year to a 40% EUI reduction of 43.98 kBTU/sf/year, making significant progress toward the 2030 goal of being carbon neutral. The team presented options to the client that ranged from 50–60% effective energy savings. By optimizing the facades and using photovoltaic panels, among other strategies, the team was able to achieve a 40% carbon reduction as compared to other buildings of this size and type. Hitting the carbon efficiency target was equivalent to avoiding 27 truckloads of ice melted per year.

Even better for the firm, the design team was able to save time and cost typically associated with design revisions after getting the detailed energy model. By understanding and refining the energy strategy as part of the design process, the team stayed closely aligned with the energy model. All of this was achieved as part of an integrated design process where the design concept was validated and refined by data generated by the design team themselves.

Benchmark energy savings model calculated in cove.tool.

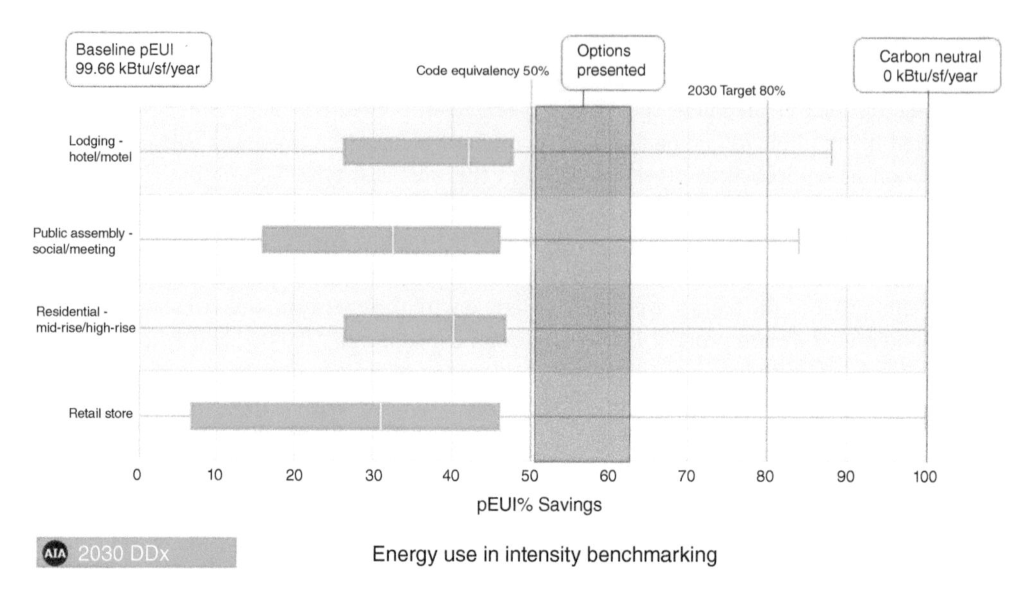

Energy use intensity benchmarking from AIA with additional options presented from cove.tool.

6.5 Harnessing the Microclimate

6.5.1 How the Conditions Facilitated by the Physical Elements Around a Building can Influence its Energy Use

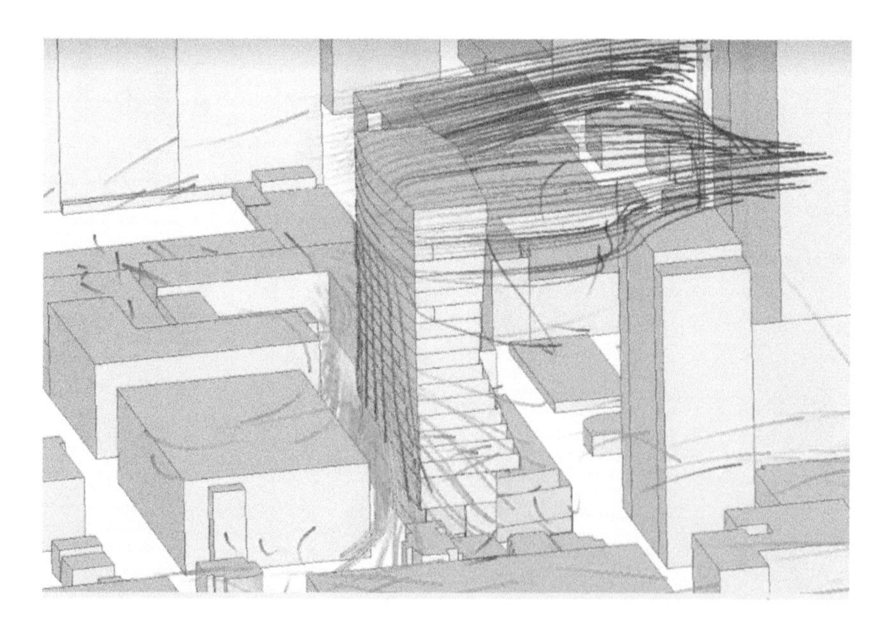

Unique environmental and topographic content.

Microclimates are the intricate tapestry of climate conditions present within a small geographic area, often influenced by the interplay of natural and artificial elements. These microclimatic conditions can vary significantly over short distances and are shaped by variations in topography, vegetation, water bodies, and built structures. The strategic harnessing of these microclimate conditions through architectural design holds profound potential for reducing a building's carbon footprint by diminishing the reliance on energy-intensive mechanical systems.

When we consider natural shade, it's not just about the immediate comfort it provides. Strategically placed vegetation and thoughtfully oriented structures cast shadows that reduce the thermal load on a building. By doing so, they significantly lower the need for artificial cooling, directly cutting down on energy consumption and associated carbon emissions.

Solar access is another critical piece of the puzzle. A building that captures the low winter sun through its windows can reduce its heating requirements dramatically. Conversely, architectural elements designed to shield interiors from the high summer sun help keep buildings cool, thereby reducing the cooling load. This dance with the sun, when choreographed correctly, minimizes the building's carbon emissions by leveraging the sun's natural energy cycle.

The concept of breezeways utilizes the knowledge of local wind patterns to create pathways for natural ventilation. By allowing air to flow through a building, it reduces the dependency on air conditioning units and fans. This natural form of temperature regulation not only conserves energy but also reduces GHG emissions by relying less on electricity generated from fossil fuels.

Proximity to water can be seen as a geothermal gift in the microclimate realm of concepts. Water bodies have a moderating effect on temperature, and buildings sited near them can leverage this for natural heating and cooling. The thermal mass of water can be a significant ally in maintaining consistent indoor temperatures, reducing the need for mechanical temperature control and, thus, the carbon footprint.

Then, there are frost pockets and topography that, when recognized and utilized correctly, contribute to natural temperature regulation. Frost pockets can trap cooler air, reducing heating needs, while a building's elevated positioning can expose it to warming sunlight and breezes, cutting down the requirement for heating systems.

Weaving these microclimatic strategies into the fabric of building design requires a nuanced understanding of the site-specific conditions and an appreciation for the natural environment's ability to provide comfort. This is not merely about building in harmony with nature but actively engaging with it to craft spaces that are inherently sustainable.

To build upon this, architects and designers can create structures that respond dynamically to the changing conditions throughout the day and seasons. For instance, adjustable facades that respond to the sun's path or operable windows that open to allow a cool breeze can reduce the need for mechanical intervention.

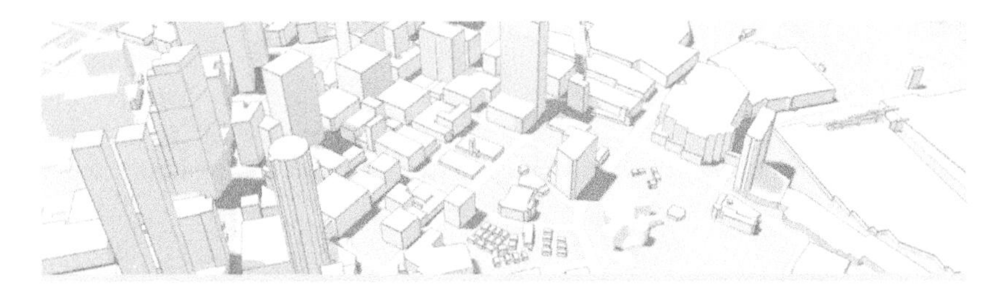

The combination of these approaches forms the backbone of a design philosophy that seeks to reduce carbon emissions not through add-on technologies but by embedding energy efficiency into the very essence of architectural design. By starting with the climate—both macro and micro—as a fundamental design tool, architects can create buildings that are less of a burden on our planet's resources, actively contributing to a future where our built environment exists in a state of symbiosis with the world around us. This is a future where our cities breathe with the rhythms of the earth, reducing carbon emissions by design rather than as an afterthought.

6.5.2 Exploring Strategies Such as Landscaping and Vegetation to Retain Moisture and Facilitate Shading, Depending on the Specific Climate and Microclimate

Landscaping and vegetation are more than just ornaments to the built environment; they are vital instruments in the ecological picture that can temper climates, conserve water, and provide natural shading. The strategic use of plant life is a sophisticated approach that, when executed with precision, marries the specific needs of the microclimate to the overarching goals of sustainable design and occupant comfort.

The deliberate positioning of deciduous trees is one such strategy that exemplifies the synergy between form and function. In the summer, their lush canopies cast cooling shadows, reducing the demand for air conditioning by shielding structures from intense solar gain. In the cooler months, as these trees stand bare, they allow the sun's warming rays to penetrate, diminishing the need for artificial heating. For instance, shade trees, when properly placed around a building, can reduce exterior temperatures outside the building leading to reductions of energy consumption for cooling.

Evergreen trees, with their year-round foliage, provide a constant barrier against thermal transfer, proving particularly effective when sited on the north side of buildings where they can shield against cold winds. By using evergreens such as pines or firs in this strategic manner, buildings can experience significantly reduced wind chill effects, potentially lowering heating requirements.

When it comes to plant placement, it's a game of inches and degrees. Groundcovers, for example, excel in cooling the immediate microclimate, creating a buffer zone that can substantially lower the temperature of the ground and the air above it. Groundcover plants, which are often drought-resistant, contribute to cooler ambient air temperatures through the process of evapotranspiration. For example, research has indicated that a well-maintained grass area can be 10–14°F cooler than exposed soil and more than 30 degrees cooler than concrete or asphalt.

The judicious selection of plant forms maximizes the potential for shade. Vining plants, when trained over pergolas, can turn a sun-drenched patio into a cool retreat. A classic example of this can be found in Mediterranean countries, where grapevines are often grown over pergolas to create shaded, cool spaces in hot climates.

Drought-tolerant plants are not just survivors; they are thrivers in their habitat. Xeriscaping with such species as lavender or Russian sage can create landscapes that are resilient to heat and water scarcity. For instance, in arid regions like the American Southwest, xeriscaping has become a fundamental practice in sustainable landscaping, allowing for beautiful gardens without excessive water use.

Profound benefits of moisture retention point to why cities should prioritize vegetation. A green roof covered with sedum plants can absorb rainwater, which then cools the building as it evaporates. Cities like Toronto have implemented green roof bylaws to take advantage of these benefits, recognizing their role in reducing urban heat islands and managing stormwater.

Landform buffers, too, play an indispensable role. By creating berms and mounds that are strategically positioned and planted, we can deflect harsh winds and nestle buildings into more temperate zones. In colder climates, these earthworks can act as windbreaks, reducing heating costs by cutting down on wind penetration.

The use of native plants is based on the principle that they are already adapted to local conditions. For instance, in the Pacific Northwest, the use of native ferns and shrubs can help maintain soil moisture and provide excellent shade due to their dense foliage. On the other hand, adapted plants, while not native, can provide similar benefits if chosen carefully. Plants like the crepe myrtle, native to Asia but adapted to parts of the United States, can offer summer blooms and shading capabilities while requiring minimal irrigation.

The physical properties of leaves can affect their performance in the landscape. Broad, heavyweight leaves can store and then slowly release moisture, contributing to cooler air through the process of evapotranspiration. Conversely, plants with reflective, light-colored foliage can help to bounce back the sun's rays, maintaining cooler ground temperatures.

Each element of landscaping and vegetation offers a specific benefit to the microclimate, working in concert to reduce dependence on mechanical systems and enhance occupant comfort. When design teams harmonize with landscape architects, they create sustainable, energy-efficient environments that gracefully withstand the challenges of climate change boosting resilience and increasing human comfort.

6.6 Refining Solar Heat Utilization: Thermal Mass and Window Placement for Climate-Responsive Design

6.6.1 The Considered Integration of Thermal Mass

In the realm of sustainable architecture, thermal mass serves as a foundation stone for energy efficiency. This is not about the overall shape and configuration, often referred to as "massing" by architects, but the very substance of the materials of a building—its capacity to absorb, store, and redistribute heat derived from solar energy. Thermal mass, the heavyweight champion in the arena of passive design, is the strategic use of materials like concrete, brick, or stone that can store significant amounts of solar energy.

However, as architects craft buildings for thermal regulation, they tread the tightrope of ecological responsibility. Each choice of material comes with a shadow—embodied carbon—the GHGs emitted during its manufacture and transportation. The sustainable selection of thermal mass materials is a complex equation, balancing the immediate thermal benefits against the longer-term carbon footprint. Just like every decision in design and construction, it is vital to ensure that the thermal inertia of strategy is not balanced out by embodied carbon.

6.6.2 Local Climate's Role in Decision Making

Larger building mass, the number of heating degree days—a metric for how much and how often additional heating might be needed in a specific location—becomes a guiding star for architects. In a city like Atlanta, where the nickname "Hotlanta" belies its true climatic nature, there are indeed more heating than cooling degree days. This subtle climatic nuance demands a design that slightly tips the scales toward heat gain over shading.

The importance of this metric cannot be overstressed. In Atlanta, the balance of window placement must favor solar gains during the colder months, ensuring that the building can harness the sun's benevolence to reduce heating demands. Yet, this strategy does not overshadow the need for thoughtful shading; it simply calibrates it to the unique rhythm of Atlanta's temperature swings.

6.6.3 The Imperative of Simulation-Driven Design

No rule of thumb can help navigate the intricate dance between thermal mass, window placement, and the specificities of a locale's climate. Here, the power of simulation-based software shines, allowing architects to model and predict how buildings will perform across a spectrum of metrics. Running simulations that calculate heating loads, embodied carbon, daylight penetration, and thermal comfort metrics in several different scenarios illuminates the path to a tailored, sustainable design strategy.

Simulations act as a crystal ball, revealing how a building will live and breathe within its environmental context. In a city straddling the line between heating and cooling demands, such as Atlanta, simulations are crucial. They allow designers to interrogate their choices, to scrutinize the trade-offs, and ultimately to strike a balance that optimizes energy use while providing luminous, comfortable spaces for occupants.

6.6.4 The Role of Window Placement

Optimizing window placement is a meticulous task that plays a significant role in sustainable architecture, striking a balance between thermal performance and daylight access. The strategy begins with understanding the sun's path and the site's climate-specific characteristics.

In climates with more heating degree days, such as Atlanta, thoughtful window placement on the south facade is pivotal. Windows here should be designed to capture the low-angled winter sun, bringing in heat when it is most needed. Consider the incorporation of sunspaces which act as solar collectors, trapping heat in a buffer zone before it permeates into the living spaces. The use of phase-change materials within these spaces can further enhance the heat capture during the day and release it as temperatures drop.

The aspect ratio of windows—broader in width than height—on these southern facades can be optimized for the winter sun's arc, maximizing heat gain without compromising the building's thermal envelope. The addition of thermal shutters can provide nighttime insulation, sealing in the warmth accumulated during the day.

For east and west facades, the approach is more nuanced. These orientations are susceptible to overheating and glare. Hence, fenestrations here require careful sizing and, sometimes, the use of dynamic shading systems that adapt to the daily and seasonal sun angles, thereby controlling heat gain and optimizing daylight.

6.6.5 Placement for Solar Heat Gains

The coordination of solar heat gains and daylight admission must be carefully calibrated with each element in harmony with the other to achieve a performance that is both functional and aesthetically pleasing. South-facing windows, while beneficial for heat gain in winter, can introduce too much direct light, leading to glare. Here, light shelves can bounce sunlight deep into the space, illuminating ceilings and allowing light to penetrate further into the core of the building. These shelves do not double as shading devices because once the heat is inside it is already heating up the space during summer. External shading or blinds are necessary to reject heat during the summer. Again, there are tradeoffs, which is why a rule of thumb is not as useful as a simulation. What works in Los Angeles will not hold true in New York City.

For north-facing windows, the strategy differs as these should be optimized for soft, diffuse light that provides consistent, high-quality illumination without the thermal penalty. Such fenestrations can be taller, ensuring that even the deepest parts of the building can benefit from natural light, which not only reduces the reliance on artificial lighting but also enhances the well-being of occupants.

In all cases, the glazing's specifications—whether triple-paned, low-emissivity (low-e) coatings, or filled with inert gas—must be chosen based on a thorough analysis of their impact on both heat gain and light transmittance. The use of advanced or dynamic glazing technologies can offer the flexibility to adapt to varying conditions, maintaining comfort and visibility while reducing the building's energy profile.

The integration of operable windows can also enhance natural ventilation, using the thermal mass as a heat exchanger between the interior and exterior environments. This can significantly reduce the need for mechanical cooling, particularly in temperate times of the year.

Window placement is not simply about positioning openings in a wall but is a complex interplay of thermal and lighting design. The ultimate goal is to achieve a building that self-regulates, minimizes energy use, and maximizes occupant comfort and connection to the outdoors. Through detailed simulation and modeling, architects can predict and fine-tune these elements to create buildings that are not only efficient but also adaptable to the changing patterns of the environment.

The alignment of thermal mass with intelligent window placement is an artful convergence of science and aesthetics. When sustainable strategies coalesce, they create buildings that are not only responsive to their climate but also respectful of their carbon impact. In heating-dominated climates, the judicious use of solar heat gains through a well-considered assembly of thermal mass and fenestration makes a more comfortable and efficient building. Let us not forget that it is through the iterative simulation that we can uncover the optimal application of these strategies, ensuring that our buildings are true stewards of energy and environmental guardians for the generations to come.

6.7 Internal Gains and Their Role in Building Performance

6.7.1 The Concept of Internal Gains and How They Can be Harnessed to Improve Building Performance

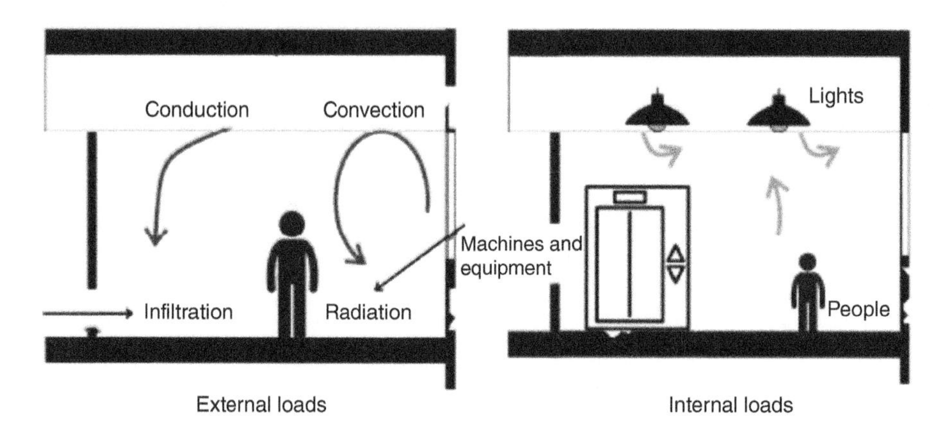

Internal vs. external loads

Internal gains—the heat produced by all sources inside a building except for the HVAC systems—are a critical yet often overlooked component in the thermal dynamics of building interiors. They stem from a variety of activities and devices such as lighting, appliances, technology, and even the occupants themselves. Effectively managed, these gains can substantially diminish the need for external energy input for heating, thereby enhancing the building's performance and energy efficiency.

6.7.2 Occupants as a Heat Source: Metabolic Contributions to Internal Gains

Occupants are surprisingly potent heat emitters. A single person can generate approximately 100–150 watts of heat just through basal metabolic processes, which can be likened to a human-sized incandescent bulb warming the surrounding space. In densely populated environments like offices or auditoriums, the collective heat from occupants can be the dominant factor in internal gains, at times accounting for over half of the heating requirements.

The heat output from occupants, however, is not a constant; it fluctuates based on the number of people and their activities. For instance, an intense workout session in a gym will significantly increase the heat emission compared to a quiet study hall. Recognizing this variability is crucial in designing adaptive thermal systems that can scale back on heating when internal gains are high.

6.7.3 Illuminating Efficiency: Lighting's Dual Role in Heat Emission and Conservation

Lighting, traditionally, has been a substantial contributor to internal gains. Older lighting technologies such as incandescent and fluorescent lamps operate at high temperatures, inadvertently heating their surroundings. Lighting, paradoxically, can account for up to a third of a building's internal heat gains, especially in commercial settings. When the authors updated the lighting in their office from florescent to LED, the system for their office was oversized leading to several months of load balancing to achieve comfort.

In response to this, there has been a shift toward highly efficient LED lighting, which not only reduces the electrical load but also minimizes unwanted heat contributions. Design strategies that couple LEDs with smart controls, like occupancy sensors and daylight-responsive systems, further capitalize on these gains, ensuring that light—and accompanying heat—is used only when and where it is needed.

6.7.4 Appliances and Equipment: Managing Heat Output Through Smart Design and Usage

The heat produced by appliances and electronic equipment is a substantial and often continuous source of internal gains. From the low but persistent warmth of a refrigerator's coils to the intermittent but intense heat of an oven, each appliance contributes to the overall thermal profile of a building.

In the design phase, considering the specification of high-efficiency, ENERGY STAR-rated appliances can substantially mitigate these heat gains. Thoughtful placement and usage can also play a role; for example, placing heat-generating appliances in areas with good ventilation can help dissipate unwanted heat, avoiding localized hotspots.

6.7.5 Mitigating Overheating: Strategies for Excess Internal Heat

The challenge arises when internal gains contribute to overheating, particularly during warmer periods. This can precipitate not just discomfort but increased energy consumption as cooling systems work harder to maintain comfortable temperatures. To navigate this, designers might incorporate high thermal mass materials that absorb excess heat during the day and release it during cooler periods. Other strategies include enhancing natural ventilation to carry away excess warmth, utilizing high-albedo materials for reflecting solar radiation, and integrating phase-change materials that absorb heat as they melt and release it as they solidify.

6.7.6 Understanding and Leveraging Internal Gains for Building Performance

A deep understanding of internal gains—how they are generated, their patterns, and their interaction with the building's fabric and external environment—allows architects and engineers to transform what could be a liability into a functional asset. It is through strategic planning and simulation that one can harmonize these internal heat sources with the building's heating and cooling systems to optimize for energy efficiency, comfort, and performance.

By modeling different scenarios, considering varied occupancy levels, appliance usage patterns, and lighting schemes, professionals can predict internal heat gains with precision. This information becomes the bedrock of a responsive design that uses less energy, costs less to operate, is low carbon, and provides a more comfortable environment for occupants.

6.8 Daylighting and Lighting Power Density: Crucial Factors in Energy Efficiency

6.8.1 The Symbiosis of Daylighting with Energy Conservation

Daylighting stands at the forefront of passive design strategies, commanding an influential role in diminishing energy requirements while promoting a healthy environment for occupants. By harnessing daylight, an abundant resource architects, and designers can significantly curtail the necessity for electric lighting, which traditionally saps energy and generates unwanted heat. The crux of daylighting design lies in its ability to channel natural light into the heart of a building by optimizing the placement and performance of windows, skylights, and reflective surfaces. When this natural illumination is seamlessly integrated with sophisticated lighting controls and concerted efforts to reduce lighting power density, the result is a substantial relief on energy systems, engaging with natural light rather than mechanical alternatives.

Daylight is not a mere substitute for artificial lighting but a superior alternative that offers a spectrum of light that artificial sources often cannot match. This full spectrum is known to support visual clarity and maintain the natural circadian rhythms that contribute to human health. By relying on daylight as the primary source of illumination, energy consumption can typically plummet by 50% to as much as 80%. This reduction is not only favorable to the environment but also eases the financial burden associated with energy costs. Moreover, with the diminished necessity for electric lighting, the internal heat it generates is also reduced, thereby easing the workload on HVAC systems and further enhancing their efficiency in cooling dominated times of the year.

6.8.2 Harnessing Technology to Optimize Natural Light

The technological intercessor in this delicate balance between natural and artificial lighting is the daylight sensor. These sensors are finely attuned to the ebb and flow of natural light, activating electric lighting only when the cloak of night or a cloud-laden sky obscures the sun's rays. Their operation prevents the unnecessary use of electric lights, reducing the lighting burden and ensuring lights are brightened only when needed to supplement daylight. When combined with occupancy sensors, these daylight sensors offer a two-pronged approach to lighting control: they ensure that lights are used only when spaces are occupied and that they operate at the optimal intensity in relation to the amount of natural light available. This judicious use of artificial lighting effectively "harvests" daylight, optimizing its use for illumination and minimizing energy expenditure.

6.9 Case Study: The Importance of Reducing Lighting Power Density

The concept of lighting power density (LPD) encapsulates the amount of electric power dedicated to lighting per unit area of a building. An effective reduction in LPD is a marker of an efficient lighting system, one that utilizes lower-wattage fixtures to achieve the desired luminosity. Astoundingly, even a modest decrease of one watt per square foot in a building's LPD can accumulate to a saving of around 10 kilowatt-hours annually per square foot. This is where the choice of lighting fixtures becomes pivotal—opting for high-efficacy options like LEDs, coupled with an integrated daylighting approach, can drive down LPDs. The resulting energy efficiency is not an isolated benefit; it ushers in a world of advantages that span economic savings, environmental protection, and regulatory compliance. With lower LPDs, businesses can look forward to reduced utility bills, an outcome that significantly impacts operational costs. Environmentally, this efficiency translates to lower carbon emissions, a critical consideration in the quest for sustainability.

Additionally, reduced energy consumption in lighting also means a corresponding decrease in the waste heat produced, which in turn reduces the cooling demands of a building—a notable advantage for spaces with dense lighting installations. From an aesthetic and comfort perspective, a space bathed in uniform light, free from the glare and shadows that accompany high-intensity fixtures, offers a more serene and productive environment for occupants. Such improvements in lighting quality can foster an ambience that encourages better performance and well-being. Compliance with building codes and standards, which often set maximum limits for LPD to promote energy efficiency, becomes less of a regulatory hoop to jump through and more of an aligned objective in the pursuit of energy conservation. Most buildings have too many lights to begin with and being judicious with electric lighting is an easy way to make a big difference.

6.10 Conclusion: Empowering Change Through Accessible Measures

As we reach the culmination of this chapter, it is clear that the path to decarbonizing our built environment is as challenging as it is imperative. The strategies discussed provide a roadmap, guiding us through the practical steps we can take to effect tangible change. We have seen how the implementation of passive design, the thoughtful selection of materials, and the shift in occupant behavior can collectively serve as a fulcrum for monumental progress.

The initiatives we have outlined are far from revolutionary; they are well within reach, requiring neither the invention of new technologies nor the commitment of considerable financial resources. Instead, they ask for a paradigm shift—a move away from short-term fixes to a holistic vision that embraces long-term sustainability. Through the optimization of microclimates, the intelligent planning of building orientation, and the integration of high-performance roofing and facade systems, we can create structures that not only serve our immediate needs but also respect the ecological boundaries of our planet.

The collective impact of these “low hanging fruits” is undeniable. They beckon us to act with the assurance that every incremental improvement is a step toward a decarbonized future. The case studies presented affirm that these interventions are not just theoretical ideals; they are practical, viable, and already in motion. They demonstrate the synergy of sustainability and economic viability, presenting an irrefutable case for widespread adoption. Let us not view these strategies as mere suggestions but as imperatives for the present and future of our global community. The onus is on us—architects, builders, policymakers, and occupants—to embrace these measures.

6.11 Dr. Pablo La Roche | Principal and Professor | CallisonRTKL, Arcadis | Cal Poly Pomona

6.11.1 Pablo La Roche | Principal and Professor | Arcadis | Cal Poly Pomona

Dr. Pablo La Roche, an accomplished architect and professor, serves as Global Sustainable Design Director at Arcadis. Holding degrees in architecture and a Ph.D. from UCLA, his extensive research focuses on passive cooling systems, low-energy carbon-neutral architecture, and affordable housing, resulting in over 150 publications, including his book, "Carbon Neutral Architectural Design."

Dr. La Roche's career includes projects worldwide, and he holds a license as an architect in Venezuela, as well as LEED BD+C accreditation in the USA. He has won awards for his projects in multiple countries and is renowned for his contributions to the field.

Dr. La Roche's active involvement includes chairing and reviewing technical sessions at conferences, co-curating exhibitions, and serving on advisory boards and associations worldwide. His leadership and expertise have earned him numerous accolades, including a Fulbright scholarship and the NCARB Grand Prize for the Department of Architecture at Cal Poly Pomona.

Dr. La Roche shared his insights on the evolving landscape of architectural decarbonization, discussing key trends, ethical responsibilities, and the pivotal role of technology in shaping the future of architecture.

6.11.2 Architectural Leadership in Decarbonization

In the pursuit of decarbonization and sustainability in the field of architecture, it is vital to consider the insights and experiences of leaders who have been actively contributing to this mission. One such leader is Dr. Pablo La Roche, a prominent figure in the field of architectural sustainability and performance. In a recent conversation, Dr. Pablo La Roche shared his thoughts on various aspects of decarbonization, the evolving role of architects, and the critical importance of technology and collaboration. This case study delves into Pablo's perspectives, highlighting key quotes and valuable insights that shed light on the trends, challenges, and opportunities in the field of architectural decarbonization.

The xylem; solar shading example.

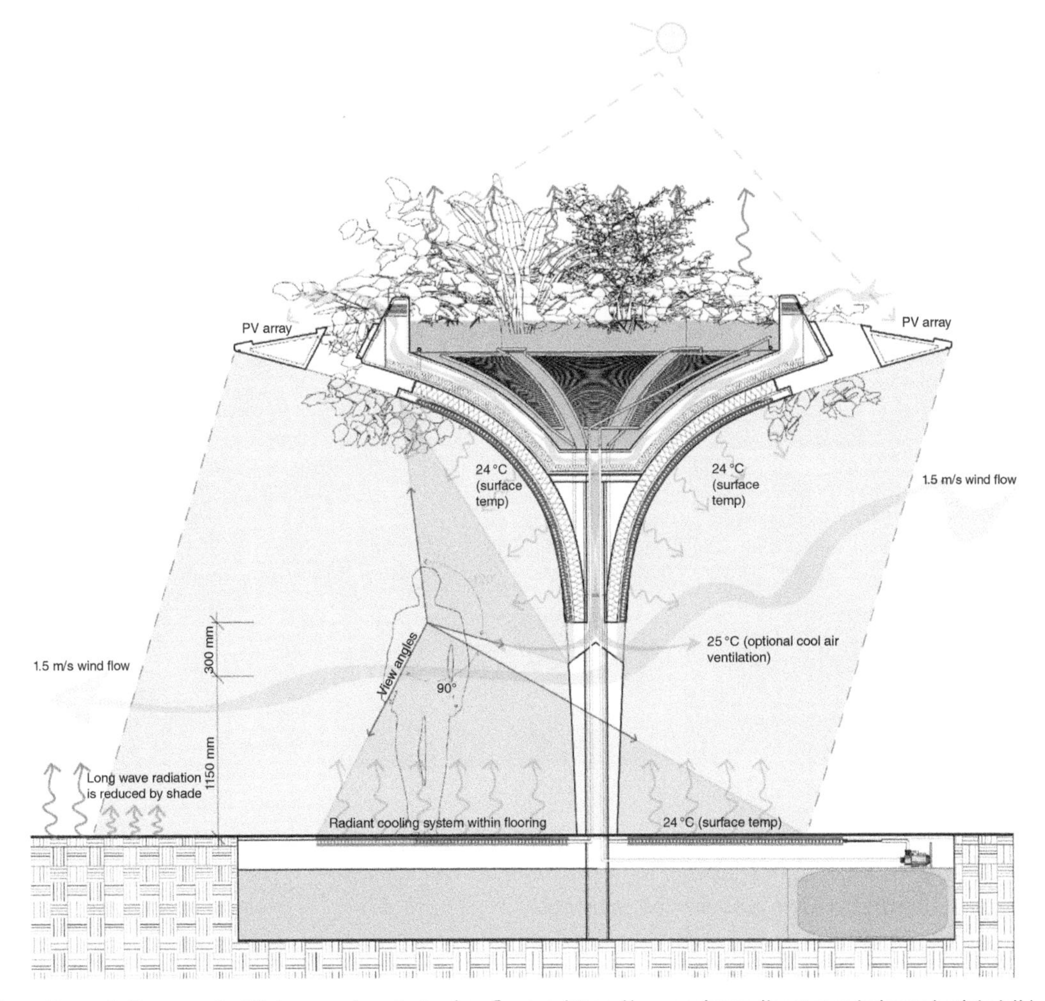

A section cut diagram of utilizing passive strategies. Source: https://www.sciencedirect.com/science/article/pii/S0378778820303078 / with permission of ELSEVIER.

6.11.3 The Ethical Responsibility of Architects

Dr. Pablo La Roche emphasizes the ethical responsibility of architects and building performance leaders in addressing decarbonization. He believes this responsibility extends beyond mere trends; it is a fundamental obligation to combat climate change. He stated, "I think we all have an ethical responsibility... It is something that we must do." This ethical stance forms the foundation for the actions and initiatives he advocates.

6.11.4 The Evolution of Sustainability in Practice and Academia

Over his extensive career, Pablo has witnessed a significant evolution in how sustainability is perceived and practiced. He recalls the earlier days when sustainable architecture was a niche concept known as "bioclimatic architecture." Today, it has become more mainstream. However, he notes that there is still work to be done in equipping architects with the knowledge and skills required to implement sustainable strategies effectively. He acknowledges the challenges and complexities of making sustainability a widespread reality, noting that "it is important to increase sustainability education in schools of architecture."

Various classes Pablo has taught.

6.11.5 Collaboration and Knowledge Sharing

One of the standout aspects of Dr. Pablo La Roche's work is his commitment to collaboration and knowledge sharing within the architectural community. He has been president of the Society of Building Science Educators and is part of sustainable leaders' groups and large firm sustainability roundtables where leaders from different firms share their experiences, challenges, and strategies. He believes in working as a collective, transcending competition for the greater goal of achieving sustainability. He noted, "We're setting up goals for all of us together... we all have the same objectives."

6.11.6 The Role of Passive Strategies in Decarbonization

Pablo places great emphasis on passive strategies as a logical starting point for achieving decarbonization. He believes architects should prioritize strategies like natural ventilation, daylighting, and energy-efficient building design. He advocates that architects must understand building physics and use software effectively to implement and test these strategies. His philosophy is summed up as "passive first."

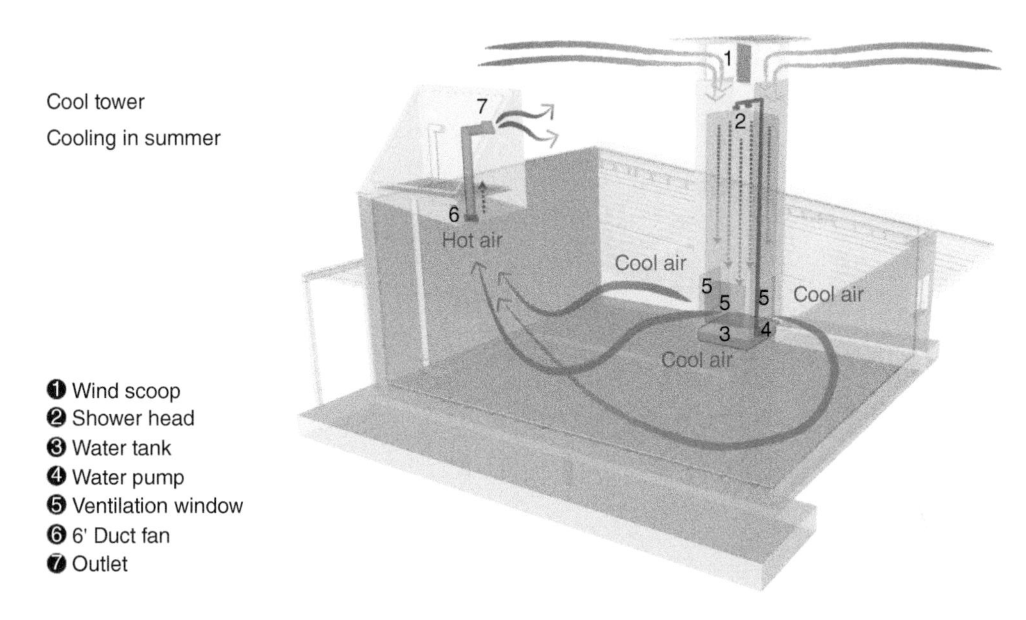

Cool tower highlighting how to passively maintain cooling in summer.

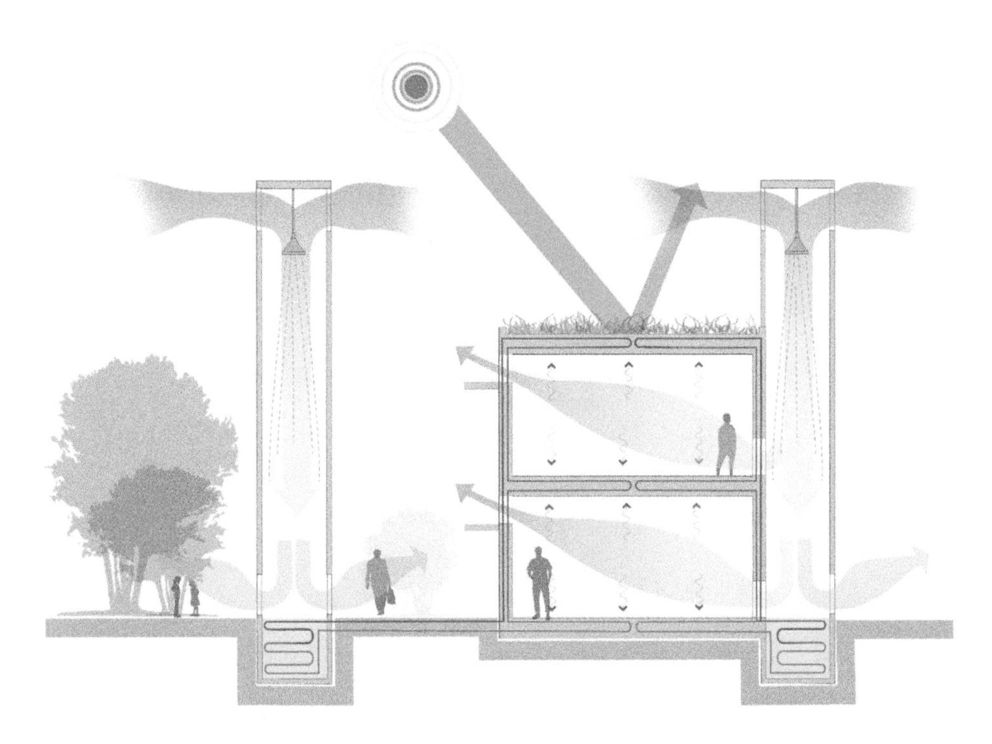

Passive strategy showing bouncing the solar radiation off with cool roofs, use of natural ventilation and more.

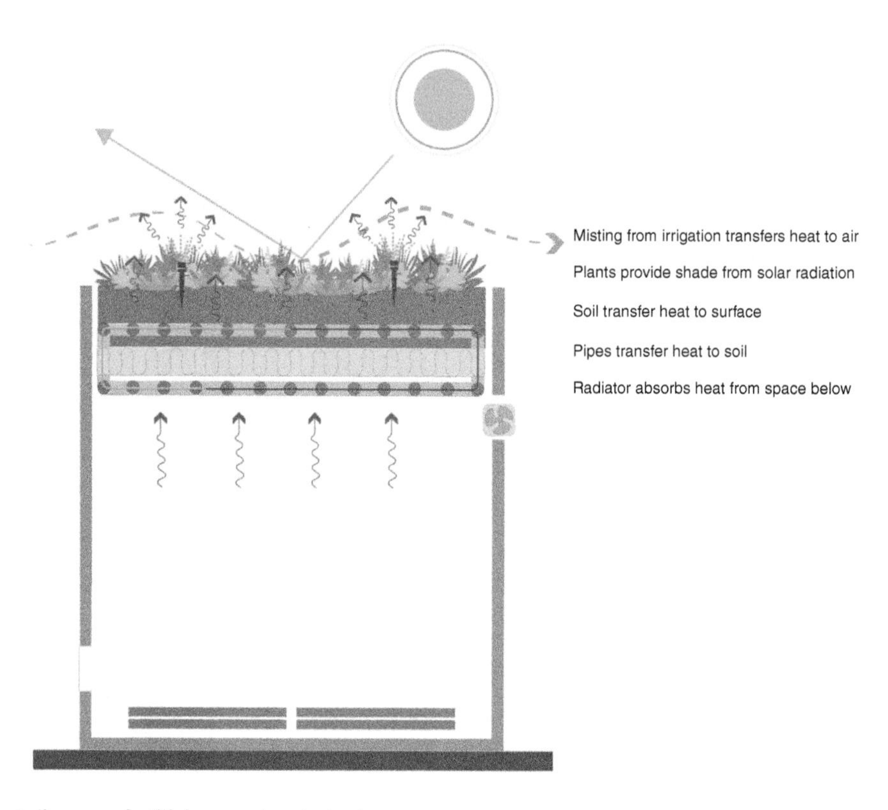

A section cut diagram of utilizing passive strategies.

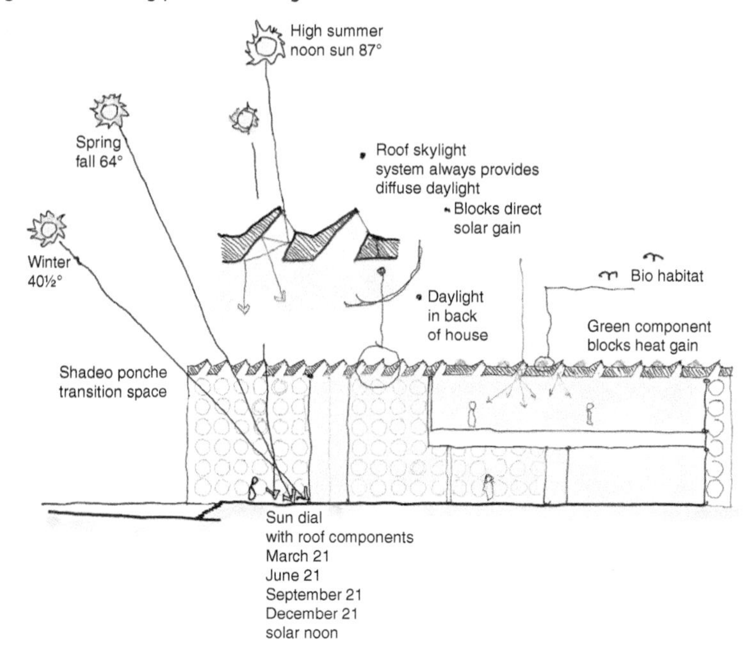

A sketch showing a holistic passive framework.

6.11.7 Advocating for Sustainability with Clients

Pablo discusses the role of architects in advocating for sustainability to clients, even when clients may not initially prioritize it. He believes that architects have a responsibility to push for sustainable strategies and make a strong case for their benefits integrating in projects. He shared instances where he convinced clients to adopt sustainable strategies without explicitly using the term "sustainability."

6.11.8 Reporting and Standardization

Collaborating with other sustainability leaders, Pablo La Roche is actively involved in standardizing reporting practices for greenhouse gas emissions within architectural firms. He believes that transparency and shared standards are crucial for evaluating and mitigating the environmental impact of architectural operations. This effort extends to CEOs, who are open to understanding and addressing environmental concerns.

6.11.9 The Future of Architectural Leadership in Decarbonization

Dr. Pablo La Roche's insights offer a glimpse into the evolving landscape of architectural leadership in decarbonization. The future requires architects to continue prioritizing sustainability as an ethical obligation and to advocate for passive strategies. Collaboration, knowledge sharing, and the alignment of goals among architectural firms will be vital in driving meaningful change toward a more sustainable built environment.

Hotel Casas Bioclimáticas, Spain.

His commitment to ethical responsibility, passion for passive strategies, and dedication to collaboration make him a prominent figure in the journey towards architectural decarbonization. His experiences and perspectives serve as valuable guidance for architects and building performance leaders seeking to make a positive impact on our environment through sustainable design and construction.

7

Shifting Decision Making Early into the Process

Envisioning a building is no simple affair; it is a tapestry woven with myriad decisions, each thread contributing to the ultimate pattern of form, function, and environmental footprint. The most critical decisions are those made at the conception of a project—the nascent stage where every choice imprints on the structure's destiny for decades. In these early sketches lie the DNA of a building's environmental stewardship, defining its relationship with the planet.

Embarking on this architectural odyssey requires a deft navigation through a labyrinth of interdependent choices: the silhouette of the structure, its inner workings, the sinews of its materials, and the ethos of its design. These decisions have traditionally been deferred until the more defined schematic phase, post-conceptual musings. However, a burgeoning body of research now heralds the merits of deep, analytical contemplation in the earliest corridors of planning. Delving into robust evaluations and a diverse array of options at this strategic juncture can unfurl a breadth of benefits, paving the way to designs that are not merely adequate, but optimal.

This chapter endeavors to map out the essential tenets that guide these preliminary choices. It will dissect the nuances differentiating early from late decision-making, the dynamics that shape these decisions, and the interplay of certainty, ambiguity, and adaptability in the embryonic stages of design. A synthesis of academic discourses will elucidate methods for sharpening these preliminary decisions, encompassing collaborative stakeholder involvement, holistic design processes, innovative prototyping, and consideration of the full life cycle.

Why is this initial phase so pivotal? It is in this crucible that the raw materials of insight, analysis, and research meld to forge high-performing environmental designs. The selection of a locale, mindful of its connectivity, topography, and climate, steers the design toward energy conservation, optimized natural lighting, and responsible material use. Early simulations that ponder orientation, volume, and the building's skin can have far-reaching effects on the end product. Choices around materials and structural frameworks, when made with foresight, solidify a commitment to reducing embodied carbon, ensuring recyclability, and enhancing thermal performance. Hence, adopting integrated design processes that pool the expertise of engineers and sustainability experts from the outset is not just beneficial—it is indispensable for aligning initial concepts with environmental ambitions.

Build Like It's the End of the World: A Practical Guide to Decarbonize Architecture, Engineering, and Construction,
First Edition. Sandeep Ahuja and Patrick Chopson.

Within the forthcoming sections, we shall delve into the principles, methodologies, and illustrative case studies that demonstrate the profound impact of information-gathering, broad exploration, and inclusive stakeholder participation at the conceptual stage. Such strategies are not merely theoretical exercises but are practical applications that have led to architectural innovations which tread lightly on the earth, bolster resilience, and even aid in ecological restoration.

7.1 The Architectural View and Decision-Making

7.1.1 The Role of Design Objectives such as Program, Construction Cost, Environmental Performance, and Aesthetics in Architectural Design

Architectural decision-making is profoundly influenced by a range of design objectives from the outset. These include the functional needs (program), budget constraints (construction cost), ecological considerations (environmental performance), and creative expression (aesthetics). Each decision, whether it's about the building's orientation, its structural mass, chosen materials, or system implementations, casts a long shadow on the project's trajectory. Early decisions heavily determine its eventual cost, effort required by the design team, and overall performance.

Imagine this scenario: a sprawling 250,000 square-foot office project is nearing the final stages of its design in downtown Raleigh. Over the past few weeks, the design team has been diligently working to complete the 100% construction drawing set by the deadline. Just as they finalize the drawings for internal review, a critical step ensues—the mechanical engineering team's energy modeling unit begins their final compliance simulation. The results show the project is still 11% short of meeting the energy reduction targets, sending both the architectural and mechanical engineering teams into a frenzy to find solutions that minimally impact the original design.

Their strategy involves subtle yet effective modifications: tweaking the glazing percentages and considering alternative materials that enhance energy efficiency without compromising aesthetics. The mechanical systems also undergo fine adjustments. After three intense weeks of iterative revisions and collaboration, they devise a plan that slightly alters the glazing dimensions and optimizes system performance, finally achieving a model that complies with energy codes.

With the new strategy in place, it takes an additional two weeks to update the construction drawings to a fully code-compliant state. This episode of rework and delay, happens far too often and underscores the crucial importance of early-stage decision-making in projects.

Invoking the spirit of MacLeamy's curve, we witness the relationship between time, influence, and cost. The curve underscores a poignant truth in architectural design: as a project matures, the latitude for impactful and cost-effective alterations narrows precipitously. Illustrated by the blue curve, the window for influencing both the fiscal and functional aspects of a building diminishes as each phase of design solidifies choices about form, fabric, and functionality.

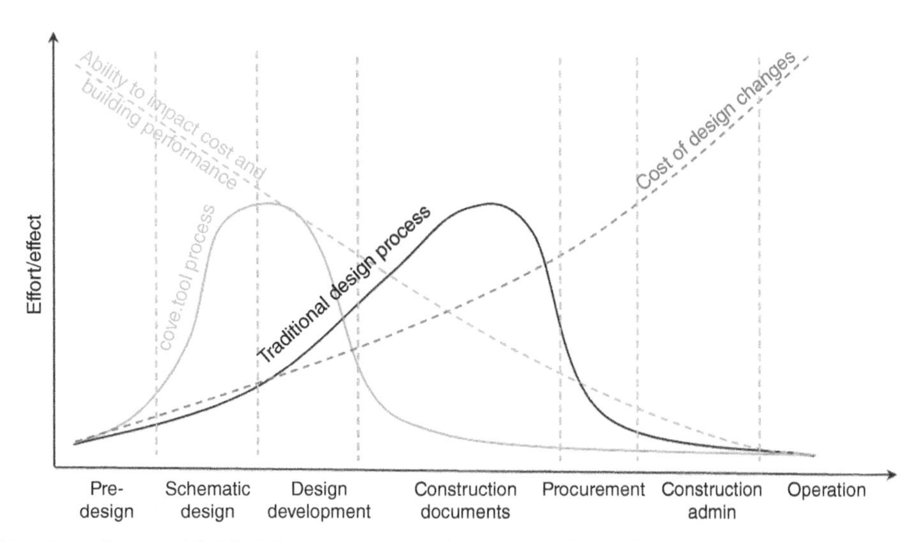

Adapted Macleamy's curve: Highlighting the cost of design changes further in the design cycles, and the low effort of making these changes early in the design process.

Consider the process of altering the window-to-wall ratio during the pre-design phase: it is a swift adjustment, taking under half an hour, in a rudimentary 3D or 2D model that requires no intricate coordination. Contrast this with a similar alteration during the construction document phase. Here, the modification becomes a herculean task, demanding extensive hours in detailed models and documents, necessitating complex interplay with a myriad of consultants and subcontractors, from HVAC design implications to structural considerations. The time investment required skyrockets.

Taking this scenario a step further, envision modifying the window-to-wall ratio during the construction administration phase. At this juncture, the cost implications are not only confined to consultancy but extend to actual construction expenses. Implementing such a change, despite unanimous agreement on its benefits for daylight quality, carbon footprint, and other performance parameters, becomes virtually impossible.

A project's program sets out essential functions, spaces, and their interrelations, which serve as benchmarks for evaluating design alternatives in terms of functionality, spatial efficacy, and life cycle expenses. Although initial massing is shaped by programmatic requirements, investigating various configurations often reveals multiple avenues to achieve cost-effectiveness. Yet, it is essential to recognize that optimizing for programmatic consideration alone can inadvertently curtail the design's potential for broader economic, environmental, and experiential advantages. Construction costs certainly delineate economic parameters, but an exclusive focus on immediate outlays neglects the substantial operational savings gleaned from energy-efficient, well-lit, and renewable interventions. Traditional emphasis on minimizing upfront costs has prevailed; however, life cycle cost analysis can recalibrate design preferences by spotlighting alternatives that diminish total ownership costs, albeit such practices are not yet commonplace.

The building's environmental performance is anchored in early-phase decisions concerning orientation, insulation, materials, and systems. Regrettably, these alternatives are typically evaluated retrospectively through simulations instead of being intrinsically woven into the design from inception. By integrating sustainability objectives earlier, a broader spectrum of high-performing options emerges, reframing the array of design choices. Prevailing practice often misconstrues optimization of an initial design as being at odds with environmental objectives, rather than envisaging design generation through an environmental prism.

Expanding our evaluation criteria to include factors affecting well-being may lead to the discovery of conceptual designs that are both aesthetically pleasing and perform better in terms of environmental and health metrics. In traditional discussions, the aesthetic appeal of architecture often does not consider its impact on well-being and the environment. These factors are essential as they significantly influence the quality of life for building occupants. By incorporating a holistic approach from the project's outset, which considers aesthetic, environmental, and health-related factors, we could develop designs that not only fulfill the client's vision but also contribute to a sustainable and health-conscious built environment. It is imperative for architects and designers to challenge the status quo and systematically explore a wider array of design possibilities early in the project planning stage.

7.1.2 How Conceptual Design Decisions About a Building's Orientation, Massing, Materials, Components, and Systems Largely Determine Lifecycle Performance

The conceptual design of a building establishes the foundation for its entire architecture and performance over the lifespan, as orientation, massing, material, and system choices at this early stage powerfully shape sustainability goals and outcomes.

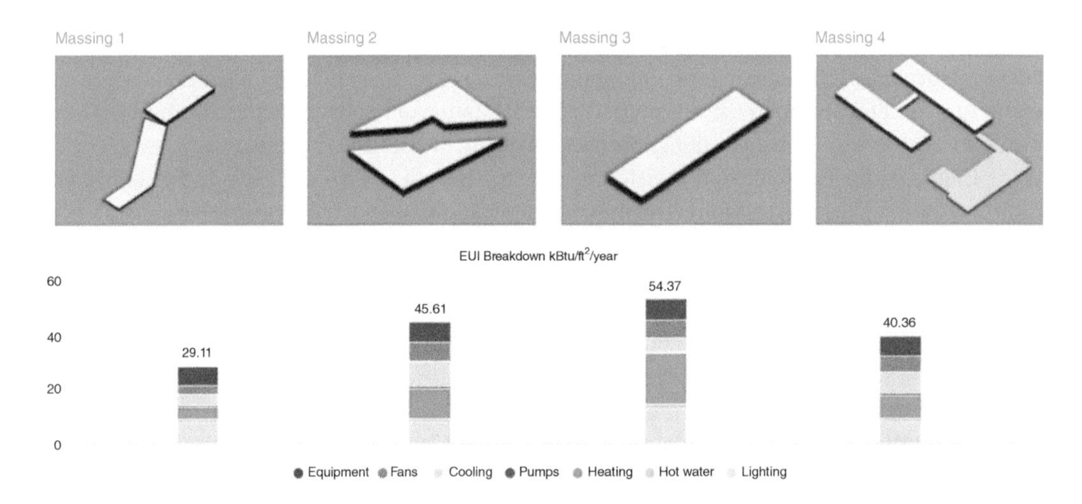

Concept and pre-design EUI breakdown.

The orientation of a building is strategic for harnessing natural light and controlling solar heat gain, which affects energy demands for artificial lighting, heating, and cooling. Massing—the building's shape and form—defines the relationship between its surface area and volume, influencing heat exchange and energy balance throughout the seasons.

The selection of materials and systems is not only a matter of structural necessity but also a determinant of the building's environmental impact. From the energy and resources consumed in their production to their performance and lifespan, these choices resonate with both the immediate carbon footprint and the long-term sustainability of the structure.

Kendeda Building at Georgia Tech. Source: KBISD/Wikimedia Commons.

While conceptual design cannot predict every detail, it sets the stage for the building's interaction with its occupants and the environment for years to come. It is a process that requires a fusion of analytical thinking and forward-looking vision, aiming to achieve an architecture that sustains and endures over time. Key elements of focus during this phase include:

- Orientation to leverage daylight, enhancing how occupants perceive and engage with interior spaces and their connection to the exterior.
- Massing to sculpt the building's interaction with its site, influencing occupant flow and interaction within its spaces.
- Material choices to optimize indoor air quality and contribute positively to the health and comfort of occupants.
- System designs that anticipate future adaptations, ensuring the building can remain functional and relevant as needs evolve.

These design considerations forge the character of the building, its purpose, and its long-term relevance. For architects, the conceptual design stage is an exhilarating juncture where the vision for a space takes shape—balancing the technical aspects of performance with the aesthetic and experiential goals, grounding environmental responsibilities in the realities of social context, and aligning immediate practicalities with aspirational futures.

Conceptual design thus becomes a powerful means to embed sustainability deeply into architecture. It is a chance to create spaces that resonate with the land and its resources, prioritize the well-being of its occupants, and champion innovation and circularity. It challenges architects to think beyond the constraints of the present, envisioning forms and functions that foster more sustainable, equitable interactions between people and the built environment.

This stage is where an architect's philosophical ethos is translated into tangible, practical decisions: from the pattern of light that dances across a room to the choice of low-emission materials and versatile systems. The transformative potential of conceptual design lies in this synergy of the tangible and intangible, combining practicality with imagination, technological precision with creative exploration, cost-effectiveness with timeless elegance.

7.1.3 The Current State of Decision-Making in Architecture and the Limitations of Tested Alternatives

Contemporary methodologies in architectural decision-making and the evaluation of design alternatives encounter several challenges that often prevent the discovery of more effective and higher performing design solutions. In the initial stages, a set of variables—including functional suitability, construction expenses, environmental considerations, and visual aesthetics—serve as the bedrock for generating design alternatives. However, the methods employed to assess these options typically suffer from a narrow focus and lack comprehensive depth. Consequently, the chosen alternatives might not fully harness the potential to deliver optimized results across all critical factors.

A fundamental problem is the prevalent practice of optimizing initial designs against isolated objectives, rather than adopting a more holistic approach from the beginning. For instance, sustainable design elements are frequently tacked on as adjustments to an existing proposal to meet environmental benchmarks, rather than being intrinsic to the creation of initial alternatives. The same goes for optimizing space usage, cost efficiency, or the execution of a specific aesthetic—these often follow a linear trajectory from an original concept, ignoring the rich possibilities that might arise from a thorough and integrative synthesis of all objectives. This piecemeal, siloed strategy can overlook the synergies and trade-offs that become apparent when multiple goals are considered in tandem from the onset.

The practice of testing a broad spectrum of alternatives is also notably deficient. Typically, only a limited range of possible designs are modeled, whether through tangible models or digital simulations, before finalizing the choice. This restricted approach, which tends to focus on slight deviations from a primary concept, inevitably means that more distinct and potentially superior performing options remain unexplored. These unexamined variations might better fulfill the complex tapestry of stakeholder needs and aspirations. The reasons for this limited exploration are twofold: resource limitations play a role, but there is also a cognitive bias toward incremental development—a path dependence that favors marginal improvements of original ideas over a bold re-conceptualization.

To surmount these barriers, a paradigm shift is required in how architectural alternatives are conceived and evaluated. A move toward systems thinking and integrated design processes that encompasses a comprehensive suite of objectives from the beginning is imperative. By adopting such an approach, architects and stakeholders can collaboratively and creatively explore a wider array of design possibilities. This not only expands the horizon of potential solutions but also fosters a more resilient and adaptable design process, one that is better suited to meet the complex and evolving demands of contemporary architecture.

7.2 The Importance of Early Understanding and Exploration

7.2.1 Research Findings that Support the Need for an Early Understanding of Design Objectives and the Ability to Explore and Analyze a Large Number of Alternatives

Several studies demonstrate the need for an early understanding of design objectives and the ability to explore and analyze a large number of alternatives during conceptual phases if our goal is to enable the discovery of higher performing architectural solutions. It is imperative to establish a clear set of design objectives at the inception of a project and to maintain the capacity to evaluate a multitude of alternative solutions during the conceptual stage. This approach is pivotal for fostering the discovery of more effective and sustainable architectural outcomes.

Research indicates that as much as 80% of a building's lifetime expenses are determined during the initial design phase when pivotal decisions regarding the building's shape, materials, and systems are made. This particular study underscores the advantage of integrating life cycle cost analysis early in the design phase, which enabled the identification of options that curtailed life cycle costs by 20–30% when juxtaposed with traditional methods. These findings advocate for a holistic and quantitative grasp of design objectives right from the start to catalyze innovation that optimizes economic efficiency over the long term. Additionally, it is estimated that approximately 60% of a building's environmental footprint is also dictated by design decisions made in these formative stages.

In the realm of generative design, investigations into the utility of digital tools for scrutinizing different massings tailored for performance metrics such as energy efficiency and carbon footprint have led to a revelation. Conceptual designs refined through these technologies exhibited a 15–40% improvement over initial drafts. These results underline the transformative potential of thoroughly investigating a broad spectrum of alternatives, propelling significant environmental improvements and integrating ecological considerations from the very conception of design ideas.

Moreover, design alternatives focused on experiential objectives are frequently neglected in the early design stages. When the criteria for aesthetic evaluation are expanded to include factors influencing well-being, alternative design variations can be discovered. These alternatives are perceived as equally or more aesthetically pleasing while also surpassing initial concepts in terms of comfort, health, and productivity measures. The potential for architects to uncover innovative conceptual designs that not only align with clients' experiential aspirations but also advance human health and well-being is high and we are just now starting to scratch the surface with data driven design thinking. However, this potential can only be realized through an initial, comprehensive framework that guides the exploration and assessment of designs.

The synthesis of these studies pinpoints the primary shortcomings in current architectural practice: the absence of an interactive, integrated perspective on multiple objectives to direct the creation of conceptual designs, coupled with a lack of systematic, computational tools to scrutinize and analyze an ample array of design alternatives. This leads to sub-optimal outcomes across economic, environmental, and experiential facets.

Incorporating expansive evaluative models and generative design methods during the conceptual stages is essential to unleashing the high-performance potential of architecture, which is currently constrained by the limited scope of conventional practices. To bridge this gap, the industry must adopt:

- A multifaceted and dynamic approach to understanding the interplay of design objectives from the beginning.
- Robust computational strategies to navigate and appraise a wide-ranging set of design alternatives.

At the crux of this transformative approach is the integration of data analytics into the conceptual design phase. Quantifying goals and performance metrics from the beginning is crucial—it enables architects to gather meaningful data, which in turn allows for the substantive evaluation of design alternatives.

This empirical foundation ensures that decisions are informed and aligned with defined objectives, rather than being speculative or solely intuition-based.

The strategy of "breadth before depth" advocates for casting a wide net to explore a variety of possibilities before narrowing focus. This approach is rooted in the principle that innovation often arises from the unexpected—it encourages architects to look beyond the conventional, thereby increasing the chances of discovering more innovative and high-performing design solutions.

An integrated perspective on project goals is essential. By considering multiple objectives concurrently, rather than in isolation, design teams can identify synergistic solutions that excel across various domains. This holistic view fosters designs that not only meet but exceed, the singular performance criteria, resulting in a harmonious blend of functionality, aesthetics, and sustainability.

Embracing life cycle thinking from the project's inception is also paramount. An understanding of a building's long-term impacts and needs allows for strategic planning that can significantly improve its lifetime performance. This forward-thinking approach equips architects to make decisions that are sustainable not just at completion but for decades to come.

Coupling the generation of design alternatives with analytical feedback creates a potent iterative process. This "generate-and-test" cycle facilitates the exploration of a broad spectrum of designs and their immediate evaluation based on how well they meet the objectives. This iterative loop can lead to a rapid convergence on optimal design solutions that might not be immediately apparent.

It is also essential to acknowledge that human biases often limit the exploration of the full design solution space. By employing computational tools and processes that systematically explore possibilities, teams can uncover options that would likely be missed due to inherent biases. These tools serve as a critical counterbalance to human tendencies that tend to favor the familiar.

Reframing design challenges to encompass a broader set of objectives and constraints from the beginning can lead to the inception of truly innovative design possibilities. This reframing can inspire teams to question the status quo and push the boundaries of what is conceivable.

Ultimately, this shift toward a more expansive, data-driven, and systematic exploration of design alternatives is not only a technical transformation but a socio-technical one. It requires a reevaluation of not just the processes and tools architects use but also the mindsets, educational frameworks, and the way collaboration occurs within the industry. Cultivating a culture that values data and prioritizes the environment and society in every design will catalyze a shift in how architecture is practiced and perceived, ushering in a new era of buildings that truly embody the goals of sustainability, performance, and human-centric design.

These tackle the scope and depth to discover substantively better performing design possibilities during conceptual phases. What does it look like when integrated into a practice? With regards to early understanding and broad exploration during conceptual architectural design, the following method can be used:

- *Dynamic generation*—Current practice tends to slightly optimize an initial design proposal, but dynamically generating alternatives from the outset based on integrated objectives can reveal higher performing options. Typically, designers propose an initial concept and iteratively refine it. But with the help of algorithms, they can systematically scan thousands of variations (not just a few), evaluating factors like energy use, material impacts, and comfort. The best performing alternatives challenge assumptions about the "best" solution, exposing trade-offs and synergies across objectives. An initial design, based on intuition, may not intrinsically perform as well against integrated priorities as dynamically uncovered designs that differ substantially. Shifting from optimizing first proposals to actively exploring possibilities through alternatives allows design innovation that realizes sustainability goals. By generating—rather than evaluating—options from the beginning, conceptual design makes a leap in performance by discovering configurations that are organically optimized.

- *Quantitative evaluation*—Meaningfully assessing design alternatives requires quantitatively measuring objective performance metrics. This requires using digital tools that can simulate and compare numeric data on factors such as energy use, comfort, material impacts, and costs. Subjective evaluation based on design aesthetics or intuition is insufficient to differentiate high-performing options from average or lower performers. Instead, data show real differences in quantified metrics like kilowatt-hours of energy used, carbon dioxide equivalent emissions, daylight illuminance levels, or construction dollars. Digital models and software simulations provide this quantitative performance data, expressing complex environmental and operational impacts in concrete numerical terms. Comparing alternatives based on diverse but digitally derived metrics reveals trade-offs that subjective evaluation could miss, helping designers identify the most suitable design considering project priorities and goals.

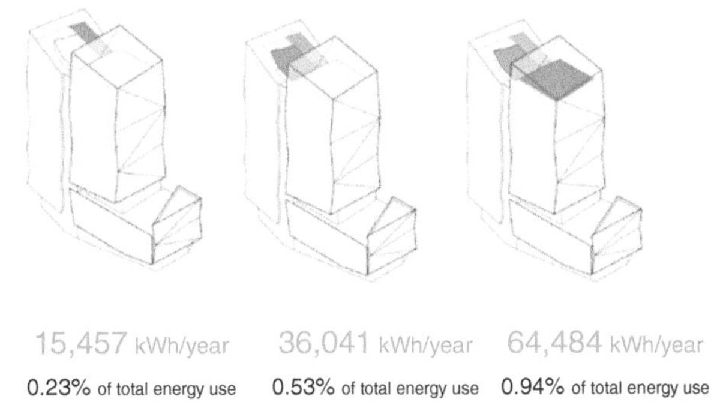

Example showing the impact of varying the amount of solar panels on the roof of the building on the energy production and the energy use intensity for the Midtown Union Building in Atlanta by Cooper Carry Architects.

- *Computational exploration*—In the light of the two previous points, manually evaluating a broad range of design alternatives at conceptual phases is infeasible, given the vast number of possible configurations and permutations. Computational methods can systematically explore the "solution space" at a scale far beyond human cognition, rapidly evaluating thousands of options against integrated objectives. Manual design exploration is limited by designers' mental models, preferences, and conventional assumptions. In contrast, algorithms can dispassionately scan the full spectrum of possibilities, uncovering high performing options that differ radically from designers' intuitions. Computational exploration techniques like multi-objective optimization and machine learning can generate and analyze design variations according to numerical performance criteria like energy use, carbon emissions, construction cost, daylighting, and thermal comfort. These techniques identify configurations that optimize objectives across factors, revealing synergies and trade-offs that expand designers' understanding of a project's potentials.

- *Life cycle thinking*—While current design practice tends to optimize for initial capital costs, exploring alternatives optimized for long-term performance metrics can potentially lower total lifetime costs. A life cycle thinking approach evaluates alternatives not just on upfront costs but on performance metrics over the full lifespan that impact operational and maintenance expenses. These key metrics include factors like energy use, maintenance requirements, material durability, flexibility for future changes, and impact on occupant health and productivity. Optimizing for high performance in these areas can lead to lower operating costs over time through reduced energy bills, less frequent repair and replacement, and a longer functional lifespan before major renewal is needed. A life cycle perspective reveals that though it begets higher upfront capital costs, these investments can pay for themselves many times over through avoided expenses and improved productivity over the full life of the building. Exploring alternatives based on life cycle metrics at conceptual phases allows designers to make high-level decisions that shape long-term value, resilience, and economic viability.
- *Multiple objectives*—Current evaluations often prioritize a single objective such as cost or program fit. Evaluating alternatives based on a single criterion like initial capital cost risks neglecting other important objectives that impact long-term value and sustainability. Considering multiple objectives together from the outset can reveal designs that perform well across diverse priorities. Optimizing for multiple objectives in an integrated manner uncovers solutions that satisfy diverse goals rather than compromising some for the sake of others, revealing innovative design options that emerge from holistic optimization rather than from prioritizing a single dimension of performance.

7.2.2 The Challenges of Managing the Design Space with Multiple Objectives and Constraints

Managing the design space during conceptual architectural phases can be challenging given the complexity of balancing multiple competing objectives and constraints that are inherent to high-performing building design. Addressing complex programs demands spanning spatial, functional, and relational requirements opens a sizable initial design space. As mentioned previously, architects still tend to narrow down and frame challenges, constraining possibilities. Optimizing for a myopic interpretation of a program risks missing opportunities revealed through broader framing encompassing life cycle perspectives. But it is understandable that some avoid it, simply because incorporating life cycle objectives amplifies dimensional complexity. Introducing performance metrics spanning capital costs, operational efficiency, end-of-life impacts, material health, and occupant well-being can be just way too much to handle, especially for "human computation" or the mind. Modeling interrelationships between these factors and evaluating tradeoffs across them through qualitative and quantitative methods is difficult and can only be properly done by using computation.

The key difference arises from a simple understanding that a decision tree is not the extent of the entire design space. As an example, if a design team follows a decision tree base framework and decides the massing first, followed by glazing percentage, followed by shading strategy, then HVAC, materiality, and so on at each decision, they are leaving behind hundreds and thousands of unexplored options that may be the right decision for the project.

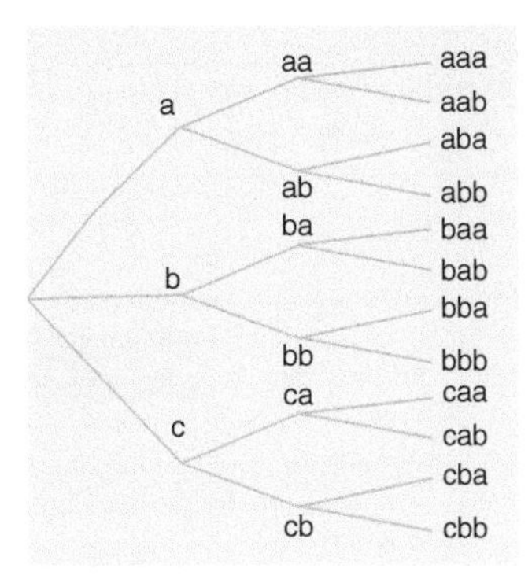

Image of a decision tree-based framework for design decisions.

Now if the same team is doing a multi-objective optimization across energy and building first cost, as an example, they can explore the full range of options before selecting the optimal one. Each dot on the graph below showcases one of the many alternatives across this optimization study, so that a design team can iterate and select the option that makes the most sense from both a first cost and energy use intensity (EUI) standpoint. The variables will vary by design, but this is the decision example at the concept design phase from the Health Sciences Research Building at Emory University.

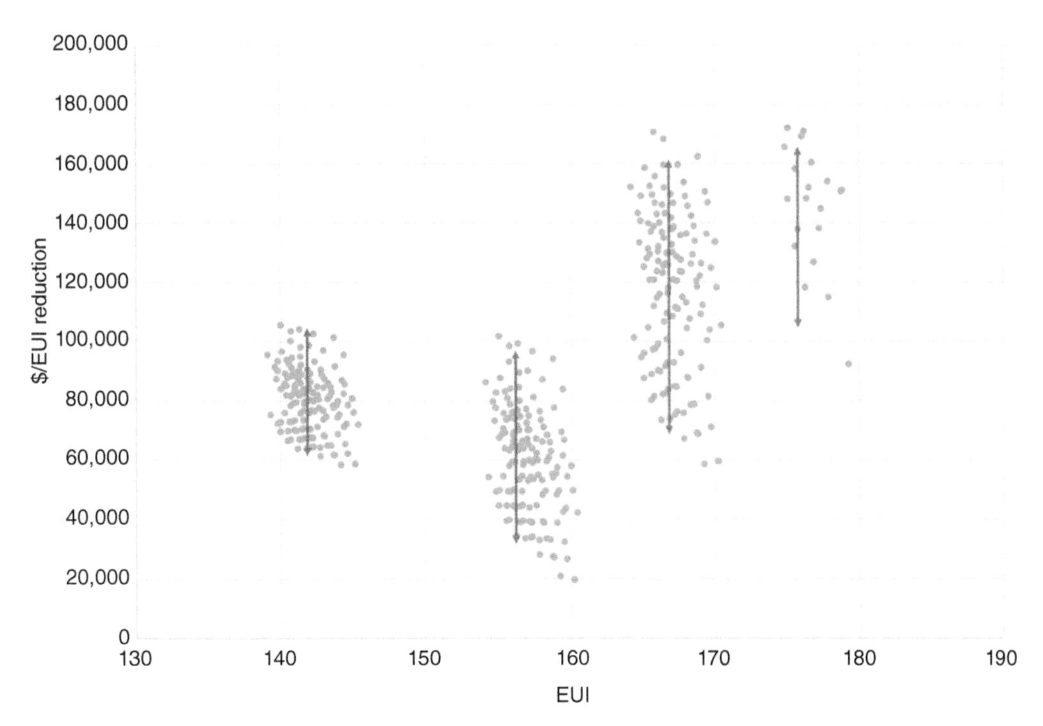

Highlighting all the possible alternatives within the design space on the decisions being considered across energy and cost.

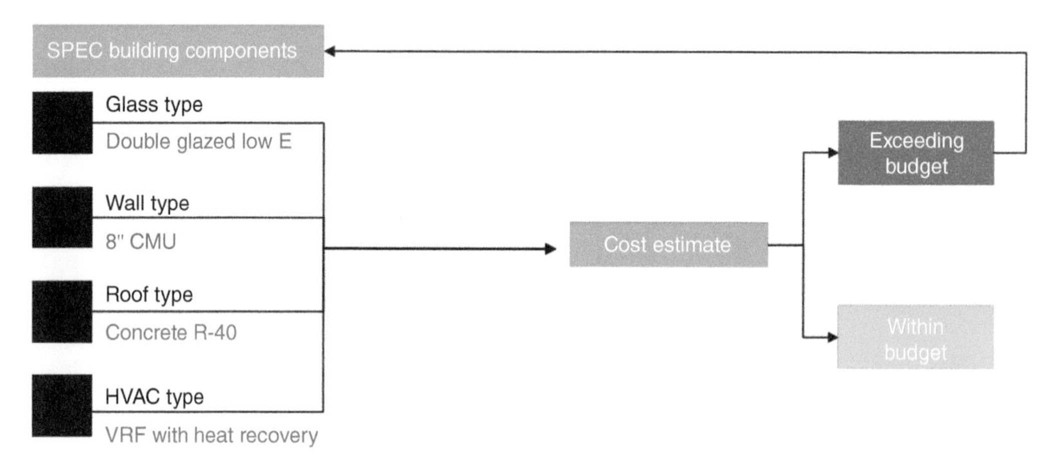

Highlighting the tree-based decision making methodology the leaves out the entire design space and leads to the design team looking at one variable at a time creating sub-optimal outcomes for energy and cost.

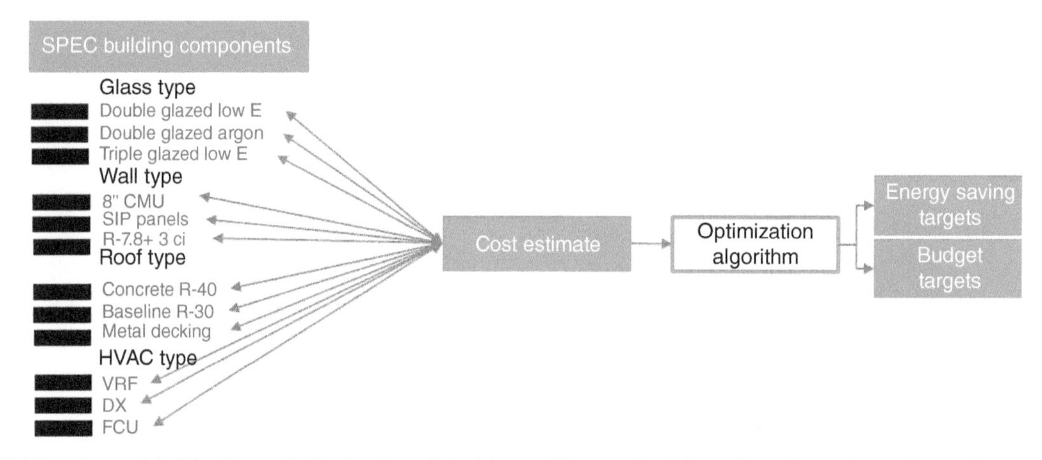

Highlighting multi objective optimization based decision making methodology that allows the entire design space to be considered and leads to the design team to the most optimal outcomes for energy and cost.

The endeavor to satisfy the multifaceted objectives of diverse stakeholders adds a layer of complexity to the architectural process. Clients come to the table with a suite of requirements that often extend beyond functional programs and fiscal constraints, venturing into the realms of aesthetics, brand identity, user experience and, increasingly, sustainability. This diversity necessitates the conversion of varied priorities into quantifiable metrics that can be employed to guide the exploration and evaluation of design spaces—a process fraught with challenges. Translating subjective desires into objective targets is inherently vulnerable to oversights and distortions.

Achieving a consensus on value amidst conflicting stakeholder interests is arduous. Decisions typically rely on subjective judgment, which inherently carries the risk of favoring certain stakeholders' values over others. The task of developing integrative value models that reconcile these divergent priorities is complex and often elusive in practice.

When technological, regulatory, and performance-based constraints are applied to the design process, they further narrow the realm of possibilities. Constraints are essential; they demarcate the boundaries within which designers operate, offering much-needed structure. Yet, transforming these constraints into generative and evaluative frameworks is not straightforward and is infused with subjective interpretations. This discrepancy is especially pronounced as architects individually translate constraints and balance them against the project objectives, often resulting in trade-offs. The current architectural practice is not optimally equipped to formalize and incorporate these constraints systematically in ways that meaningfully influence the exploration of the design space.

Navigating such complex, multidimensional design spaces is a cognitively demanding task, one that is becoming increasingly intricate as projects escalate in complexity. While computational techniques hold the promise of enabling the systematic generation and evaluation of design alternatives at a scale that surpasses human capability, their application in everyday architectural practice remains remarkably limited. These methodologies largely remain confined to experimental settings and research laboratories, rather than being integrated into mainstream architectural workflows.

The challenges of managing design spaces during the conceptual phases of architecture include a multitude of factors:

- The interplay of performance metrics such as energy usage, comfort, and constructability is convoluted and challenging to represent and assess comprehensively.
- Design options often display emergent behaviors—unexpected results that arise from the combination of objectives and constraints in nonlinear ways—making prediction based on individual factors difficult.
- Subjectivity permeates the conception of what is deemed "good" design, with valuation frameworks that vary significantly among individuals and communities, complicating the establishment of shared criteria.
- Prioritizing objectives often necessitates compromising on others, sparking value tensions where trade-offs lack clear objectivity and require nuanced negotiation.
- A partial understanding of the relationships between design attributes and their performance outcomes hampers the ability to fine-tune designs with precision.
- Despite the capacity of computational tools to explore numerous design alternatives, current algorithms often fall short when grappling with the intricacies of complex design tasks, emergent behaviors, and value tensions.
- Data gaps present a significant barrier to evaluating design alternatives with sufficient depth. More comprehensive data on performance outcomes is imperative to forge robust models for navigating design spaces.
- Beyond the technological advances needed, there is a call for a socio-technical transformation within the architecture discipline—a shift in mindsets, collaborative processes, roles, and responsibilities that supports the integration of advanced computational methods into standard practice.

The convergence of these factors underscores a pressing need for an evolution in architectural methods—one that embraces a holistic, data-driven, and computationally supported approach to designing spaces that align with the complex tapestry of human needs and environmental constraints.

7.2.3 How Limited Time and Budget can Constrain the set of Design Options Tested During Conceptual Design

In the realm of architectural design, time and budget constraints are formidable factors that directly impact the breadth and depth of conceptual exploration. The industry's prevailing business models, with a focus on maximizing billable hours, conspire to limit the investment in technology to all but the most lucrative and complex projects. This economic framework promotes a culture where immediate utilization and client billing take precedence over the potentially transformative benefits of technological advancement.

Time constraints center around limited durations afforded to schematic design phases within traditional project workflows. In many cases limited time is allotted for the schematic design phase, often just 10–15% of the total project budget, enforcing a need for rapid completion. This time allocation is scarcely sufficient for the thorough iterative processes that conceptual design demands. With such a narrow window, the rich potential for exploring diverse design options is markedly reduced. Given that the early stages of design set the trajectory for 60–80% of a building's lifecycle impacts, this time constraint is at odds with the holistic view needed for creating sustainable and high-performance buildings.

Moreover, the traditional billing structures within architecture firms impose budget constraints that deter extensive investment in computational tools and exploration methods. These tools, which include advanced algorithms and machine learning techniques, are perceived as overhead costs rather than essential investments, thus receiving scant funding. As a result, most computational efforts, which should be integral to the conceptual design phase, are relegated to overhead, with their extensive potential untapped in everyday practice.

The current framework leads to a paradox where the tools and technologies that could significantly enhance design options, reduce costs, and streamline processes in the long run are paradoxically seen as burdens on the firm's financial resources. This has created a conservative stance toward technology adoption, stifling innovation and reinforcing a cycle of underinvestment.

Design team planning for a new project. Source: Rawpixel.com/Adobe Stock.

This financial myopia hampers the shift toward more data-driven, performance-oriented design practices that could revolutionize the industry. There is an evident need for a business model revolution within architectural practices—a transformation that recognizes technology as a pivotal element for future-proofing the industry, promoting profitability, and responding to the increasing demands for sustainability and performance optimization. Only with such a recalibration will the industry harness the full spectrum of design possibilities and move toward a more innovative, efficient, and sustainable future.

7.3 Case Study: Phase 2 of Health Sciences Research Building

Health Science Research Building II Project at the Emory University Campus in Atlanta, Georgia. Source: Cove Tool, Inc.

7.3.1 Project Profile

The Health Sciences Research Building II (HSRB II) at Emory University, an impressive 300,000 square feet facility, represents the second phase of a new biomedical research center. It is designed to meet the burgeoning space requirements and to provide efficient, flexible research environments. With 60% of its space dedicated to pediatric research and the remainder to cancer, immunology, and drug discovery, the facility stands at the forefront of scientific exploration. In line with the University's sustainability commitments, HSRB II aims to achieve high-performance targets.

7.3.2 Understanding the Climate

The team began with a detailed analysis of the local climate, an essential step in performance-based design. This analysis influenced decisions on building massing, orientation, and glazing to ensure energy efficiency without additional cost burdens.

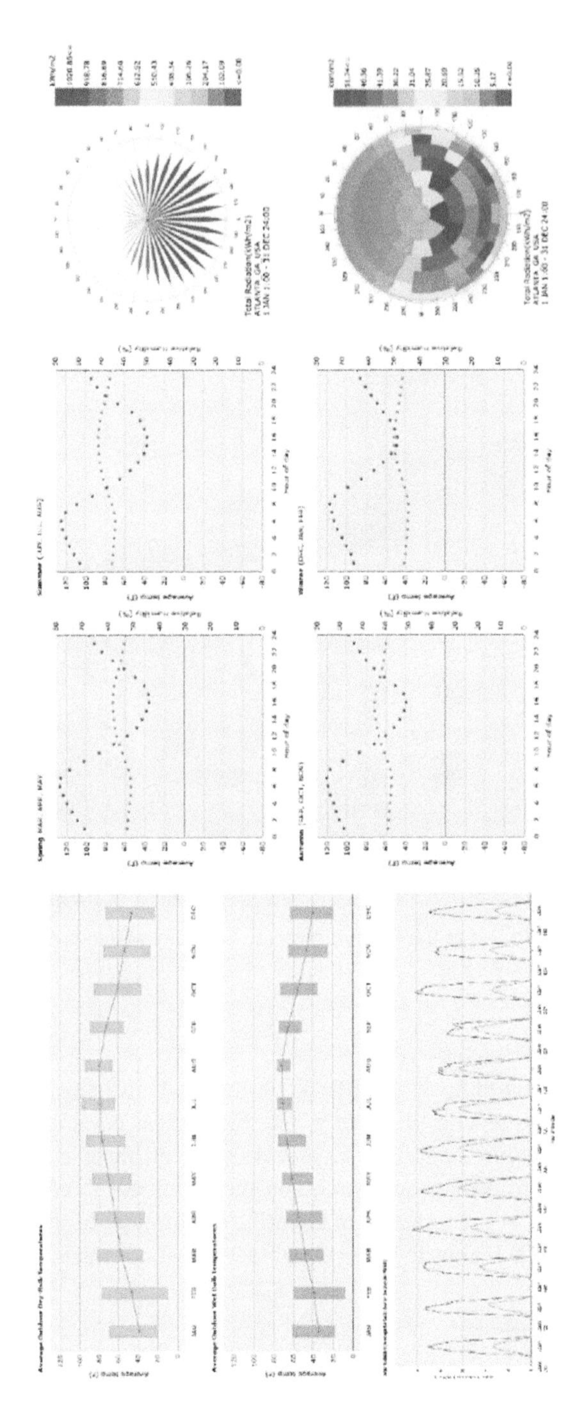

Detailed local climate analysis for the Health Sciences Research Building II project at Emory University.

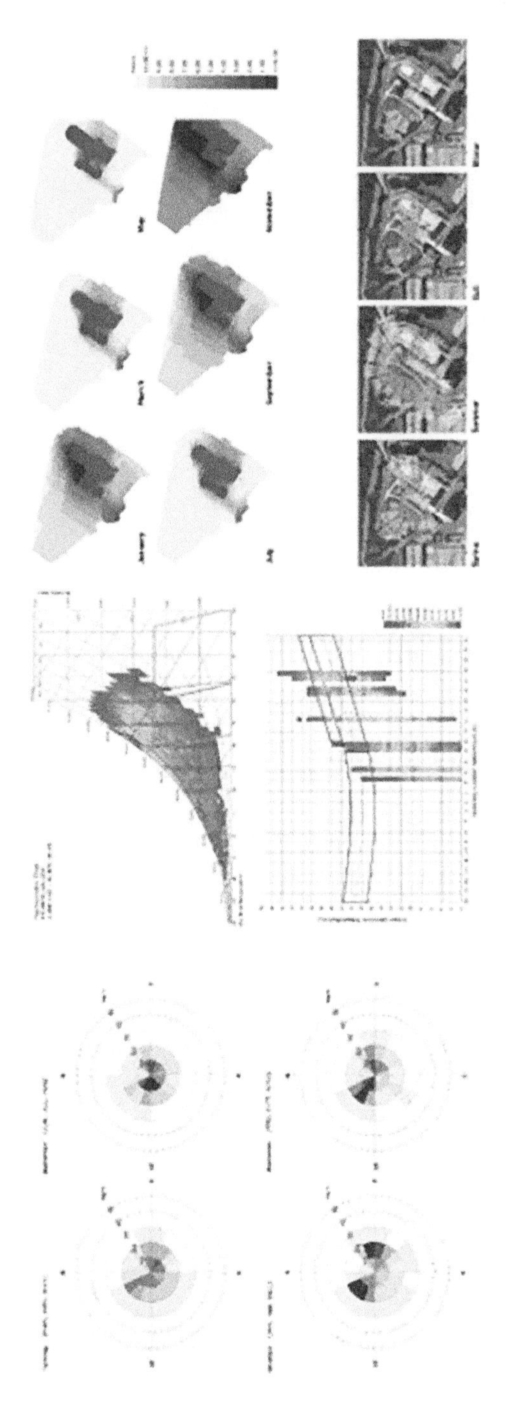

Detailed local climate analysis for the Health Sciences Research Building II project at Emory University.

7.3.3 Massing Study

Performance Based Decision Making—The team explored 10 different massing options, assessing each for energy use intensity (EUI) and spatial daylight autonomy (sDA). By combining side-by-side comparison and parametric analysis, they efficiently narrowed down the choices to two optimal schemes.

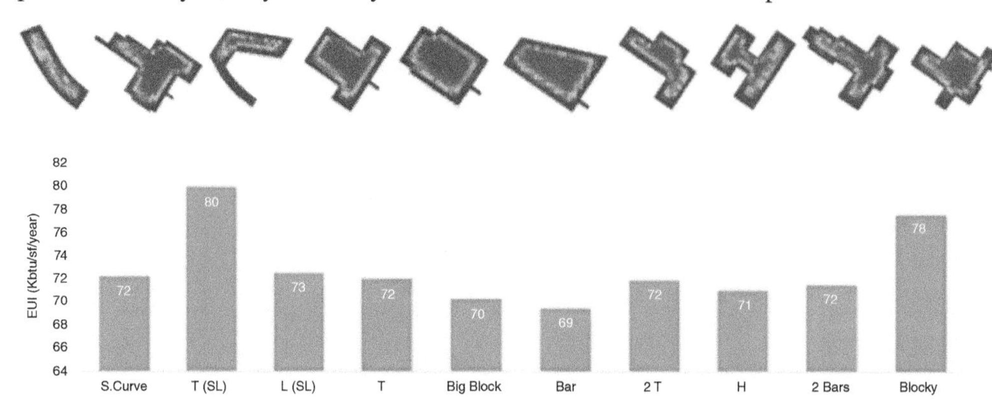

Comparing massing schemes at the concept design stages across the Energy Use Intensity and Spatial Daylight Autonomy metric.

7.3.4 Massing—Parametric Comparison

The design team concentrated on a parametric-driven decision-making approach, leveraging tools like cove.tool to interlink various design parameters. This method allowed them to comprehensively compare different architectural options, visualizing the range and impact of each choice. A key component of this approach was the use of a parallel coordinate graph, which mapped the 10 massing options. This graphical representation was instrumental in providing a clear, visual understanding of how each option performed across various parameters, facilitating informed and holistic design decisions.

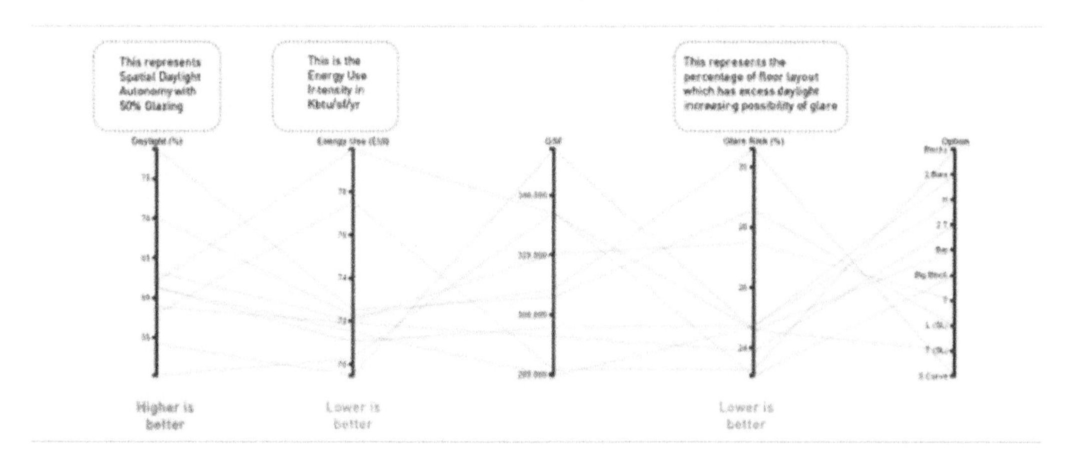

Highlighting an early stage multi-objective optimization across the variable of 10 different massing schemes and decision metric of energy, daylight and glare.

With the use of a parallel coordinate graph, users are able to weed out certain bundles which do not hit the users' bottom-line goals. Here, two from the ten remain after the design team calibrated the graph to ensure the massing options would have a low EUI, an in-between gross square foot area, and low glare risk.

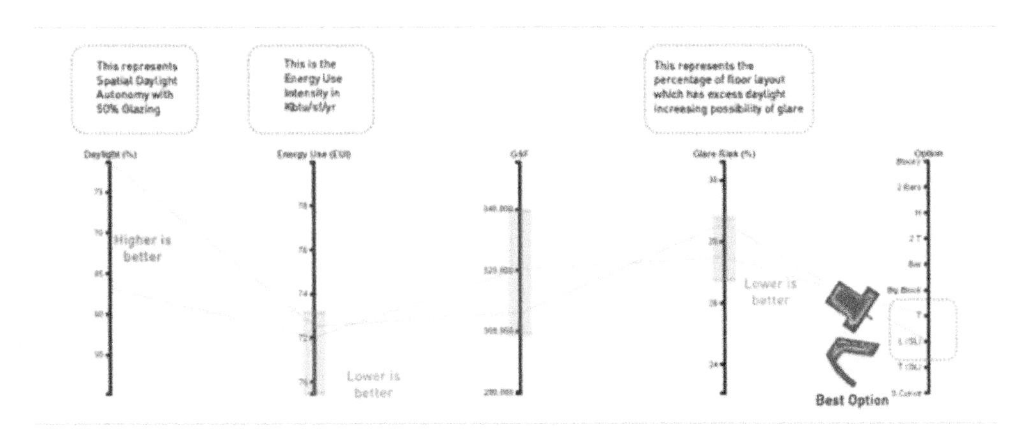

Utilizing the filtering technique in the optimization to set the decision criteria and aligning at the optimal outcomes.

Schemes which performed the best include Massing—L-Scheme.

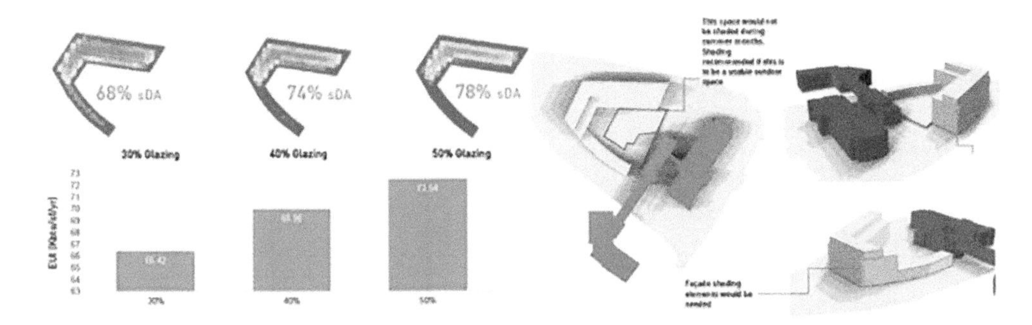

and massing T-Scheme.

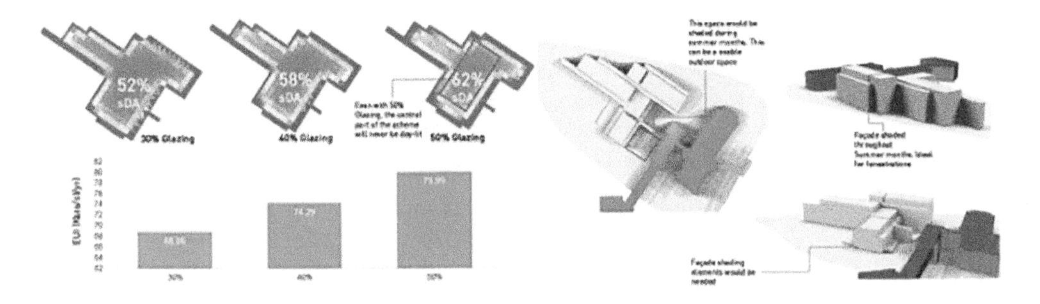

A look at the shadow, energy and daylight performance of the shortlisted scheme in cove.tool.

7.3.5 Setting up Benchmarks

Establishing benchmark data was vital for informed decision-making, guiding the team in setting realistic and achievable performance targets.

7.3.6 Cost Versus Energy and Carbon Optimization

The team undertook a detailed sustainability review, focusing on optimizing cost and energy usage. They explored various building technologies and strategies, evaluating their impact on the overall energy performance of the building.

Initial optimization graph showcasing the 60,000 bundles.

Here (below), the platform recommends a top bundle before any additional options are changed.

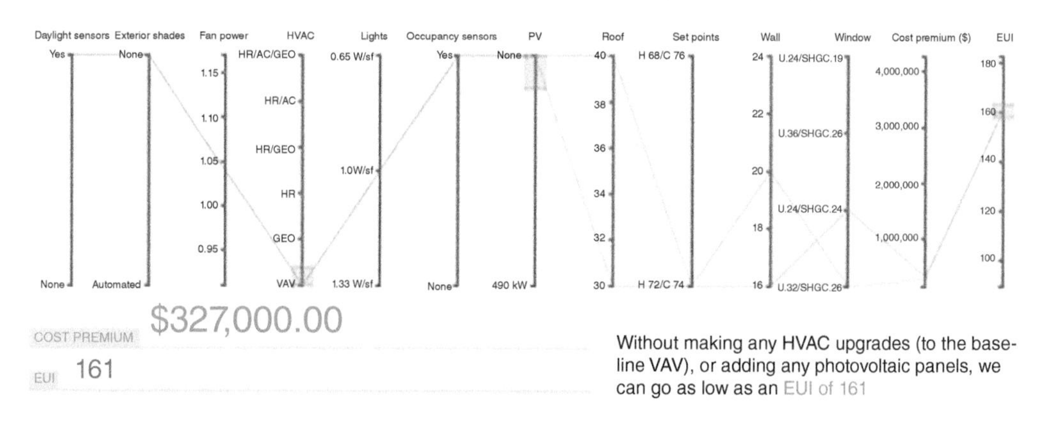

A cost vs energy optimization in cove.tool: Each vertical axis is a potential decision that needs to be made, this includes: Daylight and occupancy sensors, exterior shading devices, varying the fan power, HVAC type, lighting power density, solar panels, the roof assembly performance, the internal set-points, the wall and window assembly performance. These variables are measured for optimal energy use intensity and upfront cost investment. The filtering is utilized to arrive at optimal outcomes based on design needs and project goals.

One parameter, HVAC system, is selected and new optimal bundle arises.

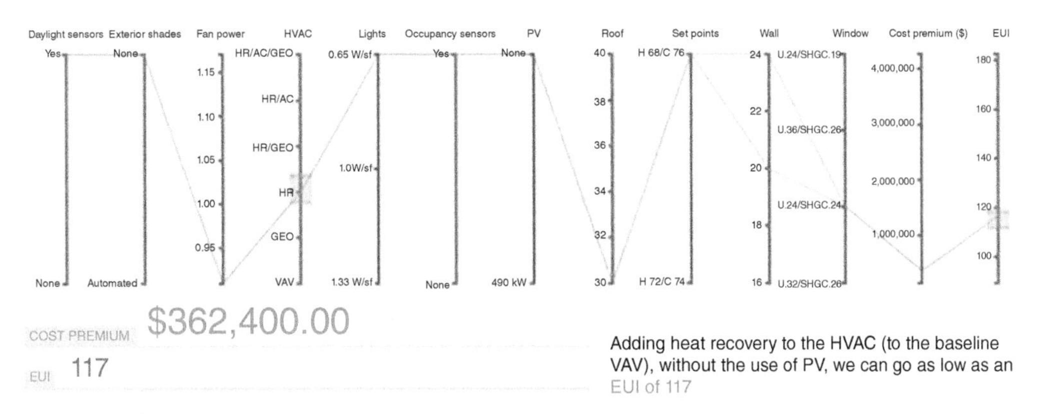

Adding heat recovery to the HVAC (to the baseline VAV), without the use of PV, we can go as low as an EUI of 117

Continued calibration of parameters and design team requests show how low the possible bundles could go.

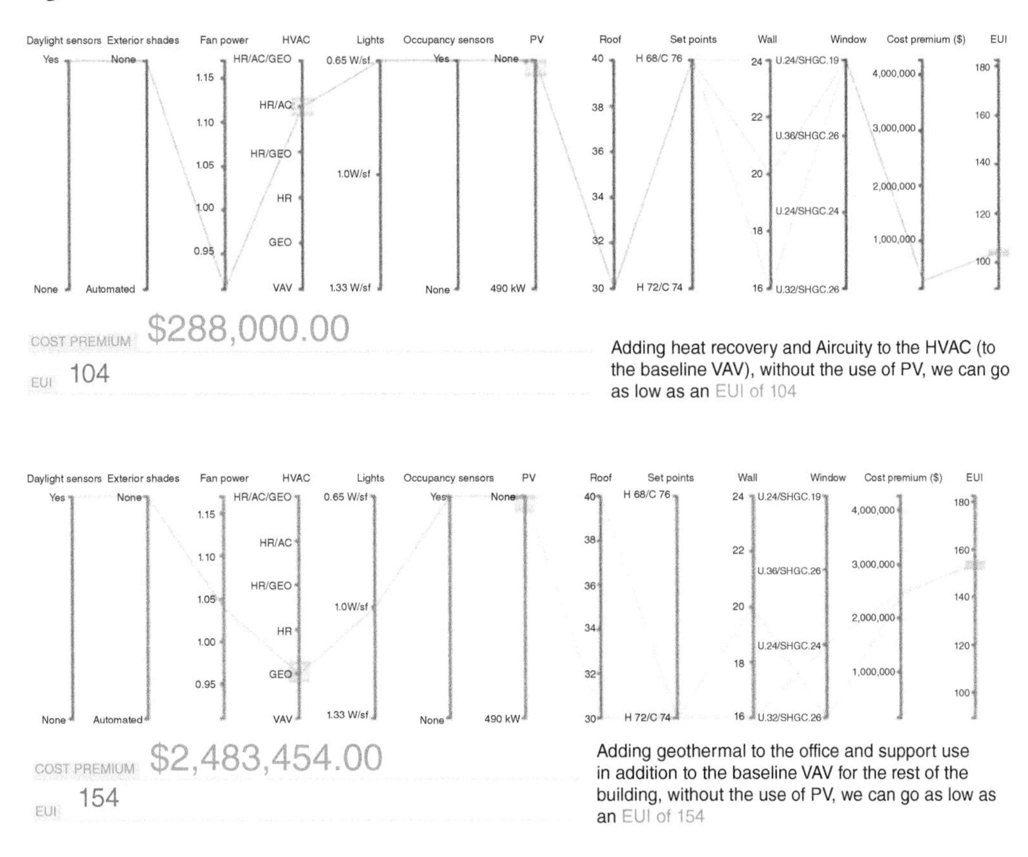

Adding heat recovery and Aircuity to the HVAC (to the baseline VAV), without the use of PV, we can go as low as an EUI of 104

Adding geothermal to the office and support use in addition to the baseline VAV for the rest of the building, without the use of PV, we can go as low as an EUI of 154

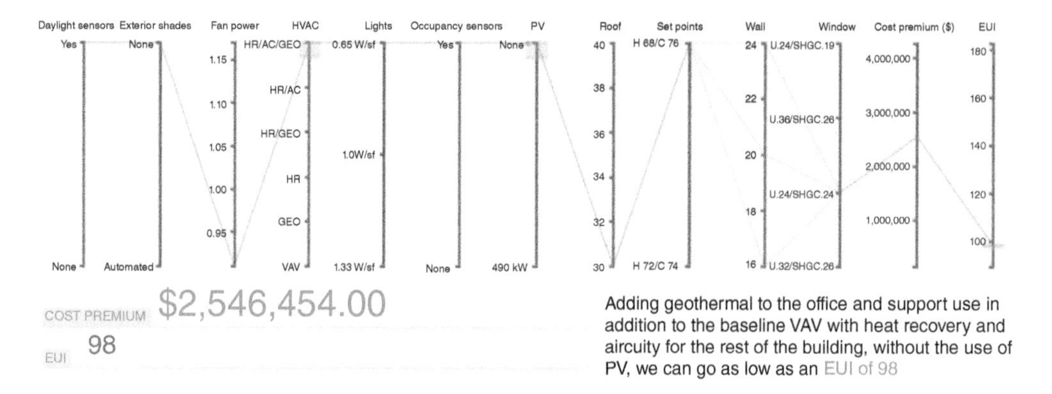

7.3.7 Comparing Best Bundles at Each EUI Segment

The design team further conducted an analysis to evaluate the relationship between cost and system/strategy combinations across different EUI segments. The team generated a series of graphs that illustrated the impact of various combinations on specific EUI targets. This analysis enabled the team to effectively shortlist potential strategies for each EUI segment, streamlining the decision-making process.

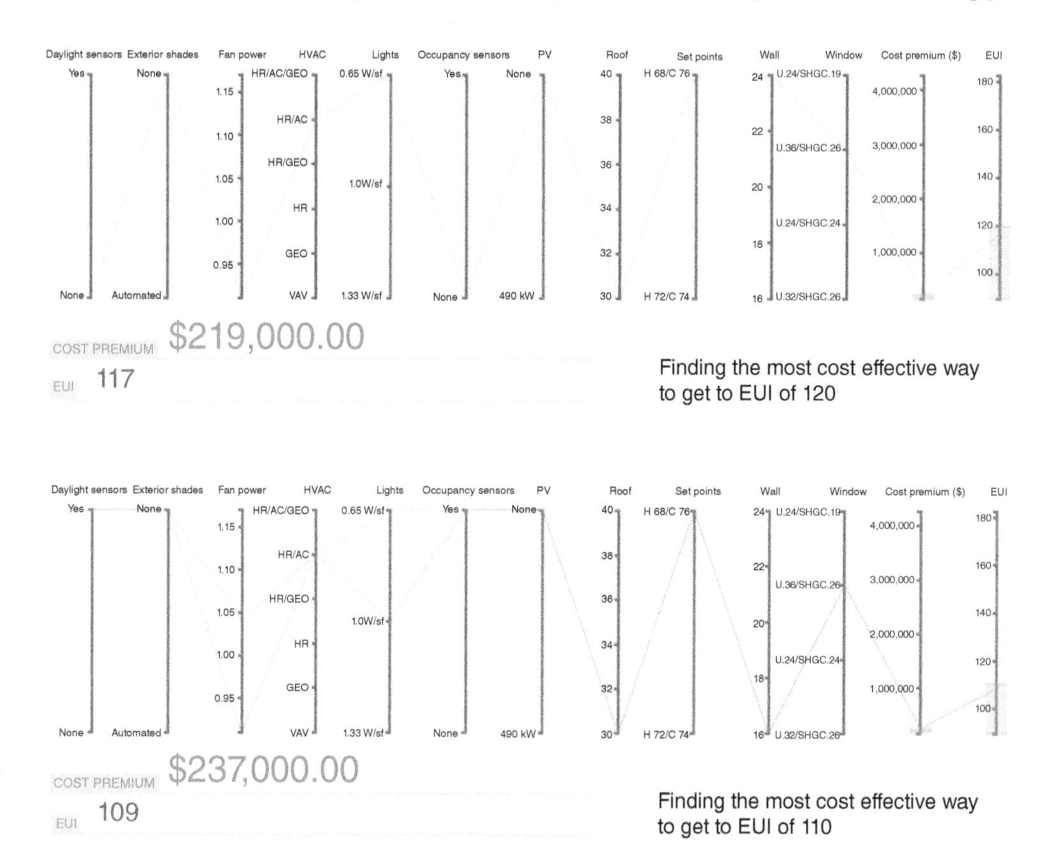

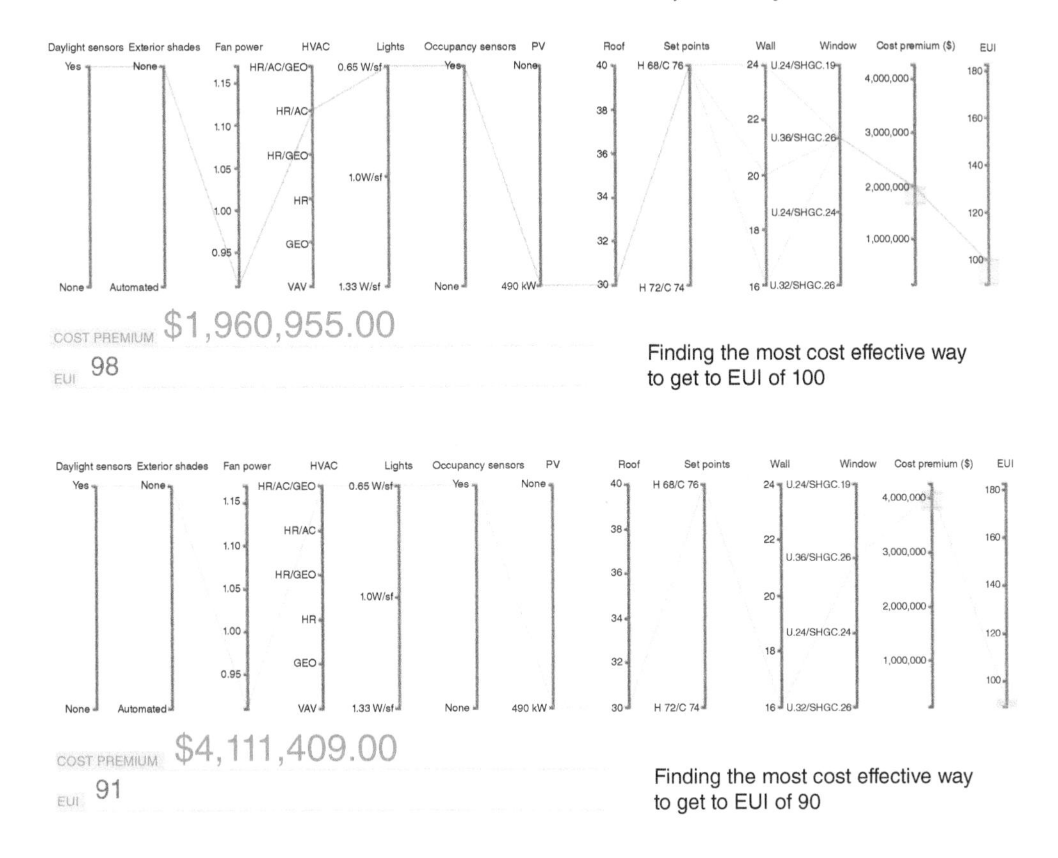

7.3.8 Key Observations

1. Automated shades are not selected in any bundle.
2. Getting to an EUI of 109 is US$1,720,000 cheaper than EUI of 98 Kbtu/ft^2/year.
3. Wall *R*-Value of 20 ft^2·°F·h/BTU is optimal.
4. Roof *R*-Value of 30 ft^2·°F·h/BTU is optimal.
5. Lowering fan power is a cost-effective strategy to lower EUI.
6. Variable air volume (VAV) with heat recovery and Aircuity are cost-effective strategies to lower EUI.
7. Photovoltaic panels are more effective than geothermal.
8. Geothermal may not be a permissible strategy per the site constraints.
9. Lighting power density of 0.65 w/sf should be the target.
10. Daylight and occupancy sensors are cost effective strategies to lower EUI.

Analyzing all the different strategies and system types proposed, and verifying their effect on the EUI of the designed building suggests a careful and comprehensive methodology used by the design team to assess these strategies. The suggestions focusing on the system type to be used is dependent strongly on the overall EUI goal. Based on the established benchmark, a low energy building would have a EUI of 100 Kbtu/ft^2/year.

7.3.9 Suggested Bundle

The team utilized a conservative shoebox energy model which did not fully account for the actual usage patterns of the building's occupants. The comparison between this standard model and a more detailed, usage-based schedule revealed a notable discrepancy in EUI—as much as 10%. This variance highlighted the importance of aligning the model more closely with real-world usage to achieve an EUI of 100 Kbtu/ft^2/year or lower.

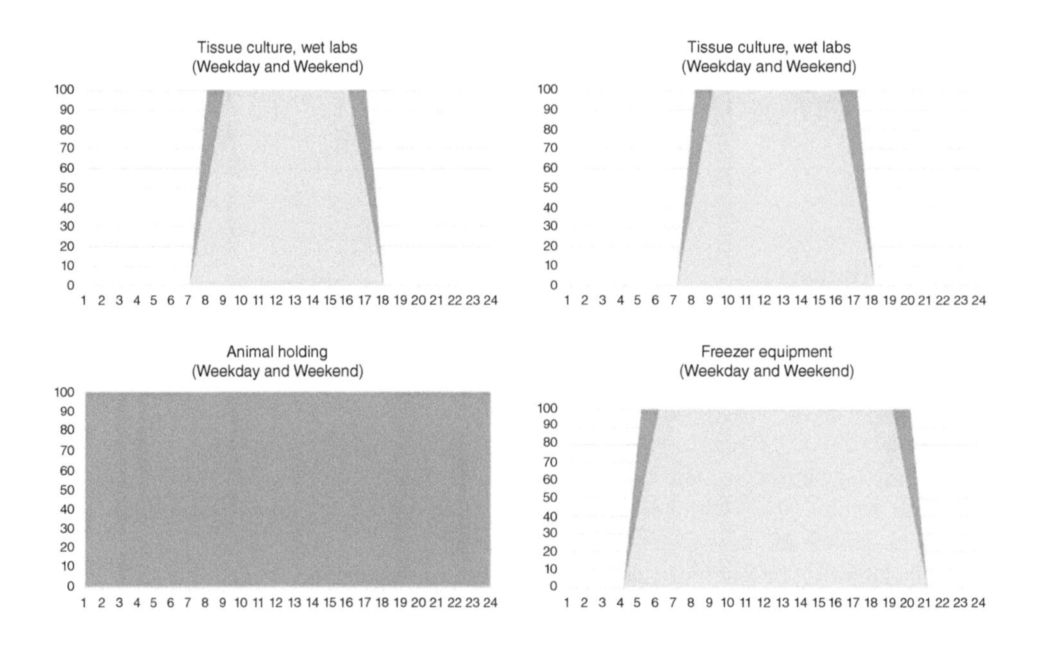

To further refine the energy model, the team considered examining usage data from the first phase of the HSRB I. Questions arose about the building's occupancy patterns, such as periods of low usage or months with fewer researchers. By adjusting variables like air changes, fume hoods, lighting, appliances, and set points during these less active periods, the team recognized the potential for significant energy savings.

7.3.10 Data-driven Decision-making

The final design proposal integrated a combination of mechanical systems and energy-saving strategies to achieve the targeted EUI of 100 Kbtu/ft^2/year. Additionally, the team made recommendations for daylight and glare improvements to meet LEED V4 standards.

In conclusion, this project exemplifies a comprehensive, data-driven approach to sustainable design, balancing the complex interplay of energy efficiency, cost optimization, and environmental considerations to create a state-of-the-art research facility.

7.4 Chirag Mistry AIA, LEED AP | Senior Principal, Regional Leader of Science + Technology | HOK | Atlanta, GA

Chirag Mistry, a seasoned principal and leader in HOK's Science + Technology practice based in Atlanta, holds a pivotal role. He is responsible for orchestrating the programming and design of cutting-edge science, research, and education facilities in the southeastern United States, encompassing both new projects and renovations. Mistry brings specialized proficiency in crafting engineering laboratories, intricate bioscience and biomedical research laboratories, as well as complex high containment facilities to the table.

Notably, Mistry has played a vital role in shaping HOK's standards for building information modeling (BIM). Additionally, he possesses a knack for developing business intelligence tools tailored for lab programming, planning, and design. His expertise extends to being a sought-after speaker, sharing insights on leveraging advanced technological resources to enhance the design process.

In the world of decarbonization, Chirag Mistry, a seasoned architect with two decades of experience at HOK, stands as a visionary. His journey through the evolving landscape of sustainability in the built environment has been marked by a significant shift in client awareness and the advent of financial incentives. In a recent interview, Chirag shared his insights into the high-level trends and opportunities he envisions for decarbonization in real projects in the years to come.

7.4.1 A Shifting Landscape: Client Awareness and Responsibility

Chirag's career at HOK has spanned a period of profound transformation in the architecture and construction industry. He recalls the days when LEED certification was a novel concept, and sustainability was a challenge to convey to clients. However, today, Chirag observes a remarkable shift in client attitudes and priorities. Clients are now acutely socially aware of their responsibilities in constructing sustainable buildings. He notes, "Now, more than ever, I've noticed that clients are socially aware of their responsibilities and what they're doing from a building perspective."

Chirag emphasizes that this change is not merely a surface-level shift; it is a fundamental reorientation of client objectives. Sustainability is no longer a stretch goal; it is a specific and non-negotiable requirement. He adds, "I see that being more and more, where it's more a specific goal and requirement from the client side."

7.4.2 The Comprehensive Approach: Beyond Design and Construction

Decarbonization is not confined to the design and construction phases; it extends to building operation. Chirag sees this as a pivotal aspect of the evolving landscape. Corporate clients with environmental, social, and governance (ESG) goals, as well as higher education and public sector entities, are all joining the sustainability movement. Chirag states, "I definitely see a big, big push in decarbonization moving forward than it ever has been on projects."

7.4.3 Technological Advancements: A Catalyst for Change

Technology is playing a transformative role in the journey of decarbonization. Chirag is enthusiastic about the analytics-driven insights that are now available during the design process. He stresses, "It's extremely helpful to have abilities and tools at our disposal so that we can really see in real-time the choices we make and what impact it can have on the project."

7.4.4 Structural Focus: A Targeted Approach

Chirag believes that structural components have the most significant impact on a building's embodied carbon. HOK has been actively researching optimized structural solutions, exploring options like sustainable concrete and mass timber. He cites the example of an Emory project, where HOK reduced embodied carbon in the structure by approximately 15% and says, "We were able to reduce our embodied carbon on the structure... it was a big push to get to that point."

7.4.5 Client Education: A Crucial Role for Designers

Educating clients about the implications of their choices is another key aspect of Chirag's vision for decarbonization. He notes, "I think there are also newer technologies that are coming to the marketplace that can help push the decarbonization effort further. It is a designer's responsibility to recommend and educate clients on healthy material choices or efficient structural design. No matter the scale, every carbon impact can go a long way in the overall building scale." Chirag emphasizes the role of technology in educating clients. Visualization tools and analytics can help clients understand the environmental and financial implications of their decisions.

7.4.6 A Bright Future for Decarbonization

Chirag Mistry's insights, accompanied by his quotes, paint a picture of a dynamic and promising future for decarbonization in the built environment. The evolving landscape is driven by socially responsible clients, financial incentives, technological advancements, and a commitment to transparency. As architects and designers continue to embrace these changes, the industry is poised for a sustainable and low-carbon future.

7.5 Consequences of Process Deficiencies

7.5.1 How Process Deficiencies Can Lead to Design Solutions with Poor Lifecycle Performance, Focusing on Environmental Performance

The current design process is fraught with inefficiencies that significantly impinge on the environmental performance of buildings throughout their lifecycle. The foundational issue is the inadequacy of early-stage evaluations that are comprehensive enough to encompass a wide array of environmental impacts. This deficiency leads to design outcomes that are often resource-intensive and environmentally burdensome over their operational life.

Despite the design industry's professed commitment to sustainability, this ambition frequently fails to transcend the realm of theory. While sustainability is championed in the rhetoric of branding and industry trends, the transition to tangible actions and measurable environmental improvements is lacking. The architectural sector's practices of retrofitting green strategies onto preconceived design concepts undermine the potential for embedding sustainability at the genesis of the design process. Life cycle assessments and rigorous energy simulations are seldom used as guiding lights in the schematic phase, resulting in design solutions that miss the mark on sustainable optimization of orientation, massing, envelope, and systems.

In most projects up to 60%—of a building's environmental impact is determined by the early design decisions related to form, material selection, and system implementation. These critical choices are often made without the necessary foresight into their long-term resource and emissions implications.

Moreover, the industry's approach to testing design alternatives is often myopic, focusing on slight iterations of a primary proposal instead of casting a wider net over fundamentally different concepts. Digital simulations have shown that when massing options are generated based on parameters such as embodied carbon and energy efficiency, the resulting conceptual designs can outperform original proposals in terms of environmental metrics. This highlights a stark misalignment between the potential for environmental impact reduction through broad-ranging exploration and optimization and the current practice of narrowly scoped conceptual testing.

The prevailing approach in the architectural process, favoring incremental development over basic innovation, inherently limits the scope of sustainable design. The linear trajectory of the design process typically favors familiar and perceived cost-effective solutions over those that may offer substantial environmental and profit increasing benefits but require a departure from conventional methods. There is a pressing need for a paradigm shift in the conceptual phase that encourages the generation and evaluation of a diverse array of designs from an integrated environmental standpoint. Such a shift is imperative to break free from the current trajectory and to make strides toward truly sustainable building practices that prioritize long-term environmental performance over short-term thinking.

The journey to a sustainable future begins with a reimagining of our problem-solving frameworks. By expanding the initial problem framing to prioritize life cycle environmental objectives, we can foster a culture of generative exploration where sustainability becomes the cornerstone of innovation. This shift in perspective allows architects to embark on design trajectories that are environmentally conscious from the outset, ensuring that each decision is made with a long-term vision.

To steer the industry toward a sustainable horizon, we must redefine what it means to optimize. The conventional targets of spatial efficiency, cost, and aesthetics must be balanced with, if not secondary to, the imperative of environmental performance. This reorientation ensures that the ultimate measure of a design's success is its contribution to long-term sustainability.

Criteria for evaluating design alternatives need to be comprehensive, embracing the full spectrum of environmental impacts. This holistic approach to assessment would require architects to consider not just individual elements, but the integrated life cycle impacts of their designs, including material provenance, energy consumption, and emissions over the entire life span of the building.

Addressing the limitations imposed by shortsighted time horizons requires architects to extend their vision beyond the immediate phases of construction and design. By incorporating operational and end-of-life considerations into their early designs, architects can anticipate and mitigate decades of resource use and emissions, crafting buildings that are truly designed for the ages.

The power of experimentation must be harnessed more effectively. Beyond traditional analog methods, embracing advanced digital simulations and optimization techniques can uncover a broader array of high-performing design options. By leveraging technology, architects can simulate countless iterations rapidly, each evaluated against stringent environmental performance metrics, thus illuminating paths to sustainable outcomes that might otherwise remain unexplored.

Embracing complexity in conceptual design can yield sophisticated solutions that align with environmental goals. Moving beyond the allure of simplicity and visual coherence opens a world where multifaceted, high-performing designs can thrive, marrying form with function in harmony with the environment.

Integrating sustainable design principles from the very beginning is paramount. Rather than retrofitting environmental considerations onto a nearly finished design, these principles should be embedded in the initial brief, guiding the generation of alternatives and ensuring that sustainability is not an afterthought but the guiding star of the entire architectural process.

7.6 Leveraging Technology for Early Decision-Making

7.6.1 Exploring the Transformative Impact of Technology on Decision-making at the Early Stages of Design

The landscape of building design has undergone a significant metamorphosis over the past three decades, thanks to technological advancements. The advent of 3D modeling and BIM software marked the beginning of this transformation, and today, we stand on the brink of a new era defined by the synergistic power of artificial intelligence (AI), advanced building information network (BIN), and multifaceted simulation software. These tools empower architects to make informed, precise decisions early in the design phase, which is crucial for the successful outcome of any project. Their use extends beyond enhancing precision and visual realism; they provoke a rethinking of our approach to architecture and its role within the environment. This shift not only streamlines the design process but also elevates the quality of the final construct.

BIN concepts, serving as a real-time digital twin of the proposed structure, encapsulate detailed insights into the building's physical and functional aspects. This method enables analysis for structural integrity, energy efficiency, and economic feasibility at the early stages of design. The accessibility of user-friendly simulation software marks a pivotal step forward, allowing us to simulate the building's performance under a variety of scenarios, including extreme weather events or emergencies. This capability to observe dynamic, real-time changes supersedes the previous limitations of static visualization, enhancing our predictive understanding of a building's safety and operational behavior.

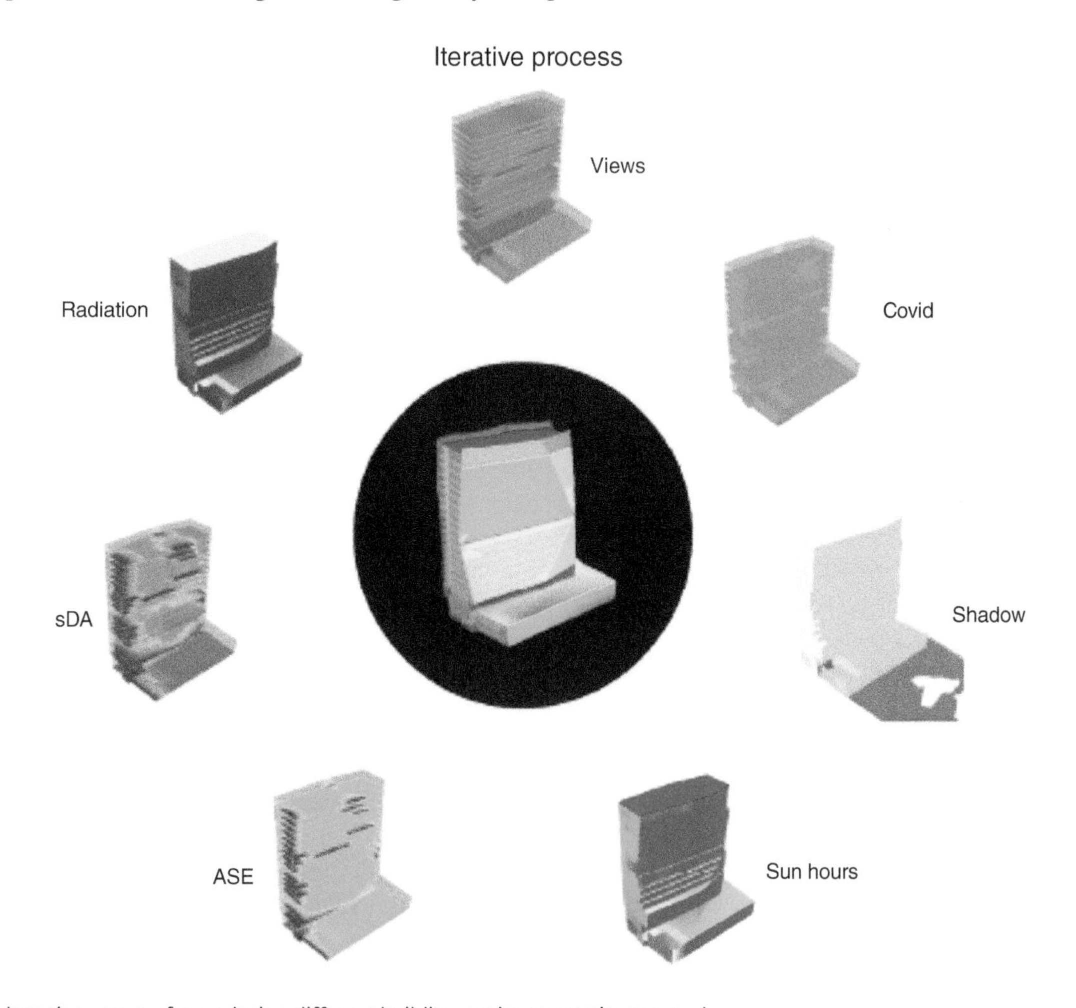

Iterative process for analyzing different building environments in cove.tool.

Being able to *analyze the environmental impact* of the designs elevates concern for sustainability in the building industry. And the more architects need to consider the environmental impact of their designs—which there is a push for in the world of policy—the more technology makes it possible to determine energy efficiency as well as the building's carbon footprint. This is key for making decisions about the materials and systems used in the building, that will be having a significant environmental impact.

Real-time collaboration on projects across geographical boundaries can be done via cloud-based software. Design teams can work side by side on a project from anywhere in the world, creating a diverse group of professionals who may bring different perspectives and expertise to the project, leading to more innovative and creative designs that are better suited to the needs of the community. Gathering data about how people use the building and then using this data on how people move through and use the building will help optimize the layout and improve traffic flow.

Leveraging machine learning for data analysis, architects can now sift through vast data sets to uncover patterns and trends that might otherwise elude human detection. This brings us to the cutting edge of parametric design, where designs are not static but dynamic, responding in real-time to site conditions, environmental variables, and user needs. Through computational algorithms, this approach optimizes building forms, material utilization, and energy consumption. Parametric design still carries the spirit of experimentation, birthing buildings that are resource-efficient and environmentally attuned, promoting a seamless integration of our built spaces with the natural environment.

To summarize:

1. *3D Modeling Software*—A revolution in the way architects approach design, allowing the creation of accurate and realistic representations of designs in a virtual environment and in real time, which can help make informed decisions about the design early on in the process.
2. *Virtual Reality* (*VR*)—Virtual reality brings immersive visualizations of the designs, which helps gain a better understanding of how space will be experienced and identify potential design flaws. Additionally, VR visualizations are interactive, which helps with presenting design to clients with a lesser ability to imagine designs in real-life scenarios.
3. *Building Information Network* (*BIN*)—BIN is a real-time digital representation of the building that includes information about its physical and functional characteristics in contrast to BIM files maintained by individual teams. This technology enables analysis of the design for structural integrity, energy efficiency, and cost-effectiveness early on in the process. BIN also makes it possible for all parties involved in the project to view and edit the same design model, which can eliminate the need for multiple versions of the design and reduce the risk of errors.
4. *Simulation Software*—Simulation software allows simulating the behavior of the building in different scenarios, identifying potential design flaws and making the necessary changes to ensure the safety and functionality of the building.
5. *Environmental Analysis Tools*—Analyzes the energy efficiency of the designs, as well as the building's carbon footprint, giving out necessary information about materials and systems used in the building to reduce its environmental impact.
6. *Parametric Design*—Parametric design software allows the creation of designs that are tailored to the specific needs of the client or community. The design software can be used to create designs that are optimized for energy efficiency or that respond to the specific needs of a particular user group.

7.6.2 How Tools like Building Simulation Software, AI, and Data Analytics can Improve the Exploration and Analysis of Design Alternatives

The architecture, engineering, and construction industry is embracing an increasing array of digital innovations to augment productivity and enhance project outcomes. Delving into the specifics of building simulation software, it's important to understand the pivotal roles of AI and data analytics in augmenting these tools.

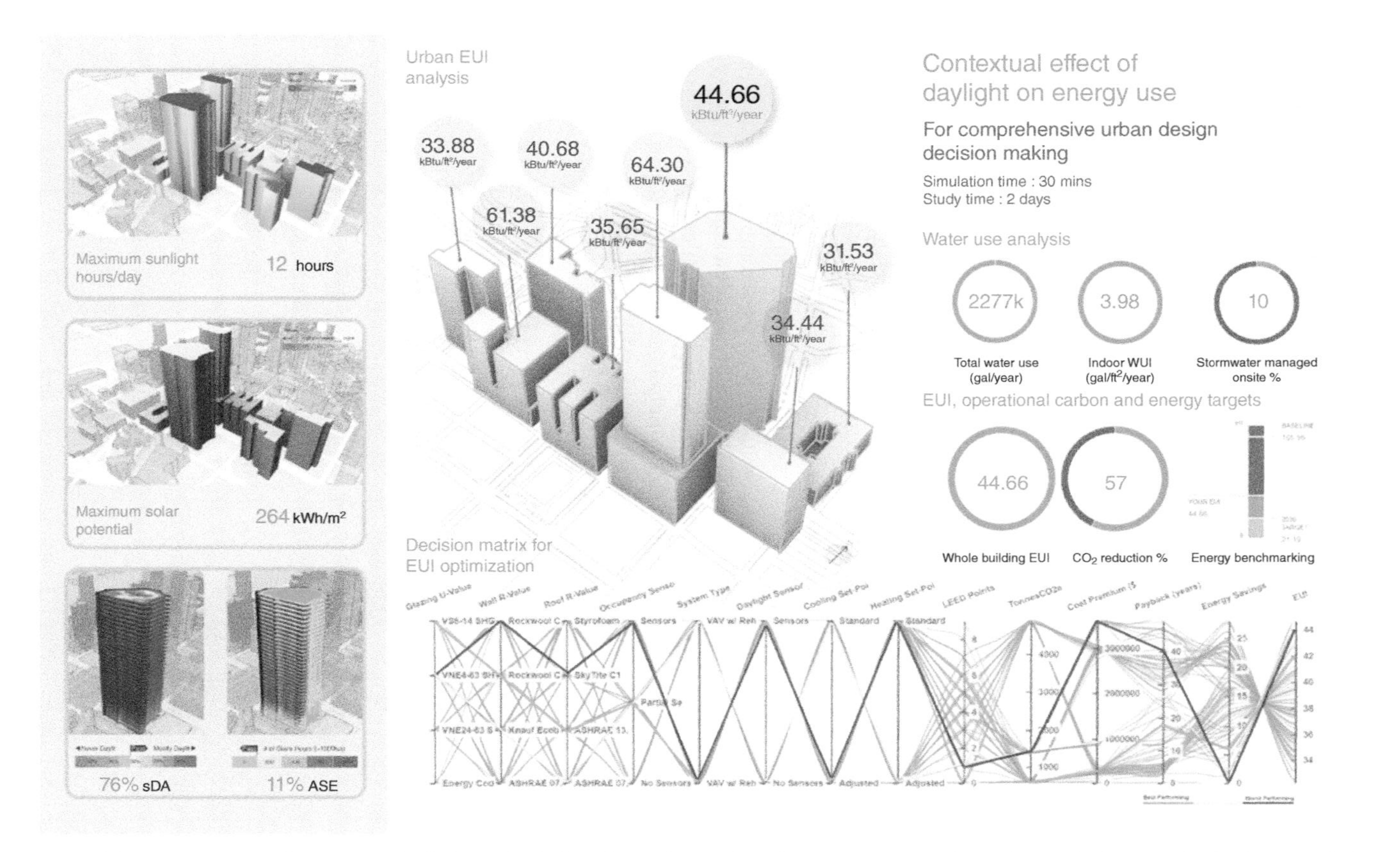

From left to right: Sun hours, radiation and daylight analysis for a high-rise project. Urban energy use intensity analysis to study the impact of the urban fabric on each other's energy use, studying the energy and water use performance along with benchmarking, at the bottom: The multi objective optimization to find the optimal solution. All analysis from cove.tool.

Building simulation software serves as a critical investigative medium, allowing designers to probe various performance metrics of potential designs within a controlled virtual space. This preemptive examination can cover a myriad of factors, from energy consumption and daylight optimization to natural ventilation and thermal comfort, steering us toward the most promising design avenues. Prioritizing these evaluations early in the design process is a strategic move, as it curtails the excessive costs and labor tied to post-design alterations.

These simulation tools excel at generating "what if" scenarios, providing a sandbox where alternative designs can be assessed and refined iteratively to approach the best possible solution. However, their effectiveness hinges on the quality and quantity of input data; without accurate data, simulations could yield unreliable predictions. This is where data analytics and AI become indispensable—these technologies excel at distilling and applying data to enhance the relevance and accuracy of simulations.

AI and machine learning algorithms act as the analytical core in the design and construction process, processing extensive datasets about building performance, material characteristics, economic considerations, and timelines. Such analysis begets new perspectives and aids in navigating the delicate balance between various design goals. For example, emerging AI solutions are integrating automated energy optimization into design platforms through the use of evolutionary algorithms, merging computational efficiency with human creativity. AI not only assesses but also ranks design alternatives created by designers against established performance benchmarks, advocating for those that promise superior performance.

This technology broadens the horizons for design exploration, expediting the discovery of valuable solutions. The true utility of AI, however, lies in its ability to decipher complex design intentions and formulate effective strategies, thereby acting as both a guide and a tool in the pursuit of excellence in design.

7.7 Policies, Regulations, and Early Decision-Making

7.7.1 How Building Codes and Regulations can Influence Early Decision-Making

Building codes and construction regulations have started playing a crucial role as codes to establish minimum mandatory requirements for structural safety, fire safety, accessibility, energy efficiency, indoor air quality, and other factors related to building performance and occupant health. By requesting compliance and setting up certain design criteria, building codes have a direct influence on key design and system choices and architectural features on a massive scale. Oftentimes, the rigidity or flexibility of codes, how up to date they are, and how well they incentivize high performance can variably impact early decisions and the ability of practitioners to innovate. To make sure to get the best of it, the code-related factors need to be examined first, providing insights into how regulations allow or hinder optimal decision-making during building design conceptualization.

With energy codes shifting drastically and moving closer to energy and carbon neutral options as the baseline, the design teams are largely struggling to meet these codes while keeping within the project budget. This can also be seen in the data from McKinsey that over 80% of projects getting built are over budget. These aggressive energy codes, paired with other sustainable certifications like LEED, BREEAM, Green Star, Living Building Challenge, Green Globes, and others, only add more challenges for ill-prepared design teams that are still utilizing archaic workflows that are neither tech-enabled nor set up for iterative decision making and optimization.

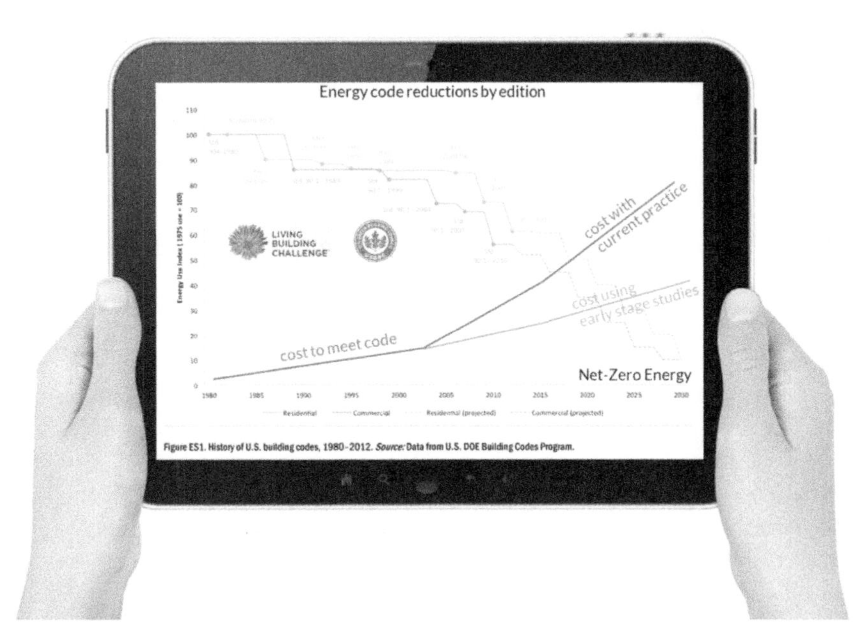

A look at the maximum accepted by code the energy use intensity over the years, with a step down every few years highlighting the impact of new more efficient regulations.

Rigid versus flexible codes differentially impact designers' options. Rigid, prescriptive codes that mandate specific design solutions provide less latitude in initial choices, restricting creative freedom and innovative solutions that may achieve or exceed code objectives in alternate ways. Conversely, performance-based codes that specify outcomes or performance targets allow practitioners to pursue a wider range of design options as long as the desired performance is achieved. For example, performance-based energy codes set mandatory energy efficiency targets but provide greater flexibility in the design features or technologies used to meet such targets. This allows designers to pursue higher performance pathways unavailable when rigidly following prescriptive solutions specified in traditional codes. That renders flexible, performance-oriented codes that emphasize outcomes rather than specific solutions able to generate broader exploration of design alternatives.

The extent to which codes are up to date influences the available options. Outdated codes cannot really be helpful in optimization and generate suboptimal solutions. For instance, energy codes that have not kept pace with recent high efficiency equipment, envelope materials or design strategies prescribe lower performance targets that designers must adhere to. Updated codes that integrate current best practices can expose practitioners to new, higher value options during early design conceptualization that would have otherwise remained unknown. Therefore, regularly updating codes based on evolving performance standards, research and technical progress ensures that the policy continues to incentivize and mandate the best currently available solutions.

Codes also shape early decision-making based on the degree to which they incentivize or simply mandate minimum compliance. Codes that establish bare minimum performance thresholds reflect a compliance-based approach. While mandating a base level of safety and quality, these usually do little to encourage higher performance beyond what is required. But codes that incorporate incentives for exceeding minimums—such as performance credits, tax deductions, or expedited permits—are more likely to succeed and inspire optimization-oriented approaches that nudge early decisions toward superior options. Such incentives can encourage designers to explore a wider spectrum of higher performing alternatives rather than focusing on minimal code compliance. Policies that reward and encourage high performance can push the development forward.

7.7.2 The Role of Sustainable Building Certification Programs in Guiding Early Design Decisions

Building certification programs aim to independently evaluate and recognize buildings that meet rigorous sustainability and performance standards. They attempt to "raise the bar" beyond minimum code compliance and instead establish a comprehensive set of environmental, social, and economic criteria that high performing green buildings should strive for. In doing so, *certification frameworks act as meaningful guidance* and outline objectives that can shape design decisions early on. They formulate rather ambitious performance targets across multiple impact areas aiming to exceed conventional codes. The energy efficiency targets that are 40–50% above code compel design teams to consider aggressive efficiency strategies, high performance HVAC and envelope systems, and design features like solar-oriented building form and natural ventilation. This is an optimization-driven framing of early decisions oriented around meeting the certification's sustainability objectives rather than minimal code compliance.

Certification criteria incentivize specific design strategies and features known to enable more sustainable outcomes. While performance targets establish the "what," the criteria detail recommended "how" through preferred solutions explicitly endorsed as meeting the certification's sustainability ethos. That provides concrete guidance.

Transparency around data-driven benchmarks within certifications inform quantitative visioning during initial design conceptualization. Certifications make public the vast database of projects, performance metrics, and best practices that form the core evidence base defining the program, which can be leveraged via prevalence of highly sustainable solutions within the databank. Instead of arbitrarily choosing generalized sustainability goals at inception, it is better to rely on real-world, data-informed baselines to define quantitatively realistic yet ambitious design targets. Subsequent decision-making can be directed toward pragmatic and high value solutions proven viable within the certification's framework.

Pursuing these certifications is an external motivator that pushes to optimize decisions for sustainability from the start. It is necessary to create a need to explore more innovative options during initial decision-making in order to succeed. The lure of a well-recognized sustainability credential based on rigorous third-party evaluation in many cases incentivizes convergence of all industry-leading solutions.

7.7.3 Case Study of a Multidisciplinary Learning Center

The 53,000 square feet multidisciplinary building provides students and faculty modern general classrooms as well as health science labs. The high-performance three-story structure showcases daylit, glare-free spaces and offsets 600 tonnes of operational carbon every year.

In the design of the new classroom building, the team carefully considered the unique climate characteristics of Nebraska, known for its high relative humidity and cool mornings and evenings. Understanding these conditions was crucial for making performance-based design decisions. For instance, the psychometric analysis suggested that dehumidification could enhance indoor comfort by 6% without the need for additional heating or cooling.

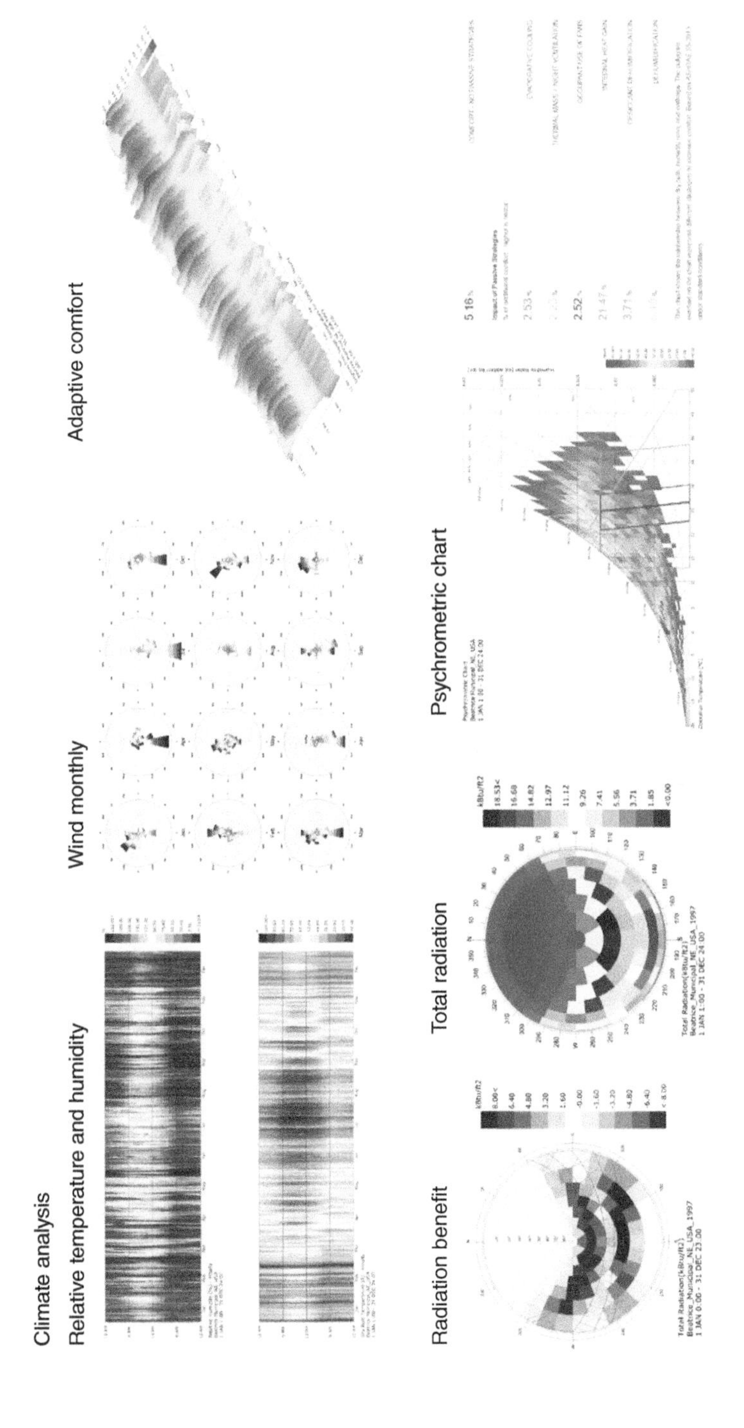

Climate analysis by cove.tool of Multidisciplinary Learning Center project.

The building was strategically oriented with its larger facades facing North and South. This design choice effectively minimized heat gain and hot spots from the East and West facades. Glare analysis indicated successful shading strategies, with less than 5% of the floor area experiencing glare potential. The team also assessed spatial daylight autonomy (sDA) and annual solar exposure (ASE) to ensure balanced daylight and glare management.

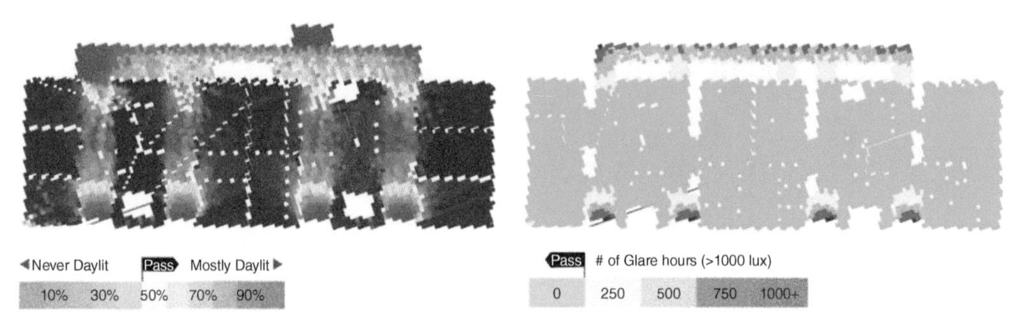

Spatial daylight autonomy (sDA/Daylight) and annual solar exposure (ASE/Glare) of the Multidisciplinary Learning Center project in cove.tool.

The West facade was designed with minimal glazing and a significant overhang on the first floor to counteract glare and heat gain.

The west facade with minimal glazing and a large overhang on the glazed first floor to offset glare and heat gain challenges.

Through integrated architectural and engineering efforts, the team achieved a predicted EUI of 50.5 Kbtu/sf/year. Massing optimization and a high-performance VRF system reduced operational carbon by 65%. The cost versus energy optimization process furthered this to a 70% carbon reduction, aligning with the AIA 2030 target. The project also aimed for 11 potential LEED v4 points for energy and a significant reduction in CO2e emissions.

While domestic hot water was not a major load, the team analyzed both indoor and outdoor water usage and stormwater management. Low-flow fixtures were identified as a key strategy, potentially contributing to LEED points.

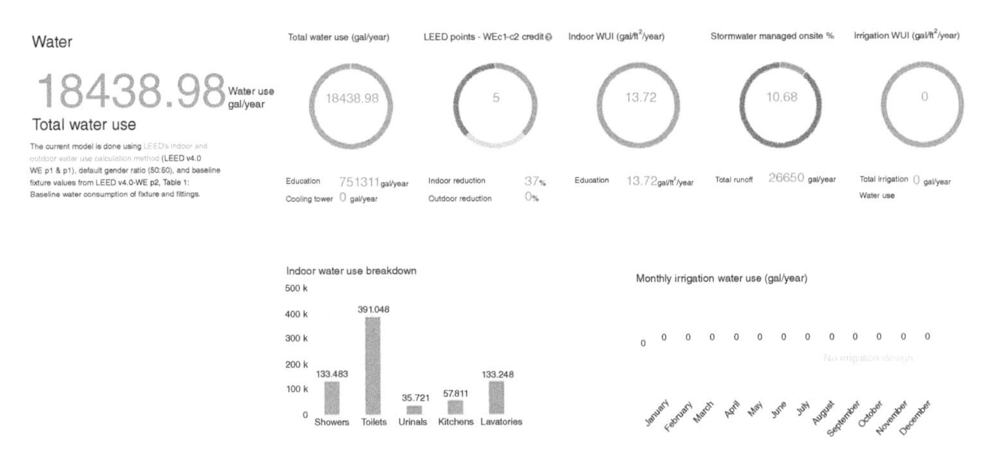

Water analysis for both indoor and outdoor water usage and stormwater management of the Multidisciplinary Learning Center.

The Morrissey Engineering team collaborated with the architects to further the building's energy performance to meet the AIA 2030 EUI target of 46.5 Kbtu/sf/year/. The team came up with a range of variables that they were comfortable parametricizing. In other words, they made a list of all the characteristics of the project that were open to being modified, for example, the HVAC type, the wall R-value, and more. The purpose of optimization modeling is to develop bundle recommendations of energy conservation measures (ECMs) to show various ways of achieving the project's energy efficiency goals in the most cost-effective manner. These variables were utilized for a parametric analysis and an ECM bundling exercise to develop the "Optimized Design" recommendations. In addition to the impact on performance, each variable also had an impact on the project's budget. The holistic cost versus energy optimization allowed the team to study all 360 combination and pick the lowest cost option that allowed them to meet the performance targets as laid out in AIA 2030. Below are all the options that were studied, along with their impact on the project's budget.

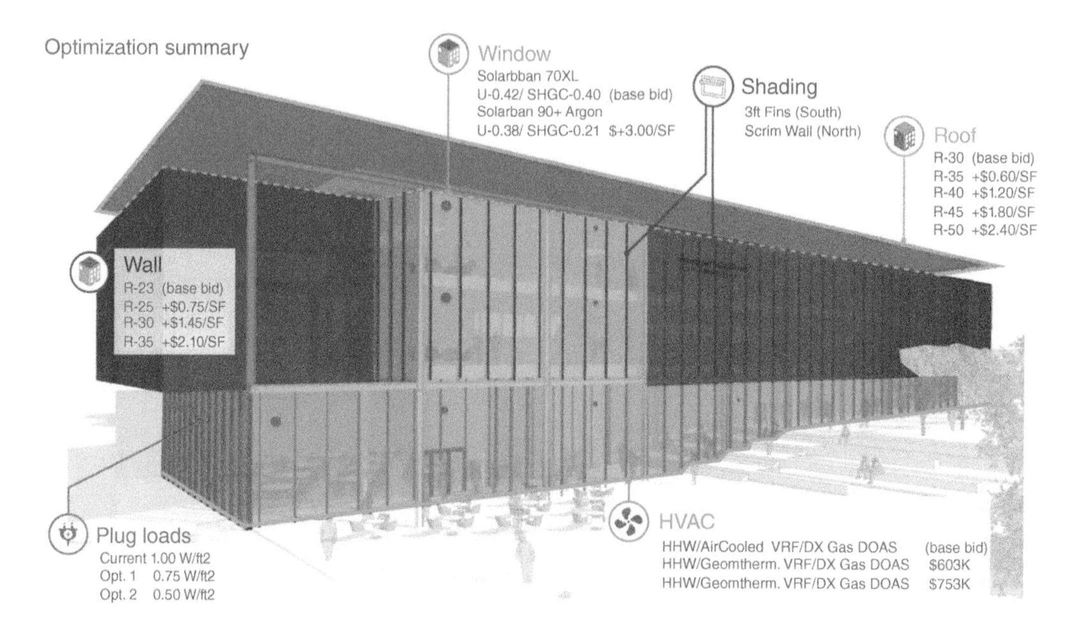

Options that were studied, along with their impact on the Multidisciplinary Learning Center project's budget.

Of all 360 options, the team utilized the dynamic slider to eliminate bundles that did not meet the project's EUI target. As can be seen from the figure below, any of the three HVAC system options could meet the 2030 Target EUI. Using the sliders to parse through the data further, eliminating cost premiums that do not pay back in the lifetime of the building (eliminates HVAC system Options 2–3), shows seven potential bundle combinations for consideration.

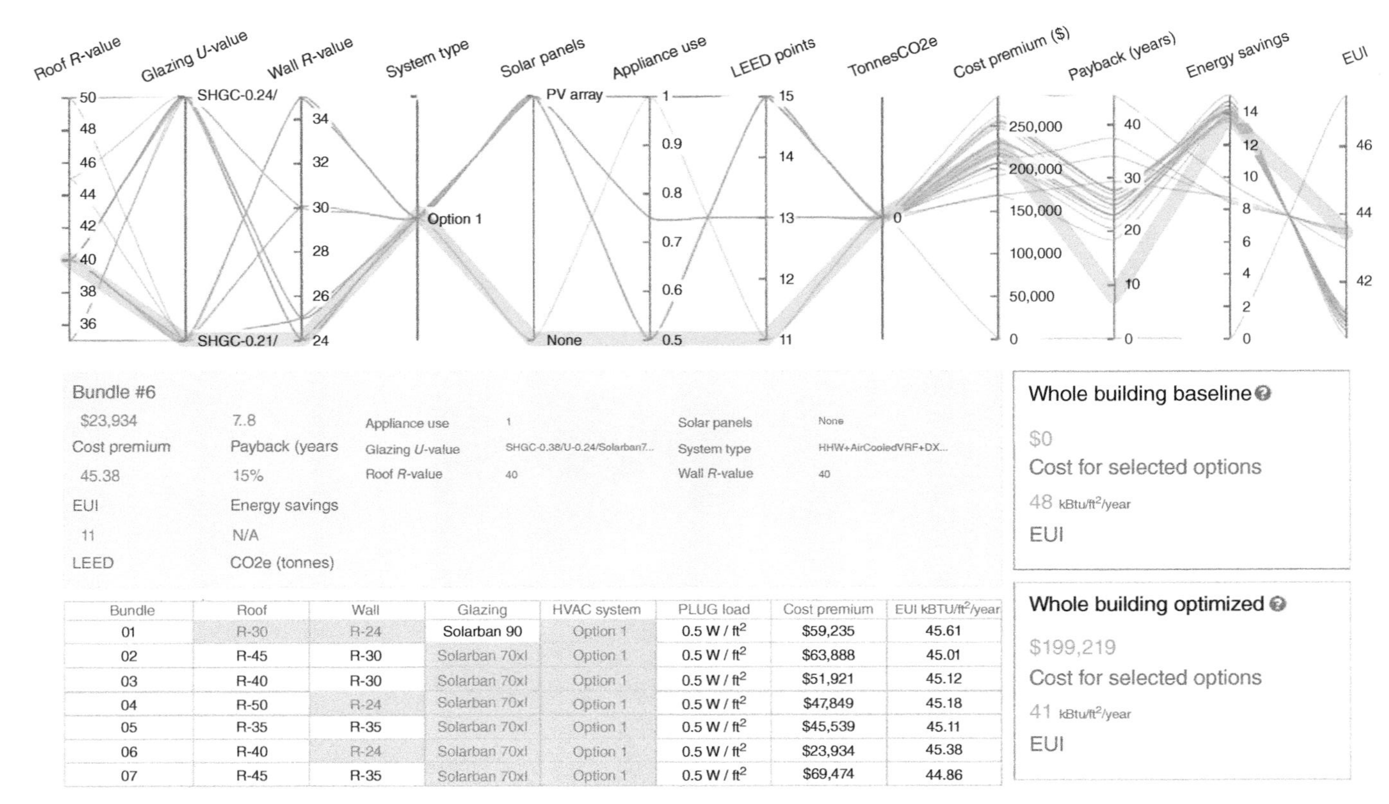

Bundle	Roof	Wall	Glazing	HVAC system	PLUG load	Cost premium	EUI kBTU/ft^2/year
01	R-30	R-24	Solarban 90	Option 1	0.5 W / ft^2	$59,235	45.61
02	R-45	R-30	Solarban 70xl	Option 1	0.5 W / ft^2	$63,888	45.01
03	R-40	R-30	Solarban 70xl	Option 1	0.5 W / ft^2	$51,921	45.12
04	R-50	R-24	Solarban 70xl	Option 1	0.5 W / ft^2	$47,849	45.18
05	R-35	R-35	Solarban 70xl	Option 1	0.5 W / ft^2	$45,539	45.11
06	R-40	R-24	Solarban 70xl	Option 1	0.5 W / ft^2	$23,934	45.38
07	R-45	R-35	Solarban 70xl	Option 1	0.5 W / ft^2	$69,474	44.86

Data slider showcasing that any of the HVAC systems chosen for the Multidisciplinary Learning Center could meet the 2030 Target EUI based off cove.tool reporting.

Variables finally included in the design are roof insulation of R-40 alongside Air-Cooled VRF with DX Gas DOAS system, wall insulation of R-24 and Solarban 70XL for glazing. The most cost-effective solution to implement allows the team to meet the AIA 2030 target with an EUI of 46 and approximate payback time of 6 years.

The holistic building performance allowed this integrated team to design a low energy, low carbon, high daylight, low glare classroom building while meeting the budget constraints.

7.8 Insights from Sarah Gudeman, Principal and Building Science Practice Lead at Branch Pattern

Sarah Gudeman, a Practice Lead at BranchPattern, is a seasoned professional mechanical engineer and building scientist who has made her mark in the realm of energy efficiency, sustainability, and occupant health within the commercial real estate sector. Her focus centers on implementing durable, resilient, and human-centered strategies right from the inception of design projects. Sarah relies on targeted simulations to predict performance outcomes and validates these predictions through commissioning and post-occupancy evaluations. Her fundamental belief is that empowering individuals to measure, understand, and manage their environmental impact is the key to creating Better Built Environments® for all.

Sarah Gudeman, Principal and Building Science Practice Lead, speaking at BranchPattern.

7.8.1 Emerging Trends and Opportunities for Mechanical Engineers

When contemplating the future of decarbonization in commercial real estate, Sarah sees a pivotal role in electrification with a specific emphasis on addressing the heating challenges faced in cold climates. Even without gas bans, the industry faces the task of providing alternative heating solutions that can rival gas-fired boilers in cold regions. Prioritizing beneficial electrification is crucial to achieving significant reductions in overall emissions. Sarah underscores the importance of enhancing regional knowledge regarding utility power generation, grid interoperability, campus energy distribution, and harnessing heat reclamation from unconventional sources. Additionally, she highlights the opportunity for engineers and analysts to incorporate embodied carbon considerations into their design decisions and to advocate for the production of environmentally friendly and healthier products.

7.8.2 Impact on the Carbon Dialogue

Sarah's substantial contributions to the carbon dialogue primarily revolve around education. She actively mentors students and young professionals, seeks opportunities to share her knowledge with colleagues and clients, and leverages regional and national peer networks, working groups, and volunteer initiatives. Sarah acknowledges that mechanical engineers, in comparison to their peers in Civil and Structural engineering, have entered the sustainability conversation at a later stage and have some catch up to do. BranchPattern is actively championing the most effective strategies to combat greenhouse gas emissions, leveraging the company's technical background in engineering design, combined with sustainability consulting and enhanced analytics to tackle both operational and embodied carbon at the portfolio scale for their clients.

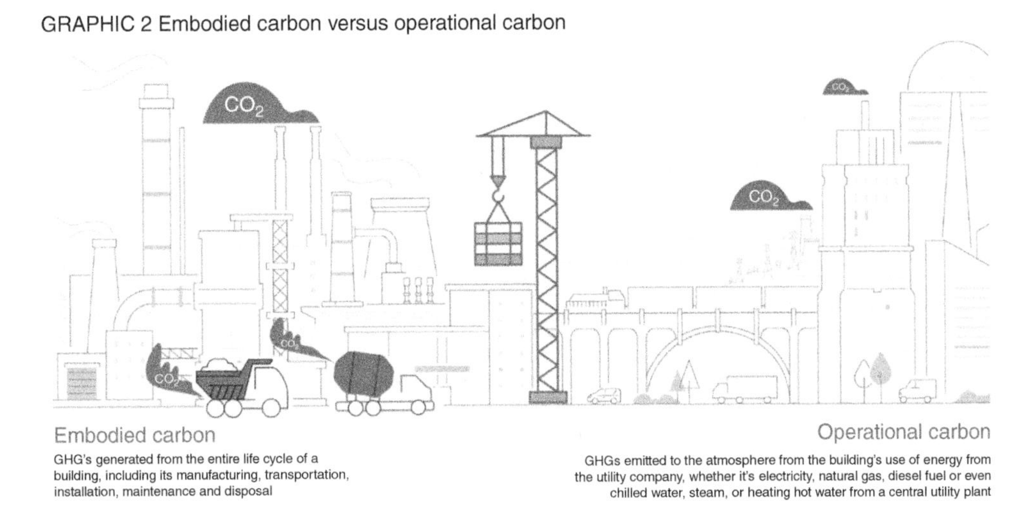

Graphic showing embodied carbon versus operational carbon.

7.8.3 BranchPattern's Carbon Achievements

BranchPattern's influence on decarbonization has grown significantly in recent years. Initially focused on reducing operational carbon (energy consumption) through energy modeling and consulting, the company's portfolio expanded to include large-scale projects and real estate portfolios. Programs like LEED® have served as catalysts for clients to pursue higher certification levels, often leading to net-zero energy and carbon status. Sarah notes that political and economic trends, such as the UN Sustainable Development Goals (SDGs), Paris Agreement, evolving regulatory landscapes across the globe, and ESG, have reshaped their approach to serving clients. These frameworks necessitate a holistic perspective, including consideration of embodied carbon in both new and existing buildings. As a result, many clients are now conducting life cycle assessments and carbon accounting for their global real estate portfolios, with a focus on reducing all carbon. BranchPattern's commitment extends to tracking and mitigating their own carbon footprint, with efforts directed toward minimizing emissions associated with travel and supporting sustainability initiatives in their local communities.

7.8.4 Mechanical Engineers and Decarbonization

Historically, mechanical, electrical, and plumbing (MEP) designers and engineers focused primarily on ongoing energy performance, with limited consideration of the broader environmental impacts of their design decisions. Today, the conversation has shifted to encompass embodied carbon, prompting MEP design firms and manufacturers to take proactive steps, especially in areas like refrigerants. The MEP 2040 Challenge sets a clear goal for engineers to advocate for and achieve net-zero operational carbon by 2030 and net-zero embodied carbon by 2040 in all projects, marking a significant transformation in the industry's approach to decarbonization.

7.8.5 Firm Leaders' Perspective on Building Decarbonization and Technology

While technology alone cannot solve the climate crisis, it plays an essential role in achieving decarbonization goals. Innovations in heat pumps, water-to-water heating, battery storage, and performance modeling tools are pivotal in delivering the all-electric buildings that clients increasingly demand and that are mandated in various jurisdictions. The path forward also involves grid interoperability, systems thinking, unique project delivery methods, and a continued focus on improving the performance of existing buildings. Firm leaders recognize the importance of these factors as they work toward the 2030, 2040, and 2050 sustainability goals.

Understanding the importance of leveraging renewable energy sources. Source: malp/Adobe Stock Photos.

7.9 Conclusion

In architectural design, the multitude of decisions made throughout the process is pivotal in determining the ultimate success of a project. Traditionally, key decisions were often deferred to later project stages, leading to commitments of time and resources that could have been better directed with earlier, more informed deliberation. The insights presented in this chapter underscore the transformative potential of frontloading the decision-making process, whereby initial choices are made with a full understanding of their implications.

This chapter highlighted the benefits of integrating digital tools, adhering to performance-based regulations, and aiming for ambitious sustainability certifications right from the conceptual phase. These strategies empower architects to make decisions that are not just timely but are optimized for design excellence. Simulation technologies, AI, and data analytics significantly enhance our ability to rapidly evaluate a wide array of alternatives, thereby pinpointing the most advantageous solutions at the outset. Performance-based regulations and associated incentives foster an environment where a spectrum of compliant and high-performing options can be considered from the very beginning. Meanwhile, sustainability certifications provide clear objectives, vetted approaches, and tangible benchmarks, guiding early-stage decisions toward outcomes that are in sync with overarching environmental goals.

The systematic and data-driven approach to decision-making outlined here marks a departure from the conventional reliance on intuition and experience. It advocates for a structured assessment of design options against measurable performance objectives, leveraging empirical data. This proactive orientation toward identifying and adopting the best possible designs from the project's inception greatly enhances the likelihood of realizing outcomes that align with sustainability, efficiency, health, and user experience goals.

Thus, the strategies and principles delineated act as a guide to be tailored and implemented within our design methodologies. Digital instruments must be employed not merely for generation but also for the critical evaluation of initial designs against established objectives. Regulations and certifications should be viewed not as mere hurdles but as catalysts for excellence, helping aim higher from the very beginning. In this rigorous and iterative process, each design alternative is to be scrutinized through simulation and optimization, with early selections being refined in light of emerging insights on performance.

8

Ensuring Carbon Positive Decisions Make it to Final Construction Documents

Good intentions do not always translate into realized outcomes if proper processes and oversight are lacking. This chapter focuses on a crucial but often overlooked aspect of sustainable design—ensuring that carbon-conscious decisions make it from concept design all the way through to final construction documents and implementation on site. As professionals in this industry, we are acutely aware of sustainability issues, from the earliest stages of schematic design, teams work to incorporate low-carbon solutions. Sophisticated building simulations and life cycle assessments help quantify a design's projected carbon and environmental footprint. But even the most well-intentioned concepts can fall by the wayside if they are not properly tracked and verified throughout the entire design and construction process.

All too often, carbon-positive decisions are value-engineered out to save costs or sacrificed for other priorities that emerge later in a project. Materials may get substituted, envelope details simplified, or efficiency measures de-scoped without realizing the cumulative impact on carbon emissions. Construction waste and improper installation can further undermine optimized designs. If genuine carbon reductions are the goal, simply focusing on concept design is not enough—the structural engineer's drawings and specifications must match the sustainability consultant's recommendations on paper and in practice. This means having processes to ensure that low-carbon solutions are not lost in translation between design teams or value engineering rounds. The consequences of failing to carry through initial intentions can be significant. Every ton of carbon that ends up emitted due to changes from approved designs is a real missed opportunity that compounds the climate crisis. In an era where emissions reductions must happen rapidly to avoid worst-case warming scenarios, we cannot afford such preventable carbon leaks. As professionals with direct influence over built outcomes, it is our responsibility to close these leaks through diligent oversight and accountability measures throughout the entire design and construction lifecycle.

This chapter provides guidance on processes and best practices to successfully shepherd carbon-conscious decisions from concept to completion. We will first discuss why traditional design and construction practices can undermine sustainability goals if left unchecked. We will identify key weaknesses where carbon positive strategies are most at risk of being de-prioritized or substituted out. Then we will present a framework for formal carbon tracking protocols to maintain oversight from early design through construction closeout. Case studies will show real-world applications of carbon accounting systems that have successfully minimized emissions through diligent documentation and verification of low-carbon commitments.

With the knowledge of formal processes, we can position our goals front and center from start to finish and orient the design and construction teams toward outcomes that meet both economic and

Build Like It's the End of the World: A Practical Guide to Decarbonize Architecture, Engineering, and Construction,
First Edition. Sandeep Ahuja and Patrick Chopson.

environmental priorities. This is a "closed-loop" approach to sustainable building that delivers genuine emissions reductions with integrity—moving the needle on the global transition to net zero carbon construction. The framework in this chapter aims to demonstrate how even imperfectly optimized designs can still achieve significant emissions savings through diligent documentation and constructability reviews.

8.1 The Continuous Effort in Meeting Performance Objectives

8.1.1 Ensuring the Project Meets all the Owner's Functional and Sustainability Objectives

Meeting the performance objectives of a project, particularly those pertaining to functionality and sustainability as envisioned by the owner, is an intricate and ongoing endeavor that requires consistent and concerted effort from conception to operation. This process is not simply a milestone achieved the moment a building is constructed or commissioned but is an enduring commitment to decision-making excellence throughout the project's lifecycle. At the very genesis of a project, a collaborative pact is forged among all stakeholders to create a blueprint for success through strategic predesign and early decisions that lay a robust foundation.

Goal setting workshop to help achieve project milestones. Source: David Pereiras/Adobe Stock Photos.

In the programming phase, a series of deliberate and interconnected steps ensure that the project remains true to its goals. The iterative process of goal-setting establishes the project's sustainability compass, with owners and designers coalescing around core objectives like carbon emissions reduction, energy and water efficiency, and indoor environmental quality. These aspirations are then translated into tangible energy budgets and performance indicators through comprehensive whole-building energy modeling, guiding the architectural vision.

Life cycle cost analyses add a dimension of foresight, appraising the long-term economic viability of low-carbon initiatives, thereby bolstering the case for upfront investments in sustainability. Coordinated scoping studies and basis of design narratives crystallize these parameters, ensuring they persist as a lodestar throughout the design journey.

As the project matures, continuous systems analysis rigorously tests and refines the proposed solution sets. Iterative energy modeling acts as a crucible, in which building envelope compositions, mechanical systems, and integrated controls are continually reassessed to align with the evolving sustainability targets.

The practice of value engineering and constructability reviews serves a dual purpose: protecting initial investment costs while also considering the longevity and impact of the building's operation. These assessments are instrumental in optimizing design without compromise.

To safeguard the integrity of these objectives, commissioning agents provide an essential layer of third-party oversight. Their expertise guarantees that comprehensive documentation, meticulous submittals, and effective owner training are in place, setting the stage for successful start-up and performance verification testing.

As the project transitions into construction, contract document specifications must unequivocally communicate the sustainability commitments forged early on. This is crucial to minimize disruptive last-minute alterations and to protect the energy performance post-occupancy. Meticulous material submittal tracking and vigilant construction oversight are pivotal in ensuring that the installed solutions are true to the optimized designs envisioned.

Commissioning remains the guardian of performance, with system functional testing being paramount. Acceptance is only granted once there is demonstrable and verified evidence that the performance aligns with the ambitious objectives set forth at the project's inception. In the United States, there are significant tax breaks in the Inflation Reduction Act that can benefit from commissioning as well. This comprehensive approach encapsulates not just the creation of high-performance buildings but the stewardship of their sustained excellence over time.

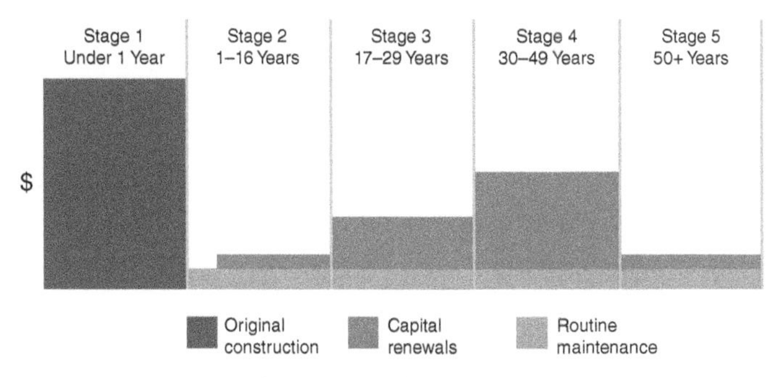

The cost of maintaining a building over its life. After the initial investment of the construction cost, the key costs coming in at age 30–49 years of the building's life with capital renewals.

8.1.2 The Role of a Well-coordinated Project Team in Achieving Carbon Positive Outcomes

From the outset, establishing an *integrated design approach* centered on sustainability fosters the communication and cooperative workflows essential to optimize performance in the long term. Early in programming and conceptual design, we should focus on an inclusive integrated design charrette that brings together owners, architects, engineers, lighting designers, commissioning agents, and other key stakeholders. Jointly mapping out performance aspirations and constraints is the fast way toward a shared understanding of overlapping systems interactions and tradeoffs. We can co-locate diverse expertise and set up the currents of creative problem-solving around carbon-conscious interventions, and by capturing these interdepartmental solutions on their implementation we can support realization down the line.

(a)

	CONCEPT DESIGN	SCHEMATIC DESIGN	DESIGN DEVELOPMENT	CONSTRUCTION DOCUMENT
DESIGN TEAM	Client and site research initial programming and massing Concept generation	Revised massing Envelope, structure and systems options	Revised design with Envelope, structure and HVAC systems	Final design with Envelope, structure and HVAC systems
ENGINEER			MEP + structure alternatives Materials alternatives study Detailed dynamic simulation Interior layout and design	Final MEP + structure design Code compliant dynamic simulation Interior layout and design
INTERIOR DESIGNER			Interior layout and design	Interior layout and design
CONTRACTOR		Initial Cost Estimate	Detailed Cost Estimate	Final cost estimate

(b)

	CONCEPT DESIGN	SCHEMATIC DESIGN	DESIGN DEVELOPMENT	CONSTRUCTION DOCUMENT
ENERGY ANALYST	Massing alternatives study	Massing alternatives study Materials alternatives study Cost Vs Energy optimization	Materials alternatives study Cost Vs Energy optimization Detailed dynamic simulation	Engineer co-ordination for code compliant dynamic simulation
DESIGN TEAM	Client and site research initial programming and massing Concept generation	Revised massing Envelope, structure and systems options	Revised design with Envelope, structure and HVAC systems	Final design with Envelope, structure and HVAC systems
ENGINEER			MEP + structure alternatives Materials alternatives study Detailed dynamic simulation Interior layout and design	Final MEP + structure design Code compliant dynamic simulation Interior layout and design
INTERIOR DESIGNER			Interior layout and design	Interior layout and design
CONTRACTOR		Initial Cost Estimate	Detailed Cost Estimate	Final cost estimate

(a) The authors' representation actors and processes and inter-relationships over time in "typical" commercial building design process. (b) Suggested variation to the current practice to achieve higher energy and cost performance.

As a project progresses through schematic design and design development, the importance of interdisciplinary collaboration becomes even more pronounced. Regular, dynamic interactions such as workshops or sessions utilizing building information modeling (BIM) are indispensable. These collaborative engagements are essential for keeping the collective eye on the goal of carbon reduction, unraveling potential challenges, and reshaping the project vision as necessary.

Central to this cooperative approach is the codification of targeted energy budgets and the endorsement of technical solutions. This is achieved through the use of comprehensive narratives and detailed models, which together forge a shared vision that guides the project through the lens of value engineering. Such documentation becomes the bedrock for cross-disciplinary engagement, proving invaluable not only during the design phase but also as the project moves into construction documentation, procurement, and on-site execution.

The benefit of this integrated approach is most evident when unforeseen challenges surface. A commitment to transparent, agile problem-solving becomes the modus operandi, allowing for modifications in strategy, reevaluation through periodic reviews, and the fine-tuning of processes to better align with the project's goals. This flexible and responsive method can occasionally impact the trajectory of project delivery, particularly when the focus is on honing long-term operational performance.

To ensure continued performance optimization, the commissioning process may evolve to include ongoing recommissioning workshops. These sessions serve as checkpoints to validate system efficiency, diagnose persistent issues, and determine refurbishment actions that uphold the building's efficiency for the future. By bringing operators into these discussions, a user-focused perspective is adopted, establishing a feedback loop that encourages incremental improvements through capital planning strategies. This occupancy-centered approach not only elevates the immediate project outcomes but also steers the building toward a trajectory of sustainable operation and efficiency over its lifetime.

8.1.3 Total Building Commissioning (TBCx) as a Quality Assurance Process That Takes all the Systems of the "Whole Building" into Account

Total Building Commissioning (*TBCx*) provides a robust quality assurance process that considers a building as an integrated whole system extending performance oversight beyond handoff. TBCx evaluates the coordinated operation of all technical and functional elements essential to meeting sustainability and programmatic objectives over both short- and long-term horizons.

Commissioning is a key part of ensuring a building's performance. Source: DC Studio/Adobe Stock Photos.

From project outset, the process establishes a comprehensive performance-based framework driving early decisions through long-tail operations. Commissioning agents facilitate integrated design charrettes and capture consensus envelopes, heating, ventilation and air conditioning (HVAC) selections, and control strategies into performance-based design narratives and whole-building energy models. These establish target budget baselines against which construction performance can be verified. Proceeding to the construction, TBCx coordinates submittal tracking, mockups, site inspections, and pre-functional testing validating optimized assemblies and startup procedures. Daylighting, thermal comfort, building pressurization—these are checked synergistically to uncover potential downstream issues with persistence.

TBCx provides extensive documentation and owner training when transitioning responsibilities during post-construction. Immediately before substantial completion, integrated system functional performance testing verifies if the construction matches the design and whether it's right-sized and balanced in all interrelated building and control systems. All is guided by comprehensive manuals and exploration diagnostics—that is, all operations and maintenance. It also expands oversight into long-term operations. There are periodic post-occupancy reviews and seasonal recommissioning that identify minor adjustments that need to be optimized. Performance baselines need to be re-verified against changing usage to support conservation. Involving facilities staff is important to cultivate ownership and to continue high performance.

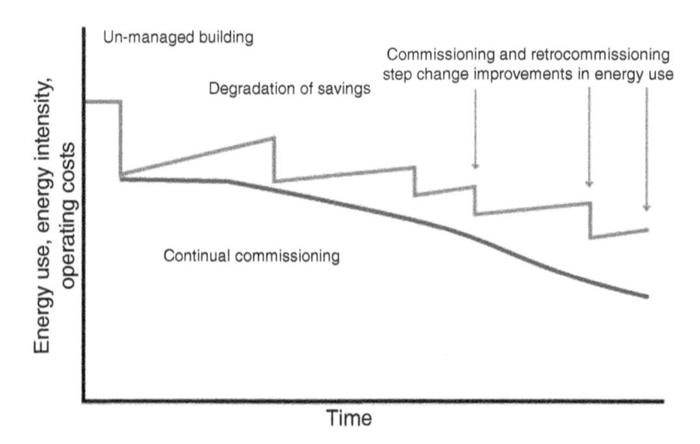

Graph of energy savings of continual commissioning. Source: Adapted from Ethan Rogers / https://www.aceee.org/blog/2016/04/intelligent-efficiency-continual/last accessed Jan 23, 2024.

A whole-building mindset should be continuous—from programming to occupancy. And TBCx delivers a process that ensures that priorities seamlessly coordinate into one vision. Meeting performance targets demands constant coordination and effort beyond a single point in time like occupancy or construction closeout. Roles evolve along with transition of responsibilities from design to construction to operations, but we should always prioritize the post-completion performance for the decades afterward, instead of how the building performs at the moment of its completion. Quality assurance processes help close the loop and ensure that owners can reliably obtain a whole system performing as promised to achieve both business and sustainability objectives.

8.2 Assuring Appropriate Programming and Establishing Design Objectives

8.2.1 The Process of Facilitating Discussions with Key Stakeholders to Establish Project Requirements and Goals

Effective project coordination hinges on skillfully guiding discussions with key stakeholders. Establishing clear requirements and goals is essential, providing a framework that directs all project activities. Effective communication is crucial, requiring careful listening and clear articulation to ensure stakeholders' insights significantly influence outcomes. This approach is similar to how scientific methods guide research, ensuring that discussions align with project management principles to seamlessly integrate functionality and design, thereby fostering a conducive environment for success.

Effective stakeholder discussion is designed to allow each representative to share their unique perspective. Source: BullRun/Adobe Stock Photos.

Effective stakeholder discussions demand preparation. At its core is identifying key stakeholders—the diverse group of voices that shape the project's outcome. The strategy is followed by an evaluation of the project's scope and context, where teams scrutinize the variables, which uncovers the main resonance points. Similar to the initial calibration of experimental parameters, defining appropriate programming and design objectives lays the groundwork, establishing a clear vision for the discourse that follows. This discourse results in delineation of goals, setting a direction toward concrete resolutions. This preparation mirrors the fusion of analytical precision and imaginative insight—qualities that are emblematic of both formal scientific inquiry and the nuanced discipline of management. It contains elements of strategic finesse and genuine interpersonal exchange. The initial stride involves cultivating rapport and establishing trust among peers as a foundation that paves the way for *active listening*. Coupled with this is empathy that characterizes understanding the challenges and motivations of fellows, allowing project leaders to embrace stakeholders' viewpoints. Adapting communication to suit stakeholders' diverse expertise levels and varying technical adeptness ensures that discussions resonate with all participants. This culminates in the creation of an inclusive and collaborative multidisciplinary environment.

Stakeholder discussions need an adept balance of formal structuring and approachable engagement. The preliminary step involves an introduction and agenda overview—a verbal roadmap that acquaints participants with the discourse's trajectory. The next segment entails elaborating on the fundamental tenets, drawing parallels to core principles to underscore their significance within the project context, which primes the atmosphere for an open dialogue. This is an interactive phase for stakeholder inquiries to align all sides to effective communication and dynamic engagement. Engaging and eliciting stakeholder input is a multifaceted process. Employing open-ended questioning techniques enables a comprehensive exploration of stakeholders' insights and unveils a panoramic view of perspectives that bring us closer to specific requirements and goals. Accurate capturing of inputs hinges on attentive listening and note-taking, ensuring that nuances are represented. The stride involves the translation of these discussions into clear project requirements that synthesizes information and contextualizes it. Ultimately, it is about seeking stakeholder validation on documented information.

Navigating project discussions inherently involves addressing challenges and concerns. The task of addressing resistance to change requires a blend of empathy and persuasive communication to foster a climate of receptivity. Similarly, mitigating unrealistic expectations calls for candid dialogue to recalibrate anticipation. The possibility of scope creep necessitates attention to ensure that discussions remain aligned with predefined objectives and stay within established boundaries. Managing disagreements among stakeholders stands as a quintessential skill—addressing challenges paves the way for constructive resolutions while upholding the integrity of the project's course. Promoting interdisciplinary discussions often results in innovative solutions. Bridging gaps between technical and nontechnical stakeholders should preserve clarity and relevance. The facet of ensuring accountability and ownership is about clearly defined roles. Assigning responsibilities for different requirements and goals helps build a cohesive framework for execution. Tracking progress and milestones makes sure that the project remains on course and deviations are promptly addressed. Holding stakeholders accountable for their commitments is a matter of diligence and responsibility. Aligning this accountability reinforces a symbiotic relationship wherein project ownership is intertwined with overarching aims.

8.2.2 Organization of Working Sessions

Prior to the *working session*, commissioning agents conduct in-depth discovery interviews impartially documenting stakeholder priorities regarding function, operations, maintenance budget, and sustainability values. A review of gleaned insights can refine core goals and then brainstorm creative options at a high conceptual level that is unrestricted by early assumptions.

During the charrette, a neutral facilitator should expertly moderate the interactive session:

1. First, soliciting individual inputs and then guiding structured ideation. Iterative dot-voting and feedback rounds surface synergies and tensions across viewpoints and break down silos toward consensus-based opportunities. Discussions should remain open and positive-minded, with the facilitator paraphrasing to capture all perspectives accurately.
2. To maximize productivity, the facilitator then applies proven techniques prompting new associations and evaluating proposals factually. Breakouts and report-backs fuel deeper exploration of priority

subjects, returning findings for refinement by the collective. Throughout, a live working document captures parameters, objectives, performance targets, and streamlined conclusions ratified together.

3. Follow-up sessions continuing discussions online extend participation, iterate priorities against delivery methods, and further develop performance benchmarks transforming general agreement into specific, actionable language.
4. Final charrette outcomes form the definitive guide establishing design objectives, from envelope targets to success metrics. Continuous reassessment maintains focus on owners' evolving needs through construction and into long-term operations.

An interactive charette to help guide the design objectives.

To ensure that it stays consistent, we should instill an ongoing validation that maintains focus on evolving needs through all subsequent phases. We can employ an intensive, collaborative programming approach that ensures that design objectives align with long-term strategic visions that can reliably guide high-performance building delivery:

- Discovery interviews last one to two hours, using open-ended questions to understand priorities without preconceptions, while facilitators take thorough notes.
- The charrette is a full-day event with 25–50 participants with an agenda that balances focused discussion and breaks to encourage networking.
- In breakouts, 8–10 people discuss 2–3 topics under a facilitator, and the feedback shares key insights and questions to answer.
- Online follow-ups use collaborative software for two-week iterative commenting on charrette outputs.
- Periodic interviews revisit priorities against market changes or new technology.
- Status meetings keep stakeholders engaged throughout design, while facilitators ensure consensus guides 100% documents handed to contractors.
- Charrette recordings and detailed facilitator notes provide institutional memory for future programming efforts and handoffs.
- Evaluation surveys give feedback on facilitation and outcome usefulness. The data improves future programming processes.

The level of rigor, inclusiveness, and documentation embedded in programming charrettes and follow-ups helps maximize stakeholder buy-in essential to delivering optimized building performance for decades.

8.2.3 The Importance of a High Level of Communication Between Project Team Members During Programming and Throughout the Facility Development Process

Effective communication in project management is especially pronounced in the area of facility development. The construction of facilities thrives on a high level of information exchange to ensure alignment with design goals, manage challenges, and navigate changing requirements. Digging into the specifics of programming, it becomes evident that it plays a multifaceted role. Maintaining a high level of communication in these circumstances—and between all project stakeholders—is imperative to appropriately program a facility and establish design objectives that can be reliably realized through completion. From project outset, collaborative discussions lay the groundwork for cooperation extending over subsequent conception, construction, and operational phases.

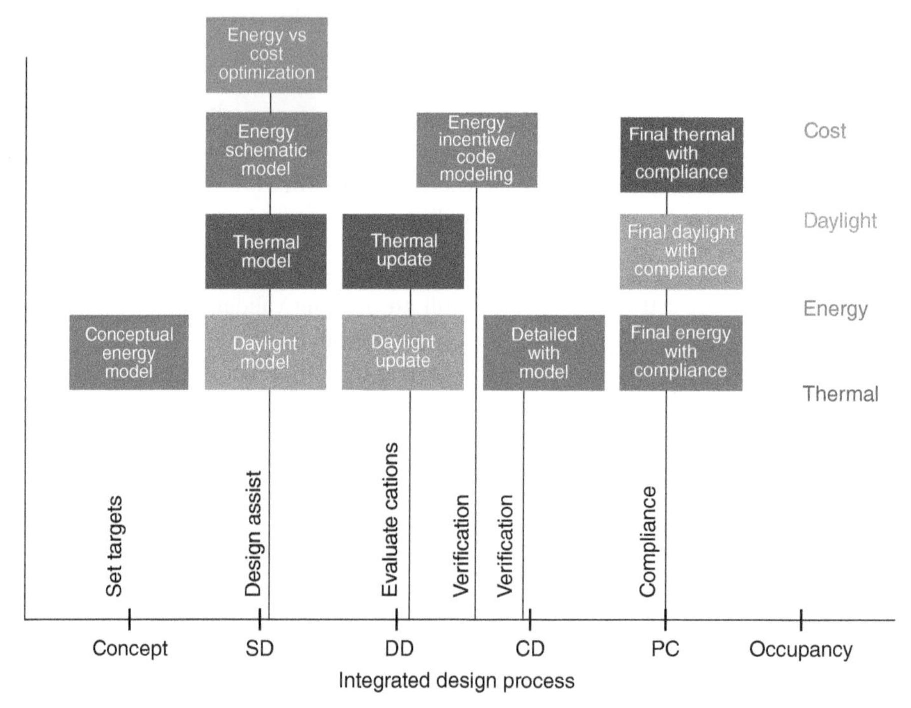

Highlighting the various types of analysis that are necessary for different stages of design, from concept to occupancy.

Inception begins with inclusive programming and exchange sessions. Here, active listening across specialties surface synergies and creative, integrated solutions amongst owners, designers, and future operators. Questions clarifying perspectives and assumptions debunk misconceptions important for developing shared understanding. Documenting all viewpoints builds consensus while respecting varied priorities. Programming solidifies objectives through workshops, further cultivating trust between participants. Web conferences extending engagement allow distributed teams to comfortably contribute ideas evaluated factually by neutral facilitators. Continuous feedback ensures evolving priorities find their way into living briefs guiding design evolution. As schematics materialize, frequent interdisciplinary coordination exchanges validate compatibility between mechanical selections, daylighting schemes, and structural envelopes. Cross-departmental buy-in facilitates minor adjustments maintaining optimal synergies, headed off by open resolutions accessible to all stakeholders.

Transitioning to construction, transparency remains key as general contractors and construction managers invite commentary safeguarding sustainability targets and quality. Accessible documentation shares the rationale behind value decisions for input strengthening optimal outcomes. Facilities staff involvement continues through commissioning as operators receive comprehensive systems training. Maintaining collaboration cultivates ownership, assisting facility optimization over decades of service. Periodic recommissioning extends partnership, identifying minor adjustments, enhancing efficiency or comfort long term. Cultivating a cooperative communication culture from project inception improves understanding across roles. Inclusive participation assures all perspectives find voice, informing design compatibility and delivering balanced priorities into the future.

What we can do to assure appropriate programming and establishing design objectives:

- Open dialogue is key to aligning expectations between owners, designers, and builders. Discussing assumptions, constraints, and priorities upfront prevents costly reworks down the road.
- Addressing important issues like building uses, maintenance needs, budget parameters, and sustainability goals requires interactive conversations where all viewpoints are respectfully considered.
- Facilitating discussions demonstrates listening skills that foster continued participation. Participants are more likely to share ideas openly when they feel heard.
- Communicating regularly maintains focus on consensus-driven solutions as projects evolve from concept to construction documents. Revisiting objectives ensures any changes still satisfy original goals.
- Frequent, transparent information sharing builds trust between specialists. Teams are more willing to compromise knowing rationales behind other design elements.
- Documentation of discussion outcomes creates institutional memory that hands off effectively to new members over the long project timeline.
- Involving future operators engages the end users' perspectives needed to develop facilities that function optimally as planned over decades.
- Establishing clear protocols for coordination and documentation exchanges streamlines collaboration, resolving conflicts before they delay schedules or budgets.
- Maintaining communication habits extends partnerships far beyond construction. This cooperative culture continues to deliver beneficiaries high performance outcomes throughout occupancy.

The continuous two-way dialogue required to develop comprehensive, balanced, and achievable objectives is as important as the objectives themselves. Open communication is truly the foundation for success.

8.3 Kate Simonen, AIA, SE | Professor | University of Washington

Kate Simonen, an AIA-certified architect and licensed structural engineer, serves as the founder and director of the Carbon Leadership Forum. Additionally, she holds the position of Professor of Architecture at the University of Washington. With extensive experience in high-performance building design and a strong background in environmental life cycle assessment, Kate is dedicated to driving collaborative efforts that will ultimately eliminate embodied carbon. Her approach involves cutting-edge research, fostering cross-sector partnerships, and nurturing innovative strategies.

In her role, Simonen oversees the research initiatives of the Carbon Leadership Forum. Under her guidance, the organization has gained global recognition for its contributions to advancing embodied carbon data, methodologies, and policy frameworks. Furthermore, it has succeeded in inspiring and empowering collective action through both online and in-person communities.

8.3.1 Integrating Decarbonization into Construction Documents

In the pursuit of decarbonization within the construction industry, it is crucial to embed this goal into the very fabric of a project from its inception. This case study examines a conversation with Simonen, an expert in the field of embodied carbon, and delves into the strategies and trends that can help make decarbonization a fundamental objective in construction projects.

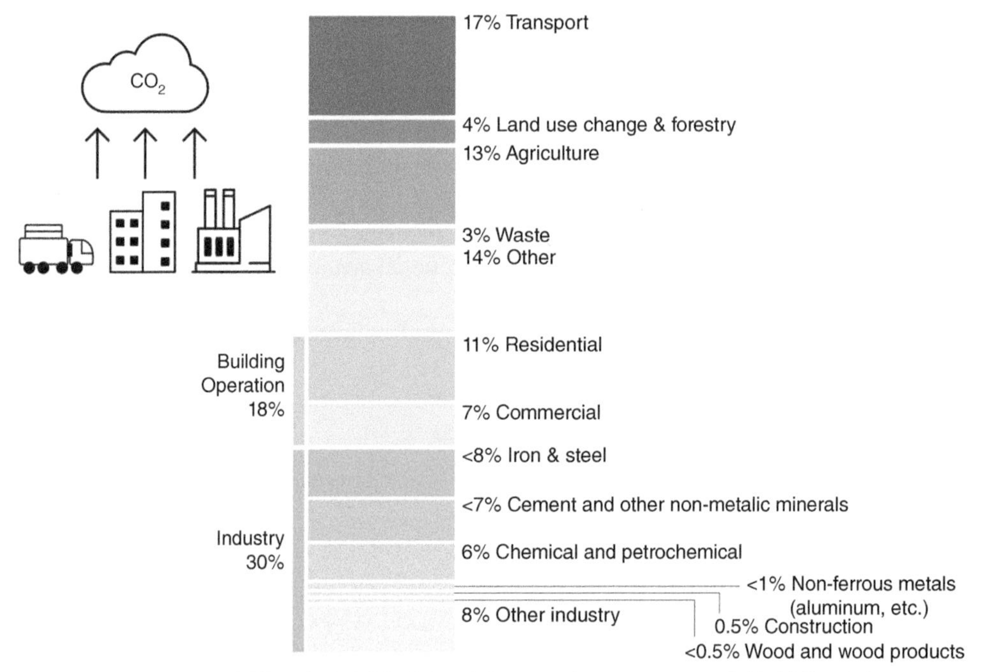

Global greenhouse gas emissions and the life cycle of buildings. Building materials are one of the largest sources of industrial emissions and, therefore, are potential solutions for reducing emissions from this sector. The top two industrial sectors alone—steel and cement—are each individually responsible for more emissions than all of commercial building energy use each year. Source: Adapted from Carbon Leadership Forum / https://carbon-leadershipforum.org/carbon-challenge//last accessed Jan 23, 2024.

In the pursuit of decarbonization in the construction industry, Kate Simonen, a renowned architect and structural engineer, stands as a beacon of insight and inspiration. Her journey into the world of embodied carbon began with a simple question: Can we truly create sustainable buildings without considering the carbon footprint of materials and construction processes? This section dives into Kate's conversation, illuminating her perspectives and highlighting her contributions to the ongoing decarbonization movement. Through her insights, we gain a deeper understanding of how the industry is evolving and how professionals at all levels can play a role in this transformative journey.

8.3.2 Setting Decarbonization as a Primary Goal

To achieve decarbonization in construction, Kate emphasizes the need to establish it as a primary project goal from the outset. "In order to have decarbonization get all the way into construction documents, it clearly has to be one of the primary project goals," she explains. Early buy-in from project owners and the entire construction team is crucial to ensure that decarbonization remains a top priority throughout the project's lifecycle.

8.3.3 Incorporating Embodied Carbon in Specifications

Kate highlights the significance of specifications during the construction document stage. This is where the rubber meets the road, and decisions that impact a building's carbon footprint are made. Kate recommends using tools like the EC3 tool to identify low-carbon material options within the available range of choices. Specifications can include requirements for material suppliers to disclose their carbon footprints, laying the foundation for informed decision-making. "You could set targets or caps, ensuring that certain materials do not exceed a specified carbon threshold," Kate suggests, illustrating the potential for innovative approaches within material specifications.

Optimize	Project	System	Procurement
Strategies	- Build less, reuse more - Design to reduce embodied carbon and increase material/structural efficiency	- Choose low-carbon system and assemblies - Use alternate, low-carbon materials	- Select the lowest carbon version of the selected product - Clean manufacturing (efficiency, fuel switching)
Tools	Early design calculators, rules of thumb	While building life cycle assessment (WBLCA)	Environmental product declarations (EPDs) / EC3 tool

Optimize project system procurement workflow; what matching policy measures and embodied carbon reduction strategies.

8.3.4 The Growing Momentum for Change

When asked about the trend in embodied carbon integration, Kate observes a remarkable surge in demand and curiosity about reducing embodied carbon. "We see architects, engineers, building owners, and policymakers recognizing the importance of addressing embodied carbon and conducting environmental life cycle assessments to evaluate design options," she notes. The federal government plans for significant investments in low-carbon materials along with increasing embodied carbon policies at city and state levels to underscore the growing intersection of building decarbonization and global climate goals. Buildings and infrastructure are now seen as key drivers for industrial decarbonization, leading to a groundswell of action.

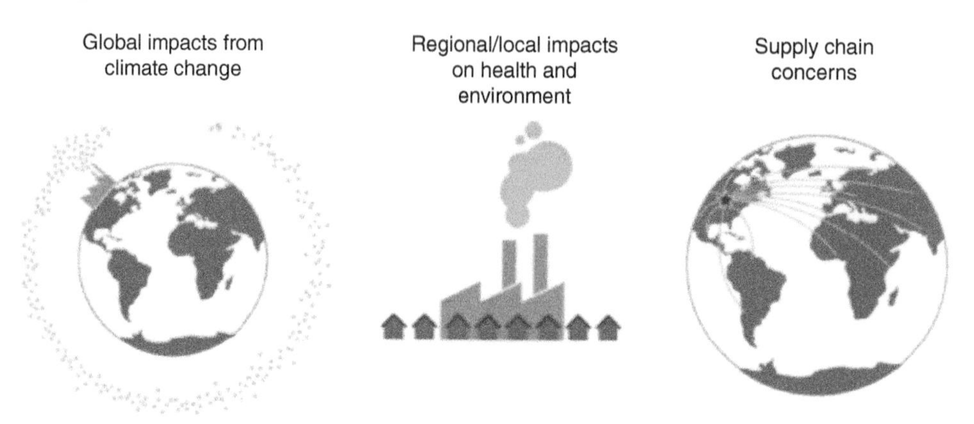

The local, regional, and global impact of embodied carbon leading to a rise in carbon emissions. Source: Adapted from Meghan Lewis & Megan Kalsman, 2023/Carbon Leadership Forum.

8.3.5 The Role of Building Product Manufacturers

Kate sheds light on the evolving role of building product manufacturers in the decarbonization movement. Some manufacturers have been pioneers in low-carbon solutions for over a decade, driven by a sense of responsibility and the desire to differentiate themselves. "Manufacturers that have taken ownership of their product carbon footprint see this as an opportunity to do good and stand out," Kate remarks. Simultaneously, there is a surge in research and development of new materials generated through the extraction of carbon dioxide from the atmosphere and storing it in long life building materials and products, promising a more sustainable future.

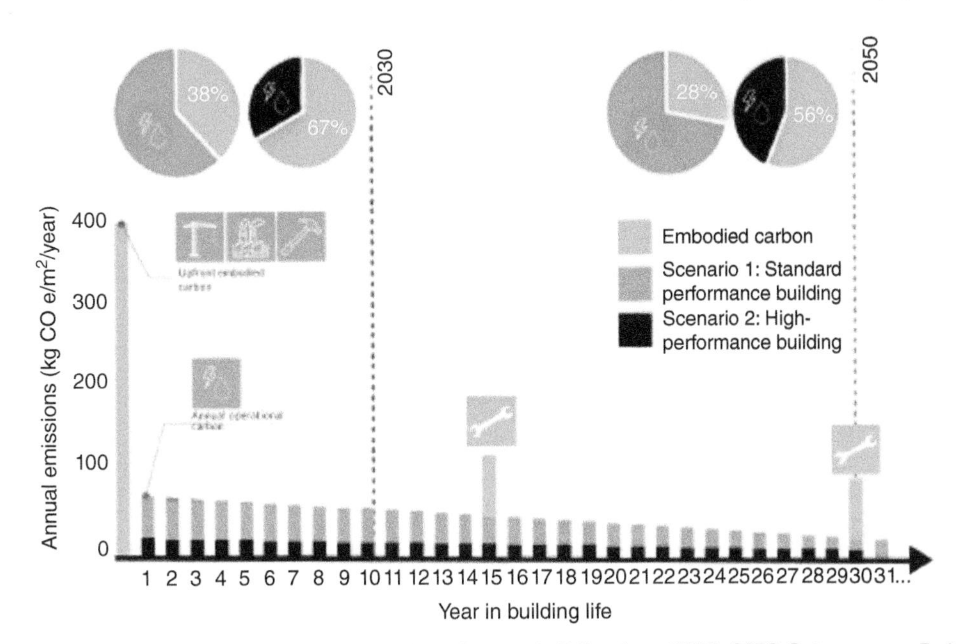

Relative impact of embodied and operational carbon of a new building from 2030–2050. Data sources: Embodied Carbon Benchmark Study and Commercial Buildings Energy Consumption Survey (CBECS), assuming a medium-sized commercial office building. Assumes gradual grid decarbonization to zero by 2050. Source: Adapted from Carbon Leadership Forum / https://carbonleadershipforum.org/carbon-challenge//last accessed Jan 23, 2024.

8.3.6 Empowering Change Through Action

Kate's journey into embodied carbon began as a practicing architect and structural engineer. Her desire to answer questions and make a difference led her to look into life cycle assessment and carbon accounting. She stresses the importance of setting ambitious goals, even when uncertainties exist. "We need to start moving forward, taking those bold steps and recognizing that we will learn both through successes and mistakes as we innovate to develop improved building practices," she states.

8.3.7 The Future of the Carbon Leadership Forum (CLF)

Kate envisions CLF's role as one that conducts critical research, provides resources to the industry, and empowers policy and practice changes. The goal is to drive embodied carbon emissions down to zero and, ideally, store more carbon in building materials than emitted during their production. She underscores the importance of research, knowledge dissemination, and building a network of individuals and organizations committed to decarbonization.

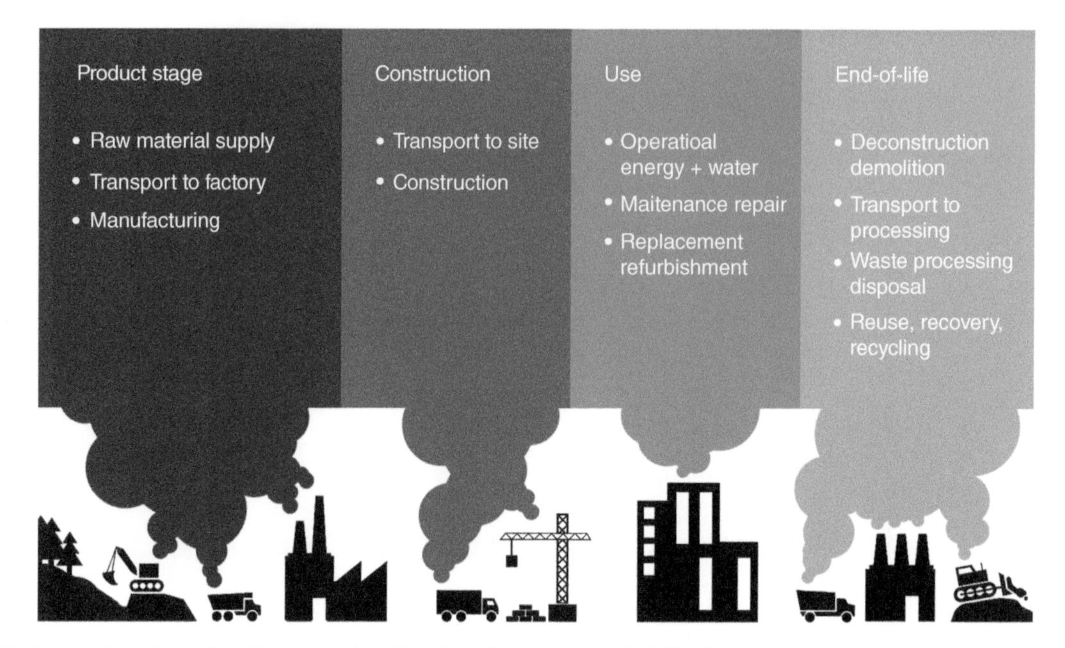

Understanding the carbon from a product from its origin to the end of life. Source: Adapted from Meghan Lewis & Megan Kalsman, 2023/Carbon Leadership Forum.

8.3.8 A Call to Action

For students and professionals alike, Kate encourages becoming domain experts in the field of embodied carbon. With the industry rapidly evolving, there is a need for professionals who understand the intricacies of decarbonization. She suggests incorporating embodied carbon considerations into project checklists and emphasizes that every role within the building sector can contribute to reducing embodied carbon, whether directly or indirectly.

Simonen's conversation underscores the transformative power of setting decarbonization as a primary goal and integrating embodied carbon considerations into construction specifications. Her insights reveal a growing momentum for change and highlight the pivotal role of building product manufacturers in the decarbonization movement. Empowering individuals to take bold steps and act in the face of uncertainty is crucial, and the CLF continues to be a driving force in advancing research and knowledge dissemination. As the industry evolves, her call to action reminds us that each of us can contribute to lowering embodied carbon and ushering in a more sustainable future for the construction industry.

8.4 Leveraging Lessons Learned and Instituting Quality Assurance

8.4.1 How to Leverage Corporate Knowledge and Avoid Repeating Past Mistakes by Reviewing "Lessons Learned"

"Lessons Learned" stands as a cornerstone *quality assurance practice* for corporations developing complex capital projects. What this means is that captured knowledge is being viewed and reviewed through structures of past experiences. The results shed light on successes and challenges on prior endeavors, illuminating strategies for continuous improvement, safeguarding against recurring pitfalls. A standardized process extracts valuable insights methodically. Post-occupancy commissioning agents administer confidential interviews with project principals—documenting performance metrics, anecdotes of complicated issues resolved, ideas to streamline processes, areas where expectations fell short, and more. Third-party facilitation elicits candid revelations that are free of political sensitivities, capturing subtle perceptions. This is an organized capture that separates discrete lessons by phase such as programming, design development, or construction closeout. It deals with relevant qualitative themes that emerge through analytical coding of transcripts. Quantitatively, metrics such as energy use, construction costs, or change order percentages are compared against targets highlighting over- and underperforming aspects that warrant corrective focus.

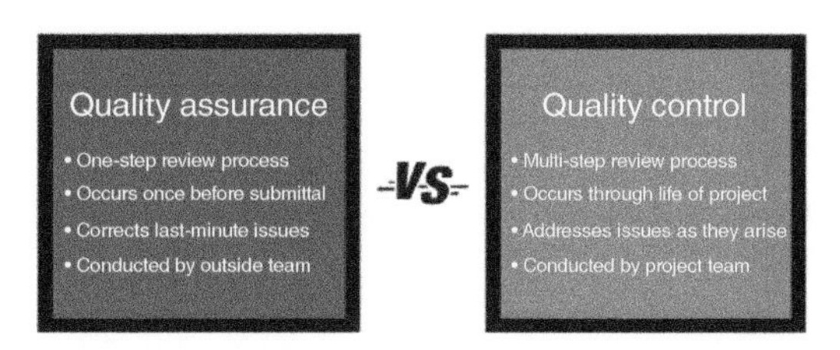

Highlighting the key different between quality assurance and quality control. Source: Adapted from https://meadhunt.com/quality-assurance-quality-control-2/.

Organizing regular "Lessons Learned" *workshops* convene multiproject stakeholders. It gives an opportunity to review anonymized cases for group discussions and unpack root causes behind positive lessons and deficiencies. We can use interactive voting to prioritize action items for targeted implementation or piloting. Documentation of action plans indicates responsible parties and timelines. This way, we are basically building a knowledge repository that systematizes insights of everyone. Collections organized by topic allow universal searchability, which is of great support to the research teams. When we use standardized templates, it is easier to maintain structure and ensure compiled lessons remain straightforward to retrieve years hence. Continuously populating these information libraries strengthens the living knowledge asset over time and cultivates an atmosphere embracing transparency and accountability. By proactively learning from history, corporations cyclically improve quality and fortify against risks—enhancing reputation, optimizing resource use, and delivering maximum sustainable value with each successive project.

Suggestions on what to do step by step and what to think about:

- Lessons capture both tangible metrics and softer insights around team dynamics, communication workflows, and so on that statistics do not reveal.
- Confidential individual interviews should put participants at ease to share candidly without fear of repercussion.
- Analytical coding of transcripts by multiple analysts should identify common themes across diverse projects.
- Lessons workshops at best incorporate real-time polling/surveys to engage participants and capture perspective shifts.
- Lessons filtered by project type/scale before review to focus discussion on most relevant cases.

- The knowledge repository uses tagging and rich descriptive fields to enhance searchability and group related lessons.
- Periodic review of past action item progress should encourage accountability and continuous learning over time.
- Case studies in the repository should be anonymous and linked to aggregated metrics for statistical trend tracking.
- Lessons should be socialized across business units to optimize performance at portfolio scale beyond any one project.
- Facilitators must probe discussions for implications both within and beyond the industry to spark new innovations.

Tips on how to capture the learning even more effectively:

- Record interviews and transcribe them verbatim to accurately preserve all perspectives shared.
- Use multi-vote democracy techniques at workshops to surface priorities by quantifying qualitative themes.
- Bundle lessons into reusable modular content like training modules, specification language, and digital playbooks.
- Use storytelling case studies to bring lessons to life for external conferences and industry publications.
- Track KPI progress toward goals, informing strategy, and resource allocation by metric benchmarks and dashboards.
- Integrate lessons into project lifecycle documentation like manuals, commissioning reports.
- Give ownership of the knowledge platform and outreach to dedicated staff to champion continuous culture of learning.
- Filter lessons by time to reveal evolving priorities and allow revisiting past action items in changing contexts.
- Value lessons from failures and unrealized goals to avoid complacency from past wins alone.
- Maintain relationships with high-performing alumni to maintain access to their evolving perspectives over career trajectories.
- Quantifying return on investment of applied lessons to incentivize commitment to the program.

8.4.2 The Process of Instituting a Project Delivery Quality Assurance (QA) Program

Making robust and comprehensive project work relies on delivery of a *quality assurance program*. Those are ideally anchored via the lessons learned to continuously improve process integrity and deliver optimized outcomes across projects. Proven methods, tailored to the organization's culture and priorities, lay the groundwork for success.

Executive leadership commitment establishes a productive learning environment where transparency and accountability should flourish. Streamlined systems extract value from experiences through cross-project examination of quantifiable metrics and softer insights alike. When it comes to hard data, that needs to be contextualized by qualitative themes that transform purely anecdotal information into actionable intelligence. The project should be assessed periodically and administered impartially by qualified third parties in order to validate conformance with procedures and high-performance targets. Including on-site field observations, document audits and stakeholder interviews impart candid perspectives to unpack pervasive issues while commending exceptional practices warranting emulation.

Workshops are always a great way to include interactive components and subject matter experts. What often surfaces are implementable strategies that have the potential to enhance delivery processes, contracts, and training. To maintain living resources, we need to keep building up comprehensive digital workspaces that include corrective case studies, training modules, specification language, and playbooks exemplifying standards. Robust documentation embeds institutional memory that will guide future project launches. Anchored by stakeholder buy-in and adaptation over time, we can cultivate a proactive problem-solving culture based on history to continuously deliver sustainable value through quality-assured project outcomes. Only this way can *formal QA* strengthen portfolio resilience and reputation.

How to ensure an effective project delivery for QA program:

- Third-party assessments occur at regular milestones (30–50% design, pre-commissioning) using standard forms to ensure consistency.
- Assessment teams include commissioning agents and subject matter experts to provide multidisciplinary insights.
- Document audits examine completeness/compliance against approved programming documents and design standards.
- Interviews solicit anonymous feedback from owners, contractors, and engineers to surface any cultural issues.
- Workshops analyze assessment findings using techniques like root cause analysis to identify systemic factors.
- Corrective action plans have assigned owners and timelines tracked to closure for accountability.
- Training modules embed continuous learning into program roll-out while socializing enhancements.
- Key performance metrics are tracked over time using dashboards to quantify program impacts.
- Case studies showcase demonstrated benefits to gain organizational buy-in for resource investment.
- External audits provide impartial validation that the QA program itself meets governance objectives.
- Program evolves through periodic review cycles ensuring relevance against shifting priorities and markets.

The key performance metrics in this case that could be tracked in a project delivery quality assurance program are the following:

- *Carbon (Operational and Embodied) and Energy*—Tracked before/after project completion to validate built performance meets design targets.
- *Change order percentage*—Measures deviation from baseline budget to contain costs. Lower is better with experience.
- *Construction defect rate*—Quantifies issues per square foot found during commissioning/warranty period. Lower over time shows improvements.
- *Design schedule adherence*—Percentage of projects meeting milestones like 50% design development (DD) review. Higher adherence saves rework costs.
- *Construction schedule performance*—Earlier substantial/beneficial occupancy dates increase return on investment.

- *Procurement variance*—Tracks supply chain execution against specifications and schedules. Less variance = less disrupts.
- *Staff/contractor engagement survey scores*—Anonymized feedback improves processes and workplace culture.
- *Training participation*—Ensures personnel get continued education to maintain high quality standards.
- *Program audit findings closeout*—Percentage of prior audit action items addressed ensures continuous progress.
- *Utility rebates/incentives received*—Validates built performance met stringent program requirements.

8.5 Eric Borchers | Structural Engineer | KAI Hawaii, Inc. | Honolulu, HI

Eric Borchers, a structural engineer at KAI Hawaii, embodies the commitment to sustainable engineering. Leading the database team for the SE 2050 Commitment Program, he champions embodied carbon reduction, fostering a complete perspective on carbon accounting. With a design portfolio spanning diverse materials and contexts, Borchers epitomizes the profession's transformation. As engineers embark on this journey of innovation and change, Eric Borchers and SE 2050 are pushing structural engineering beyond strength and serviceability to consider carbon footprint.

Eric Borchers: Leading the database team for SE 2050.

Change comes slowly to structural engineering. Carbon presents the biggest opportunity to shake up traditional practice and the future is rich with possibilities. In the coming years, structural engineers will navigate a landscape where environmental concerns and sustainable practices will hold more sway. A shift toward carbon-conscious designs and the exploration of alternative materials, like mass timber and low carbon concrete, are anticipated to make waves. The clarion call for lowering carbon footprints will reshape business models and practices. As the profession embraces change, collaboration with architects, contractors, and manufacturers will be instrumental in realizing a sustainable built environment.

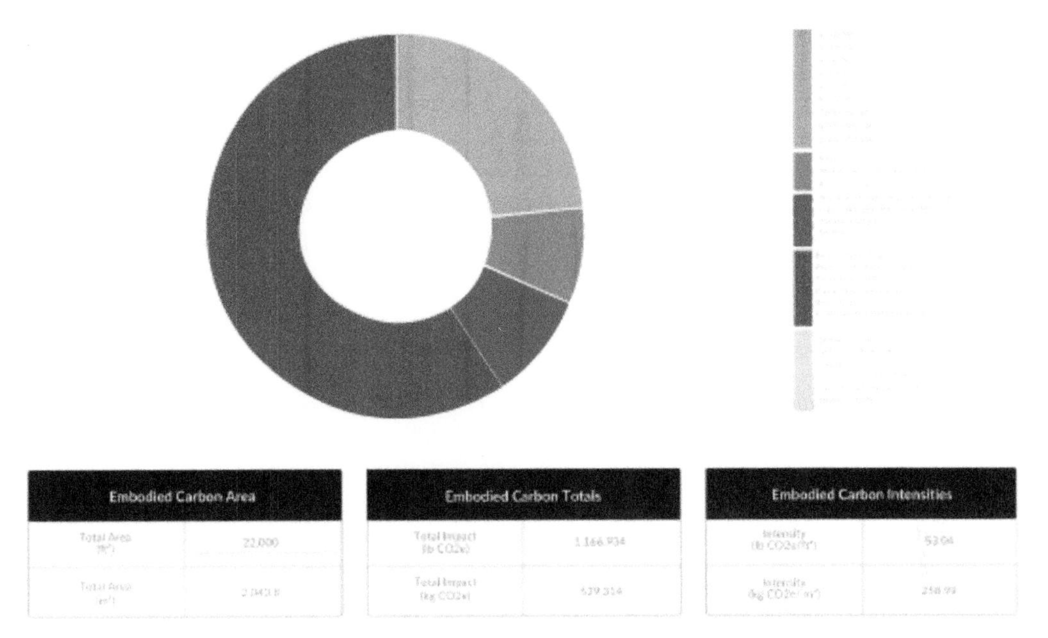

Quick calculations of embodied carbon intensities of structural materials by building area. Source: Adapted from https://www.enr.com/articles/55637-se-2050-is-in-quixotic-pursuit-of-eliminating-embodied-carbon-in-building-structures.

8.5.1 Amidst Evolving Trends and Challenges

Structural engineers of the near future will find themselves embroiled in new trends and challenges. The seismic shift toward sustainable practices will compel engineers to incorporate carbon considerations into their designs. A key driver of this transformation is the SE 2050 Commitment Program, igniting a dialogue that was once foreign in structural engineering circles. Embodied carbon, previously an enigma, is now at the forefront of discussions, with the program fostering awareness and driving transformation.

8.5.2 Shaping a Sustainable Path Forward

The SE 2050 Commitment Program has kindled a carbon dialogue, propelling structural engineers into action. The once-obscure notion of embodied carbon has permeated the industry. Firms have embraced the commitment, with over 100 signatory firms collectively representing thousands of engineers dedicated to eliminating embodied carbon from their designs. As this movement gains momentum, resources like the Embodied Carbon Order of Magnitude (ECOM) tool empower engineers with insights into their designs' impact (Fig. 1). With such tools readily at hand, daily tasks such as designing floor framing or even a temporary tower crane foundation now bear the weight of environmental responsibility, enlightening engineers about their contributions to emissions.

8.5.3 Carbon Integration and a Holistic Approach

The integration of carbon considerations into structural engineering marks a paradigm shift. Engineers accustomed to optimizing for strength, serviceability, and cost are now tasked with incorporating carbon as an integral dimension. Embodied carbon's impact becomes a key determinant in design decisions. The transition extends beyond engineering calculations, demanding an astute understanding of material choices, construction methods, and life cycle assessments. Carbon-positive solutions that also lead to cost savings appropriately garner the most attention, but the industry must recognize the need to embrace a wide variety of carbon reduction solutions, including those that have a higher upfront cost than business as usual.

8.5.4 Empowering Change and Impact

To translate carbon-positive decisions into tangible results, structural engineers must wield their influence throughout the construction process. Design documents become the blueprint for change, laying the foundation for sustainable building practices. Setting stringent requirements and thresholds for emissions and materials at the outset ensures their presence throughout the project lifecycle. With early intervention, engineers can exert a substantial impact on suppliers' practices, steering them toward carbon-friendly alternatives. Engaging suppliers and setting bold carbon limits can lead to meaningful reductions, championing sustainability at every stage.

9

The Consequences of Inaction

The consequences of our inaction in the face of climate change manifest not just in global reports and future forecasts but in the intimate spaces we hold dear—the landscapes of our childhood memories and the rhythms of the seasons we cherish. The impact is deeply personal, altering the fabric of our daily lives and eroding the experiences that define us.

Remember the summers of your youth, the enchanting flicker of fireflies at dusk, and the graceful dance of monarch butterflies in your garden. These delights, once markers of the seasons, are becoming increasingly rare, their absence a poignant reminder of the ecological upheaval wrought by our changing climate. The same skies and fields that once thrummed with life now bear witness to the dwindling of these beautiful creatures, a direct consequence of our environmental negligence.

Each day, we feel the personal toll of climate change. The surge in pollen levels, fueled by rising carbon dioxide, aggravates our allergies, transforming the simple act of breathing into a struggle. The unseasonable warmth of winter days, triggering premature blooms in trees like the redbud, disrupts the delicate balance of ecosystems. These distress signals from the natural world, once anomalies, now serve as constant reminders of the profound changes we have set in motion through our collective inaction.

The cost of this crisis is not just measured in statistics, but in the irreplaceable experiences and landscapes, we stand to lose. The places that shaped us—the swimming holes of our childhood, the gardens where we chased butterflies, the forests that taught us wonder—are being irrevocably altered, victims of drought, extreme weather, and shifting seasons. As these environments change or disappear, we lose not only biodiversity but also the touchstones of our identity and heritage.

For those in the architecture, engineering, and construction industry, this is a deeply personal matter. Every unsustainable choice we make in our professional capacities contributes to the degradation of the very environments that hold our memories and dreams. Our decisions ripple out from every BIM model, email, and request for information on the construction site, shaping the world that will be inherited by future generations.

As stewards of the built environment, we must recognize these signs and respond not just with technical solutions but with a fierce commitment to preserving the planet that sustains us. Let the changing world ignite in us a sense of urgency, compelling us to adopt practices that protect not just abstract ecosystems but the personal landscapes we hold dear. In this fight against climate change, the stakes are not just professional but profoundly personal.

Build Like It's the End of the World: A Practical Guide to Decarbonize Architecture, Engineering, and Construction,
First Edition. Sandeep Ahuja and Patrick Chopson.

9.1 What Real Commitment Looks Like

Genuine commitment to sustainability in the AEC industry transcends mere adherence to standards or the adoption of green technologies. It demands a form of moral courage and leadership that challenges conventional practices and sets new benchmarks for environmental responsibility. It requires visionaries who are willing to confront uncomfortable truths, make difficult decisions, and prioritize the long-term well-being of our planet over short-term expedience.

In this context, moral courage means having the conviction to question the status quo and advocate for transformative change, even in the face of skepticism or resistance. It is embodied by the architects, engineers, and builders who lead by example, willing to risk criticism and navigate uncharted territory because they are driven by a higher sense of purpose and accountability to society and the environment.

These leaders understand that true sustainability is not a box to be checked but a foundational principle that should inform every decision, every design, and every project. They strive not just to minimize harm but to create built environments that actively regenerate and revitalize the natural world. Their vision extends beyond the confines of individual buildings to encompass the health and resilience of entire ecosystems.

Leadership in this new paradigm is about more than project management or resource allocation. It is about inspiring a fundamental shift in mindset and empowering others to think and act with sustainability at the forefront. These leaders are the early adopters, the ones who embrace innovative practices and technologies not because they are easy or proven but because they are necessary and right. They are the trailblazers, forging the path toward a more responsible and sustainable future.

You will find these leaders at the vanguard of change, advocating for stricter environmental regulations, investing in the development of low-carbon materials, and educating the next generation of professionals to prioritize ecological integrity alongside aesthetic and functional considerations. They understand that each of these actions, while potentially challenging in the short term, is crucial for safeguarding the future of our planet.

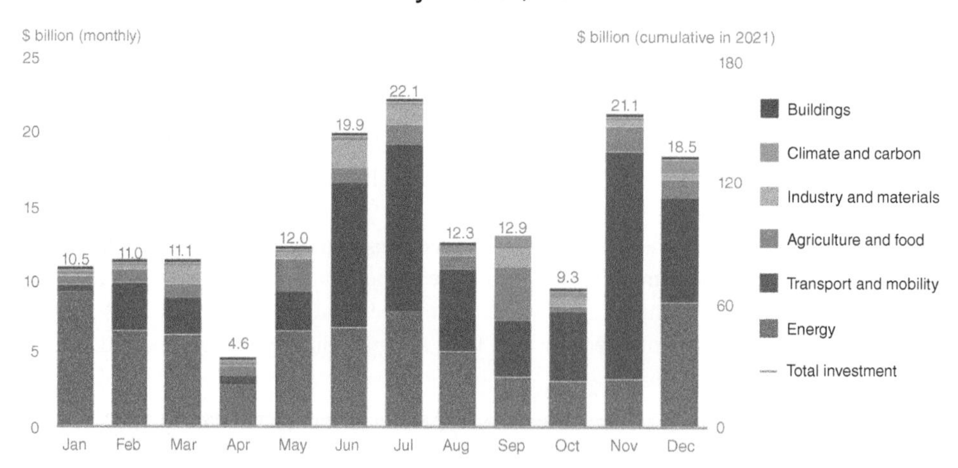

The financial commitment within climate tech is a significant step forward.

But their commitment is not just professional; it is deeply personal. These individuals see themselves not just as architects or engineers but as stewards of the environment and agents of positive change. They are motivated by a profound sense of responsibility to the communities and ecosystems that their work impacts, both locally and globally.

Their dedication is evident in every sustainable project they pioneer, every harmful practice they reject, and every policy debate they engage in to push for a greener, more resilient built environment. They recognize that these choices, while not always popular or easy, are the very essence of what it means to be a true leader in the face of the climate crisis.

As we navigate this pivotal moment in history, the story of the AEC industry must be one of courageous leadership—of individuals and organizations who dared to challenge convention and act with boldness and integrity. These are the leaders who will not just react to change but proactively drive it, guided by an unwavering commitment to the health of our planet and the well-being of future generations. In their actions and their example, we find the blueprint for a truly sustainable future.

9.2 Catalyzing Change: Strategies for Decarbonization

The path to decarbonization in the AEC industry is paved with choices that demand moral courage and a clear, unwavering vision. It requires us to challenge deeply ingrained norms and conventions to advocate for a comprehensive shift toward sustainable practices. This transformation goes beyond the mere adoption of new technologies; it necessitates a fundamental re-evaluation of our project management methodologies and design philosophies, prioritizing long-term environmental benefits over short-term expediency.

To lead this movement effectively, we must set a powerful example, demonstrating the feasibility and the imperative of sustainable practices. This often involves making decisions that may initially face resistance or skepticism—such as advocating for electric heating systems over traditional gas solutions, despite the inertia of industry habits. It means being prepared to defend these choices with robust data, rigorous research, and an unwavering conviction in the necessity of change.

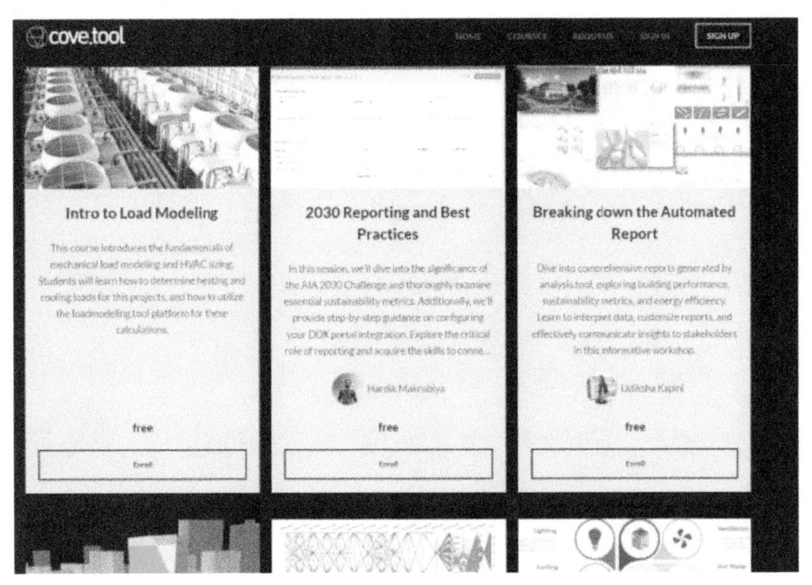

Sharing the case studies from projects can inspire others to follow.

As leaders in this movement, we will often find ourselves at the forefront of debates, pushing for changes that disrupt the status quo but ultimately lay the groundwork for a more sustainable and resilient industry. We must be the voices of reason and the catalysts of progress, even when met with resistance.

One of the primary obstacles in driving change is overcoming the early-stage design inertia—the tendency to fall back on familiar approaches rather than embracing the challenges of innovation. Decarbonization demands that we engage with better solutions from the outset, focusing our efforts on fundamental, systemic changes rather than cosmetic adjustments.

To advocate for these foundational shifts effectively, we must cultivate a deep understanding of the far-reaching impacts of our decisions, and we must educate our stakeholders about the long-term value of sustainable practices over the entire lifecycle of a project. This requires a commitment to continuous learning and knowledge sharing, both within our teams and across the broader industry.

Authors teaching the Living Building Challenge Studio at Kennesaw State University. Students presenting to the principals at Perkins&Will.

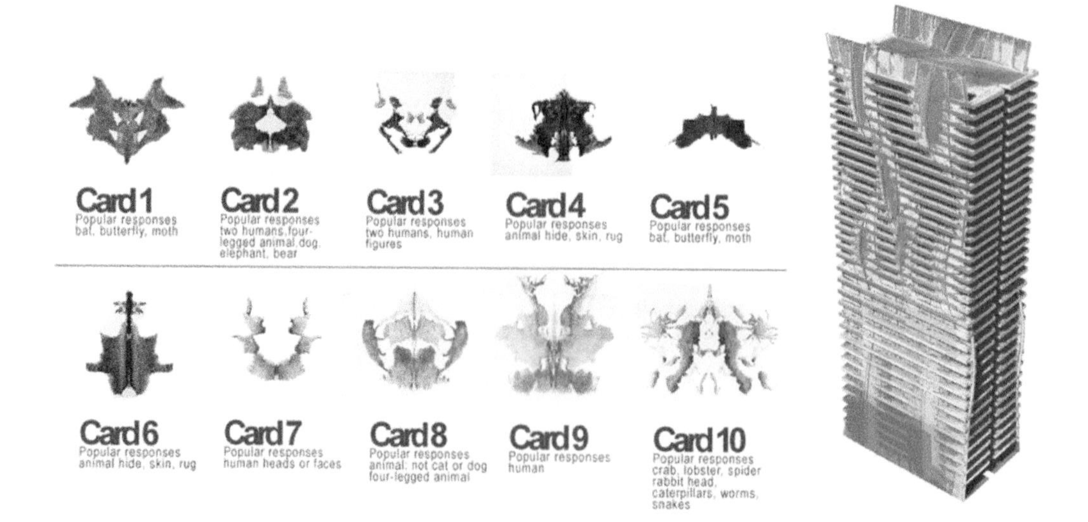

An example design scheme from a student from the studio.

However, no transformative movement can succeed in isolation. Building a robust coalition of like-minded professionals, organizations, and communities is paramount. This coalition should be united by a shared ethos of sustainability and a sense of responsibility toward future generations. Through collaborative workshops, joint research initiatives, and cross-disciplinary projects, we can foster a vibrant ecosystem of knowledge and support that amplifies our collective impact.

Yet, catalyzing lasting change requires more than initial enthusiasm; it demands sustained commitment and a willingness to adapt. It involves not just implementing new strategies but also establishing robust frameworks for monitoring, evaluating, and iteratively improving upon those strategies. This agile, responsive approach ensures that our efforts remain effective and relevant in the face of evolving challenges and opportunities.

As we embark on this journey of decarbonization, we must recognize that every decision we make, every project we undertake, is an opportunity to shape a better future. By approaching each challenge with a mix of moral courage, strategic foresight, and a commitment to collaboration, we can catalyze a profound transformation in the AEC industry—one that not only responds to the climate crisis but proactively shapes a more sustainable, resilient world.

9.3 Our Last Chance

As we conclude this urgent call to action, we find ourselves standing at a critical juncture—a point where our choices will determine not just the trajectory of our industry but the fate of our planet. The climate crisis is no longer a distant threat; it is a present reality, a test of our collective resolve, and a challenge to the very core of our professional responsibilities. In this pivotal moment, we must ask ourselves: Will we rise to meet this challenge, or will we retreat into the comfort of complacency?

The stakes could not be higher. The consequences of inaction are not abstract or remote; they are tangible and immediate. They manifest in the rising tides that threaten to engulf our coastal cities, in the wildfires that ravage our communities, and in the ecosystem collapse that undermines the very foundations of life on Earth. Can we, in good conscience, continue to contribute to this unfolding catastrophe, knowing that we possess the knowledge and the tools to chart a different course?

The legacy of our decisions extends far beyond our own lifetimes. It will be etched into the landscapes and the living conditions of future generations. Every unsustainable project we approve, every opportunity for green innovation that we forgo, adds to a tragic inheritance of environmental debt. But this is not an inevitability; it is a choice.

Envisioning a better tomorrow.

Imagine, for a moment, a future where we are called to account for our actions. How will we justify to our children and grandchildren the fact that we had the power to mitigate the worst impacts of climate change but instead chose the path of least resistance? How will we explain the wildfires that drove them from their homes, the droughts that ravaged their crops, and the species that vanished forever on our watch? These are the haunting questions that we must confront if we continue on our current trajectory.

But there is another path, one that demands courage, conviction, and a willingness to defy the status quo. It is a path that recognizes the true essence of our roles as architects, engineers, and builders—not just as designers of structures but as shapers of a sustainable future. It is a path that harnesses our skills, our creativity, and our influence to drive transformative change, lead by example, and inspire others to follow suit.

This is our moment to redefine what it means to be a professional in the AEC industry. It is our chance to channel our frustration, our fear, and our passion into concrete, meaningful action. Every line we draw, every material we select, and every project we green-light holds the potential for positive impact. Let us wield this power with intentionality and with a deep sense of responsibility to the planet we share.

Let our legacy be one of bold action in the face of crisis of unwavering commitment to a sustainable future. Let it be said that when the challenge of our lifetime arose, we met it with determination, innovation, and a resolute refusal to accept anything less than a thriving, resilient world.

This is our last chance to alter the course of history to bend the arc of our industry toward sustainability and regeneration. It is a chance that we must seize with urgency, with courage, and with a profound sense of purpose. We owe it to ourselves, to the generations that will inherit the world we shape, and to the planet that sustains us all.

The path ahead will not be easy, but it is necessary. It will demand the best of us—our ingenuity, our perseverance, and our moral fortitude. But if we rise to this challenge and embrace the opportunity to be the architects of a sustainable revolution, we can secure a future that is not just livable but vibrant and thriving.

So let us move forward with resolve and with hope, knowing that every action we take, every decision we make, is a brush stroke in the portrait of the world we will leave behind. Let us paint a picture of a planet restored, of communities resilient, and of an industry that led the charge toward a greener, more just future. The time for half-measures and hesitation has passed; the era of bold, transformative action is upon us. Let us meet this moment with all the courage and conviction it demands.

Epilogue: Crafting Our Future Landscape

As we stand at the concluding chapter of this narrative, let us engage with an essential inquiry: What shape could our world take if we accelerate the transformation of our built environment toward carbon neutrality? This is not merely about dodging a climate catastrophe; it is about proactively scripting a future of urban and architectural innovation.

Our future is not a fixed trajectory; it is a dynamic landscape shaped by our collective decisions and actions. The ideas and pathways laid out in this book are not mere theoretical musings; they are practical, actionable routes toward a sustainable and thriving future. We are envisioning a shift where our buildings and cities do more than provide shelter; they become instrumental in the healing and restoration of our planet. The adoption of renewable energy and the leap in efficiency are not just options but necessities to divert us from the path of severe climate repercussions.

Imagine the cities of 2050, transformed not just in their physicality but in their very essence. This transformation is rooted in a revolution in materials science and energy usage. These future buildings are more than just structures; they are active participants in climate change mitigation, providing affordable, carbon-neutral living spaces that promote health and well-being.

The urban landscapes of tomorrow, once homogenized in their concrete expanses, are now alive with energy-efficient, electrically powered architecture. This is not just about design; it is an intricate integration of functionality and sustainability, where buildings are interwoven into a grid powered by renewable energies.

The transformation extends beyond the buildings to the urban fabric itself. Cities have evolved into human-centric ecosystems, where the emphasis has shifted to public transportation and pedestrian-friendly spaces. The move toward shared, electric vehicles is not just a transition; it is a reinvention of urban mobility aligned with sustainable principles.

Our residential and commercial spaces have evolved from mere physical structures to embodiments of our commitment to a sustainable future. Urban planning now integrates nature as a core element, turning our cities into vibrant, green ecosystems that serve as the lungs of our urban environments.

The task of global decarbonization is a grand challenge, yet the signs of progress are undeniable and inspiring. This journey is not about incremental adjustments but about a fundamental rethinking of how we build and inhabit our urban environments. The narrative of our future is still being written, and it is up to us to author a story of a healthier, more equitable world, leaving a lasting legacy for the generations to come.

Build Like It's the End of the World: A Practical Guide to Decarbonize Architecture, Engineering, and Construction,
First Edition. Sandeep Ahuja and Patrick Chopson.

This is more than a story of survival; it is a narrative of flourishing in a world where our built environment coexists in harmony with nature. Our role as creators and custodians of this built world is clear. We are at the threshold of a new era, where our actions today will define the sustainability and vitality of our urban landscapes. The cities of the future, as envisioned here, are not static entities; they are dynamic, evolving systems, growing in symbiosis with their inhabitants and the natural world. Our decisions today have the power to shape a future where architecture and nature are not just coexistent but are intertwined in a sustainable, life-affirming dance. Let us embrace this opportunity with the vision and determination it demands, building a legacy that extends far to our children.

Appendix A

Experts, Innovators, and Thought Leaders Interviewed

Shaun Abrahamson, Co-Founder & Managing Partner at Third Sphere
Sarah Gudeman, PE, BCXP, CPHC, WELL AP, LEED FELLOW, CEM, Principal at Branch Pattern
Natalie Terrill, WELL AP, LEED Fellow, Director of Sustainability at The Beck Group
Pablo La Roche, PhD, LEED AP, Principal, Arcadis, Professor Cal Poly Pomona
Michael Beckerman, CEO, CREtech
Kate Simonen, FAIA, SE, Founder and Board Chair of Carbon Leadership Forum, Professor at University of Washington
Dennis Shelden, PhD, Director at Center for Architecture and Ecology at Rensselaer Polytechnic Institute
Chirag Mistry AIA, LEED AP, Regional Leader of Science and Technology at HOK
Eric Borchers, SE, Sr. Structural Engineer at KAI
Tristram Carfrae, FRSA, FREng, FTSE, RDI, Deputy Chair of Arup

Organizations and Universities Represented

AIA
Arcadis
Arup
Beck Group
BranchPattern
CallisonRTKL
Cooper Carry
cove.tool
CREtech
HOK
KAI
Third Sphere
US Green Building Council/Green Business Certification Inc.
Cal Poly Pomona

Build Like It's the End of the World: A Practical Guide to Decarbonize Architecture, Engineering, and Construction,
First Edition. Sandeep Ahuja and Patrick Chopson.

Emory University
TVS Architecture and Interior Design
Morrissey Engineering
Georgia Institute of Technology
Kennesaw State University
Rensselaer Polytechnic Institute
University of California, Los Angeles
University of Virginia
University of Washington
Vanderbilt University

Sources

2021 United Nations Environment Programme. (2021). Global Status Report for Buildings and Construction. Retrieved from https://globalabc.org/sites/default/files/2021-10/GABC_Buildings-GSR-2021_BOOK.pdf.

911 Security. (2019). U.S Drone Laws: Overview of Drone Rules and Regulations in USA by State. Retrieved from https://www.utsystem.edu/sites/default/files/offices/police/policies/USDroneLaws.pdf.

Andersson, H., & Mutlu, A. (2020). Digital collaboration tools. Retrieved from https://lup.lub.lu.se/luur/download?func=downloadFile&recordOId=9017193&fileOId=9017201.

Architecture 2030. (n.d.). Retrieved from https://architecture2030.org/.

BBVA Research. (2022). Economics of Climate Change: How do digitalization and decarbonization efforts interact? Retrieved from https://www.bbvaresearch.com/wp-content/uploads/2022/05/How-do-digitalization-and-decarbonization-efforts-interact_2T22.pdf.

Blue Yonder Group, Inc. (2020). The Digital Control Tower Imperative. Retrieved from https://medialibrarycdn.blueyonder.com/jp/ja/-/media/files/blue%20yonder/master/knowledge%20center%20documents/white%20paper/control%20tower%20imperative%20white%20paper.pdf?rev=-1.

Braham, W. (2015). Thermodynamic Principles for Environmental Building Design. Retrieved from https://acee.princeton.edu/wp-content/uploads/2015/12/Braham_Andlinger_111615-Slide-Presentation.pdf.

Building Green. (n.d.). The Importance of Building Science. Retrieved from https://www.greenbuilt.org/the-importance-of-building-science/.

Canada's Climate Plan. (n.d.). Retrieved from https://www.canada.ca/en/services/environment/weather/climatechange/climate-plan/net-zero-emissions-2050.html.

CDP. (2019). Climate Change Report 2019. Retrieved from https://cdn.cdp.net/cdp-production/cms/reports/documents/000/004/588/original/CDP_Climate_Change_report_2019.pdf?1562321876.

City of Toronto. (2018). Toronto Green Standard Version 3.0 Energy Efficiency Reporting and Modelling Guideline. Retrieved from https://www.toronto.ca/wp-content/uploads/2018/03/97f1-City-Planning-TGS-V3-Energy-Efficiency-Reporting-Modelling-Guideline.pdf.

Climate Bonds Initiative. (n.d.). Explaining Green Bonds. Retrieved from https://www.climatebonds.net/market/explaining-green-bonds.

cove.tool. (2019). 6 Ways to Lower Your EUI. Retrieved from https://help.covetool.com/en/articles/4533792-6-ways-to-lower-your-eui.

cove.tool. (2019). How To Use Passive Strategies In Your Design. Retrieved from https://www.covetool.com/2019/04/how-to-use-passive-strategies-in-your-design/.

Build Like It's the End of the World: A Practical Guide to Decarbonize Architecture, Engineering, and Construction,
First Edition. Sandeep Ahuja and Patrick Chopson.

cove.tool. (n.d.). Optimization and Bundles. Retrieved from https://help.covetool.com/en/articles/3452693-optimization-and-bundles.

cove.tool. (n.d.). Understanding the Automated Report. Retrieved from https://help.covetool.com/en/articles/3799650-understanding-the-automated-report.

D'Amico, F., D'Ascanio, L., De Falco, M., Ferrante, Ch., Presta, D., & Tosti, F. (2020). BIM for infrastructure: an efficient process to achieve 4D and 5D digital dimensions. Retrieved from http://www.istiee.unict.it/sites/default/files/files/2_10_ET_180.pdf.

Eldash, K. (2012). Value Engineering (Course Notes). Retrieved from https://www.researchgate.net/publication/271909862_VALUE_ENGINEERING_Course_Notes.

Energy Services Coalition. (n.d.). Performance Contracting. Retrieved from https://energyservicescoalition.org/performance-contracting.

Energy Star. (n.d.). Retrieved from https://www.energystar.gov/.

Global Alliance for Buildings and Construction. (2021). Decarbonizing the Building Sector: 10 Key Measures. Retrieved from https://globalabc.org/sites/default/files/2021-07/Decarbonizing%20the%20building%20sector.pdf.

Global Infrastructure Hub. (n.d.). Sustainable Infrastructure. Retrieved from https://www.gihub.org/sustainable-infrastructure.

Goulding, J., & Rahimian, F. (2019). Offsite Production and Manufacturing for Innovative Construction: People, Process and Technology. Retrieved from https://www.routledge.com/rsc/downloads/Ch_1_from_Offsite_Production_and_Manufacturing_for_Innovative_Construction.pdf.

Gransberg, D., & Riemer, C. (2009). Performance-Based Construction Contractor Prequalification. Retrieved from https://www.researchgate.net/publication/280610211_Performance-Based_Construction_Contractor_Prequalification.

Harvard Business Review. (2004). The Path to Corporate Responsibility. Retrieved from https://hbr.org/2004/12/the-path-to-corporate-responsibility.

Harvard Business Review. (2015). The Truth About CSR. Retrieved from https://hbr.org/2015/01/the-truth-about-csr.

Harvard Business Review. (n.d.). Corporate Social Responsibility. Retrieved from https://hbr.org/topic/subject/corporate-social-responsibility.

Hill, A. C. (2023). Climate Change and U.S. Property Insurance: A Stormy Mix. Retrieved from https://www.cfr.org/article/climate-change-and-us-property-insurance-stormy-mix.

International Energy Agency. (2021). Sustainable Recovery – Buildings. Retrieved from https://www.iea.org/reports/sustainable-recovery/buildings.

International Monetary Fund. (2022). Staff Climate Note. Retrieved from https://www.imf.org/-/media/Files/Publications/Staff-Climate-Notes/2022/English/CLNEA2022002.ashx.

Kjartansdóttir, I., Mordue, S., Nowak, P., Philp, D., & Snæbjörnsson, J. (2017). Building Information Modeling (BIM). Retrieved from https://www.ciob.org/sites/default/files/M21%20%20BUILDING%20INFORMATION%20MODELLING%20-%20BIM.pdf.

Living Future Institute. (n.d.). Living Building Challenge. Retrieved from https://living-future.org/lbc/.

McKinsey & Company. (2017). Reinventing Construction: A Route to Higher Productivity. Retrieved from https://www.mckinsey.com/~/media/mckinsey/business%20functions/operations/our%20insights/reinventing%20construction%20through%20a%20productivity%20revolution/mgi-reinventing-construction-executive-summary.pdf.

McKinsey & Company. (2019). Driving Impact at Scale from Automation to AI. Retrieved from https://www.mckinsey.com/~/media/McKinsey/Business%20Functions/McKinsey%20Digital/Our%20Insights/Driving%20impact%20at%20scale%20from%20automation%20and%20AI/Driving-impact-at-scale-from-automation-and-AI.ashx.

McKinsey & Company. (2019). The Next Normal in Construction. Retrieved from https://www.mckinsey.com/~/media/McKinsey/Industries/Capital%20Projects%20and%20Infrastructure/Our%20Insights/The%20next%20normal%20in%20construction/The-next-normal-in-construction.pdf.

ModelThinkers. (n.d.). First Principle Thinking. Retrieved from https://modelthinkers.com/mental-model/first-principle-thinking.

Munich Re. (n.d.). Natural Disasters. Retrieved from https://www.munichre.com/en/risks/natural-disasters.html.

Network for Greening the Financial System. (n.d.). Retrieved from https://www.ngfs.net/en.

Nielsen. (2022). Inaugural Brand Sustainability Report. Retrieved from https://www.nielsen.com/news-center/2022/nielsens-inaugural-brand-sustainability-report-reveals-australian-consumer-perceptions-of-the-sustainability-efforts-of-leading-brands/.

NOAA National Centers for Environmental Information (NCEI) US Billion-Dollar Weather and Climate Disasters (2023). https://www.ncei.noaa.gov/access/billions/. DOI: 10.25921/stkw-7w73.

Principals for Responsible Investment. (n.d.). Retrieved from https://www.unpri.org/download?ac=6241.

ResearchGate. (2014). 5D BIM: A Case Study of an Implementation Strategy in the Construction Industry. Retrieved from https://www.researchgate.net/publication/320391935_5D_BIM_A_Case_Study_of_an_Implementation_Strategy_in_the_Construction_Industry.

ResearchGate. (2020). Manufacture and Structural Performance of Modular Hybrid FRP-Timber Thin-Walled Columns. Retrieved from https://www.sciencedirect.com/science/article/abs/pii/S0263822320334358.

ResearchGate. (2020). Problems and Challenges in the Interactions of Design Teams of Construction Projects: A Bibliometric Study. Retrieved from https://www.researchgate.net/publication/355152428_Problems_and_Challenges_in_the_Interactions_of_Design_Teams_of_Construction_Projects_A_Bibliometric_Study.

Romanello, M., Di Napoli, C., Drummond, P., et al. (2022). The 2022 report of the Lancet Countdown on health and climate change: health at the mercy of fossil fuels. Volume 400, Issue 10363, P1619–1654. https://www.thelancet.com/journals/lancet/article/PIIS0140-6736(22)01540-9/fulltext.

Swiss Re Institute. (n.d.). Why Insurers Need to Transform Digital Distribution. Retrieved from https://www.swissre.com/institute/research/topics-and-risk-dialogues/digital-business-model-and-cyber-risk/why-insurers-need-to-transform-digital-distribution.html.

The Global Commission on the Economy and Climate. (2018). The 2018 Report of the Global Commission on the Economy and Climate. Retrieved from https://newclimateeconomy.report/2018/executive-summary/.

The Green Building Initiative. (n.d.). Retrieved from https://thegbi.org/.

The National Association of Home Builders. (n.d.). The NGBS Green Promise. Retrieved from https://www.ngbs.com/the-ngbs-green-promise.

The Passive House Institute US. (n.d.). Passive Building Principles. Retrieved from https://www.phius.org/passive-building/what-passive-building/passive-building-principles.

Toronto Green Standard. (n.d.). Retrieved from https://www.toronto.ca/wp-content/uploads/2018/03/97f1-City-Planning-TGS-V3-Energy-Efficiency-Reporting-Modelling-Guideline.pdf.

Turner, G. (2014). Is Global Collapse Imminent? An Updated Comparison of the Limits to Growth with Historical Data. Melbourne Sustainable Society Institute.
U.S. Department of Energy. (n.d.). Energy Saver: Reducing Electricity Use and Costs. Retrieved from https://www.energy.gov/energysaver/reducing-electricity-use-and-costs.
U.S. Department of Energy. (n.d.). Energy Saver: Grid-Connected Renewable Energy Systems. Retrieved from https://www.energy.gov/energysaver/grid-connected-renewable-energy-systems.
U.S. Department of Energy. (n.d.). Office of Energy Efficiency & Renewable Energy. Retrieved from https://www.energy.gov/eere/energy-efficiency.
U.S. Green Building Council. (n.d.). The Business Case for Green Building. Retrieved from https://www.usgbc.org/articles/business-case-green-building?elqTrackId=48428afea0ac4595a4d199e23a4d808a&elqaid=54&elqat=2.
Urban Land Institute. (n.d.). Developing Resilience. Retrieved from https://developingresilience.uli.org/about-resilience/.
Urban Land Institute. (n.d.). A Guide for Assessing Climate Change Risk. Retrieved from https://uli.org/wp-content/uploads/ULI-Documents/ULI-A-Guide-for-Assessesing-Climate-Change-Risk-final.pdf.
World Economic Forum. (2019). The Global Risks Report 2019. Retrieved from https://www.weforum.org/reports/the-global-risks-report-2019/.
World Economic Forum. (2022). The Cost of Net Zero: $3.5 Trillion a Year. Retrieved from https://www.weforum.org/agenda/2022/01/net-zero-cost-3-5-trillion-a-year/.
World Green Building Council. (n.d.). Retrieved from https://worldgbc.org/.
Zukunftsinstitut. (2017). The Future of Timber Construction CLT – Cross Laminated Timber. Retrieved from https://www.storaenso.com/-/media/documents/download-center/documents/product-specifications/wood-products/clt-technical/stora-enso-the-future-of-timber-construction-en.pdf.

Index

Build Like It's the End of the World: A Practical Guide to Decarbonize Architecture, Engineering, and Construction, First Edition. Sandeep Ahuja and Patrick Chopson.

f

o

p

q

r

s